Texts in Mathematics

Volume 8

A Primer of Mathematical Analysis and the Foundations of Computation

Texts in Mathematics Series Editor
Dov Gabbay   dov.gabbay@kcl.ac.uk

A Primer of Mathematical Analysis and the Foundations of Computation

Fairouz Kamareddine

Jonathan Seldin

ISBN 978-1-84890-443-9

College Publications
Scientific Director: Dov Gabbay
Managing Director: Jane Spurr

http://www.collegepublications.co.uk

Cover designed by Laraine Welch

Preface

About this book

This book discusses ideas for a different approach to teaching the foundations of mathematical analysis as well as the foundations of mathematics. The main idea is to delay the use of what Keith Devlin in [12] called "formal definitions", which are definitions that nobody can understand without working with them. These formal definitions are common in advanced mathematics, including analysis, but for students without mathematical maturity they are very difficult to learn from. This is in contrast to ordinary dictionary definitions, which the reader of a novel might consult for an unknown word and which the reader will understand immediately.

This book discusses an approach that uses the history of mathematics to first develop fundamental concepts of mathematical analysis and the foundations of mathematics and only introduces formal definitions after the concepts are understood by the students.

Many textbooks for a first course in real analysis begin with formal logic, apparently assuming that this should indicate to students what proofs are and why they are needed. Starting this way appears to place too much of a burden on the students. It would be better for the formal logic to occur later in the book when examples from earlier parts of the course can be used to illustrate how formal logic can serve to analyse them. We believe that this is the best approach for students to recognise the meaning of the logical symbols.

Furthermore, these textbooks for a first course in real analysis would introduce after formal logic, the axioms for a complete ordered field, and then proceed from there with the $\varepsilon - N$ definition of the limit of a sequence and the $\varepsilon - \delta$ definition of the limit of a function. This form of introduction to real analysis appears to consist of several formal definitions in Devlin's sense. This seems to be the wrong order for many students, since it does not help the students learn why these axioms and theorems are important and what analysis is really about.

Similar ideas apply to the foundations of mathematics and computation where students struggle between the solvable/unsolvable and find it hard to reason about paradoxes, termination, fixed points, etc. They may not even have seen cardinal arithmetic or sizes of infinite sets before and many of them struggle to construct models of the λ-calculus or the computable functions. Approaching the material in a historical order rather than a logical order may help student learning. Even if we do not have an account of the history that is beyond controversy, we think taking a historical approach to the material will make sense to students and will enable students to visualise what they will be studying.

The idea of this book is to use an approach that is more historical than the traditional logical approach. It is designed more to help students achieve mathematical maturity than to cover a fixed collection of topics. Our book aims to enable the students to take some time to help them learn what mathematical proofs are and why they are important. This book is not intended to replace a traditional course in real analysis or on the foundations of mathematics and computation, but to supplement it by preparing students to take such a course with more understanding than many of them now have. The past decades of teaching a number of mathematical topics (e.g., analysis, discrete mathematics, foundations of mathematics, theory of computation, lambda calculus, logic and type theory) have demonstrated to us that students enjoy learning the material and put in more effort if they are also told of the history and background and introduced to the detective-like search for solutions.

The audience

The audience for this approach is not only students preparing to study a course on mathematical analysis or a course on the foundations of mathematics and/or computation. There are a number of historical sides and views as well as many useful examples with an immediate appeal for students of fallacious reasoning, and how to steer clear of such pitfalls. There is a wealth of exercises, with selected solutions as well as numerous graphical illustrations which give an experienced instructor lots of possibilities to select a colourful and stimulating course in mathematical analysis or a course on the mathematical foundations of computability theory with a broader background involving for instance glimpses of cardinal arithmetic and predicate logic background. Even for just browsing by general readers beyond the student level, this book presents stories and insights, concepts and mathematical theories, covering a history window of ancient times to the present.

This book is designed for a second or third year course in an undergraduate degree in Pure Mathematics or Computer Science. Since these students often take a first course in real analysis or on the foundations of mathematics and/or computation right after a course on logic and set theory or discrete mathematics, many of them still lack a complete understanding of mathematical proofs and what they are really all about. We think that this is unnecessarily difficult for many students. The historical backgrounds and the large number of exercises provide an alternative approach which will iron out many of the difficulties these students face.

The idea of this approach for Mathematical Analysis started years back at the Department of Mathematics and Computer Science at the University of Lethbridge when the Curriculum Committee discussed the need for teaching analysis to maths students in two stages:

1. A third-year course in analysis where students would learn how to do $\varepsilon - \delta$ proofs for limits of functions and $\varepsilon - N$ proofs for limits of sequences, and

2. a fourth year advanced course on analysis where a more formal approach would be followed which would make more sense to the students now that they understand the history, background and evolution of analysis.

This approach was tested at Lethbridge by one of us (Jonathan Seldin). The tests demonstrated that almost all the students in the class, including the weaker ones, were able at the end of the course to do $\varepsilon - \delta$ and $\varepsilon - N$ proofs (although some of the weaker students and some of those whose native language was not English, did not use the correct logical words in the right places.)

Concurrently, one of us (Fairouz Kamareddine) has been teaching computability theory, mathematical foundations of Computer Science, and logic and proofs to undergraduate maths and CS students at the University of Glasgow and Heriot-Watt University and has been developing the relevant material with an eye of the evolution of the subject which has always struck a cord with students and motivated them to stick to (and actually enjoy) a mathematically laced course within a Computer Science curriculum.

This prompted us to write a detailed self-contained book that takes the idea of the informal and historical approach seriously and to apply it to mathematical analysis and the foundations of mathematics and computation with emphasis on the evolution of the subjects. We also wanted to back everything with exercises so that we could use the informal approaches to introduce the students afterwards to the formal definitions and to further accelerate the learning. The result is the present book.

In view of our desire to avoid formal definitions, we have tried not to introduce new vocabulary without first having the concepts arise naturally out of the discussion. We have also avoided dealing with foundational issues which do not arise necessarily from the subject under discussion, since these are beyond the scope of the book. We feel that these issues can easily be left for later courses. As H. B. Curry put it in [10, p. 363],

> "The foundations of mathematics are not like those of a building, which may collapse if its foundations fail; but they are rather more like the roots of a tree, which grow as the tree grows, and in due proportion."

Our book is best suited for students studying 2nd or 3rd year BSc in Mathematics or 3rd year BSc in Computer Science where it should provide a solid introductory course on mathematical analysis or the foundations of mathematics and computation. All students of a BSc in mathematics or a BSc in computer science will have studied a basic course on logic, set theory and discrete mathematics in the first year of their degree. This is the background needed before starting this book, although the book is really self sufficient and does explain the necessary background needed.

The book is self explanatory, so any staff member in the departments of mathematics or computer science can teach this course. The subjects mathematical analysis respectively mathematics of computation are standard in every undergraduate degree of mathematics respectively computer science degree. So, this course will have an international appeal.

This is a unique book and should be treated as a complementary book which should be made compulsory for UG degrees in mathematics or theoretical computer science. The book is not intended to replace the entire curriculum of mathematical analysis or the mathematics/foundation of computation. Rather, this book is there to make the courses on mathematical analysis and the mathematics/foundation of computation, easier to master by the students and as such this book will make the task of lecturers of this material a lot easier. The book will give the students the right techniques and skills to work with mathematical analysis and the theory of computation and to go on further to study more advanced courses in this area.

Just like our book should be distinguished from the traditional books in that we leave formal definitions till after the concepts have been intuitively and historically understood, our book should also be distinguished from historical books. Our book is developed as a fully pledged university course that is guided by the historical development and chronology. Its message for example is different to that of the excellent book by David M. Bressoud [6] which is focused more around the history. While Bressoud's book does not demonstrate how one can do calculus or mathematical analysis as is acknowledged by the author when he says: "This book will not show you how to do calculus. My intent is instead to show how and why it arose. (p. xi)", it remains an excellent read for any student on a calculus course and is a testimony to the importance of understanding the history of any subject especially when teaching it.

Similarly, our book is complementary to the book of Boman and Rogers [5]. This introductory real analysis text starts with the development of the real number system which it uses to give the definition of a limit which is used as a foundation for the definitions encountered thereafter. The book is more centred around the number systems and specific series limits, whereas our book is centred more around sets, cardinals and metric spaces.

Overview of this book

- The text begins in Chapter 1 with a discussion of why mathematics is about reasoning and why this reasoning requires much care as it is easy to make errors. We also briefly mention the fear of the infinite and the reasoning obstacle faced in the history of mathematics with respect to limits and infinitesimals (i.e., very very small values), and how this led to the development of calculus and analysis.

- Chapter 2 discusses briefly the development of proofs in mathematics. We start by emphasising the importance of the rhetorical style of arguments in Ancient Greece and move to discuss early Greek mathematics concentrating on the theory of odd and even numbers as developed by the Pythagorean school and documented in Euclid's *Elements*. On our way, we need to mention the motion paradoxes developed by Zeno which contributed to the Greeks fear of the infinite and their attempt to avoid infinite processes. Since Euclid's Elements developed mathematics in geometric terms and anything not expressible in such terms was excluded, proofs are taken as diagrams. We follow Knorr's suggestion [26, Chapter V, Section III] that the original proofs were proofs as diagrams using pebble diagrams and present the theory of even/odd numbers and pebble diagram proofs. We do this until we hit the point in history where proofs by diagrams or pebbles were replaced by proofs as sequences of statements. According to Knorr, this occurred at the same time as the incommensurability of the side and diagonal of a square was discovered which led to the discovery of the existence of non rational numbers.

- In Chapter 3 we reflect a bit on the history of numbers and take up proofs as statements as applied to some elementary algebra. We do this in the context of defining the positive rationals $\mathbb{Q}^+$ and the integers $\mathbb{Z}$ from nonzero natural numbers $\mathbb{N}^+$. We start by assuming the nonzero natural numbers $\mathbb{N}^+$ and the standard operations of addition and multiplication on $\mathbb{N}^+$. Then we show how we can build the positive rationals $\mathbb{Q}^+$ as values of fractions (formal quotients) of nonzero natural numbers and give the addition and multiplication operations on these newly built positive rationals. Although inverses did not exist in $\mathbb{N}^+$ (since we have no fractions and no negatives), we see that in $\mathbb{Q}^+$, multiplicative inverses do exist and that even if we start with a natural number system without a multiplicative identity (which is 1 in $\mathbb{N}^+$), the rational number system we get adds one. Afterwards, we move to build the integers as equivalence classes of accounts of credits and debits and define the addition and multiplication operations on these integers. Comparing the positive rationals and the integers, we finally build starting from a set with an operation satisfying certain "axioms", a unified theory for adding inverses and, if none is present identity elements. We will do this so that if we identify the starting set with the nonzero naturals and the operation with multiplication we will have the theory of fractions and if we identify the set with the nonzero naturals and the operation with addition we will have the theory of accounts.

- In Chapter 4 we start the discussion of the sizes of (finite and infinite) sets. Sets became important in mathematics in connection with *infinite sets*, about which some problems arose in thinking about their sizes (also known as their number of elements). In the first section we give first thoughts on sizes of sets, we define the size of a finite set, and introduce the notions *(infinitely) countable* and *uncountable*, but fall short of defining the size of infinite sets in this chapter. Instead, we focus on the notions of countable and uncountable and show that subsets of countable sets are also countable, that $\mathbb{N}$, $\mathbb{Z}$, $\mathbb{Q}$ and the set of algebraic numbers are countable whereas the interval $(0,1]$ of $\mathbb{R}$ is uncountable. In this first section, we keep everything at the informal level and don't go into the formal details of one-to-one correspondences and functions. In the second section we introduce basic operations on sets such as union, intersection, powerset, and then formally define functions and their properties (including injectivity, surjectivity and bijectivity) which will help us establish further properties of countable sets. We also establish the Cantor-Schröder-Bernstein theorem which states that there is a bijection between two sets iff there is an injection from each of the sets to the other set. In the third section we reflect further on finite and infinite sets and the sizes of finite sets as well as countability and uncountability.

- By this stage we are familiar with calculating the sizes of finite sets, but not the size of an infinite set nor the size of the union or product of infinite sets. In Chapter 5 we study the cardinal numbers which are sizes of sets (finite and infinite) and explain how to measure sizes of infinite sets and how to add or multiply infinites. We show that all infinitely countable sets have the same size and that there are infinitely many infinite cardinal numbers the smallest of which is the cardinal $\mathfrak{a}$ which is the size of $\mathbb{N}$ and of all the infinitely countable sets. The cardinal number $\mathfrak{a}$ is different from the cardinal $\mathfrak{c}$ which is the size of the real numbers. In particular we show that $\#\mathbb{N} = \#\mathbb{Z} = \#\mathbb{Q} = \mathfrak{a}$ and $\#\mathbb{R} = \#\mathcal{P}\mathbb{N} = \#\mathcal{P}\mathbb{Z} = \#\mathcal{P}\mathbb{Q} = \mathfrak{c} < \#\mathcal{P}\mathbb{R} \cdots$ where $\#S$ stands for the size of S (also called the cardinal number of S) and $\mathcal{P}S$ is the powerset of S. We then move in the second section to give the arithmetic of cardinal numbers. Although the arithmetic of the finite cardinal numbers is that of the natural numbers, we will see that the arithmetic of cardinal numbers that involve infinite cardinals has a number of surprises. Finally, in the third section we reflect on the use of the universal set and on the paradoxes and then reflect further on the so-called *Continuum Hypothesis* and discuss why we chose to use $\mathfrak{a}$ for the smallest infinite cardinal number and $\mathfrak{c}$ for the size of the real numbers instead of starting at $\aleph_0$ as is usually done in some sources.

- In Chapter 6 we are going to introduce *formal logic*, and we will see how it can be used to analyse the informal proofs we have been presenting so far. The results of these analyses of proofs will show how formalising the logic can be useful.

- During their quest for the (non-existent) unit which divides evenly and exactly into both the side and diagonal of a square, the Greeks found irrational numbers. In light of this, they faced a need to treat other quantities which are not a discrete collection of units (like the naturals or the ratios of two naturals), but which are continuous. The Greeks did not know how to handle these quantities and juggled with two notions: their notion of "numbers" (as a multitude of units) as well as the so-called *magnitudes* (which in addition to "numbers" include things like lines and areas and volumes, etc.). The Greeks developed arithmetic for their numbers, but treated their magnitudes geometrically. However, starting in the 16th century and in order to construct magnitudes (like what we call nowadays the real numbers), mathematicians used approximations. Even though the Greeks have not thought of constructing magnitudes or any new mathematical objects, they did introduce a procedure for approximating ratios that sheds light on the innovative thinking of the time. We discuss this procedure in the first section of Chapter 7. We then in the second section discuss the study on the theory of numbers after the discovery of the incommensurability of the side and diagonal of a square. Thereafter, we explain the "ruler and compass" constructions in Euclid's *Elements*.

- The calculus is based on limits. Therefore, in order to obtain a proper foundation for the calculus, we need a proper foundation for a theory of limits. In Chapter 8, we will start by looking at the basic properties we would expect for limits of functions and sequences. In this process, we revisit again Zeno's dichotomy and discuss where Zeno's conclusion is false. We move on to explain that adding infinitely many numbers need not return an infinite and then we analyse the way the ancient Greeks dealt with limits especially through the so-called *method of exhaustion*. In particular, we will look at two theorems from the ancients: Archimedes' theorem giving the area of a circle, and a theorem of Euclid which says that the areas of circles are to each other as the squares of their radii. Both theorems were proved by a method that relied on Eudoxus theory of proportions which was a geometric theory designed to overcome the difficulties expected because of the discovery of the incommensurability.

- We start Chapter 9 by using the method of exhaustion to suggest a definition of the limit of a sequence which we will use to prove the properties of limits of sequences discussed in the previous chapter. We then demonstrate how to construct proofs in the theory of limits starting with scratch work that goes backwards. We then do for limits of functions what we have already done in this chapter for limits of sequences. All this is done without specifying that we are dealing with real numbers. This corresponds to the historical order of the development of calculus and analysis in European mathematics. The ancient Greeks had distinguished between numbers and magnitudes, and Descartes had shown that both could be considered together as quantities. But a rigorous theory of these quantities had not been developed, and many questions about them had not been answered. Most of the elementary calculus was developed before the real number system was defined. In fact, it had not even been settled whether or not there were infinitely small quantities; for example, for an indication that Cauchy **thought** that there were infinitely small quantities, see [27].

- In the previous two chapters, we did not say much about the magnitudes/quantities (real numbers) that we were using in dealing with limits. This was left ambiguous on purpose: we were looking at both an intuitive approach and some results from Archimedes and Euclid. However, we cannot leave this ambiguous indefinitely. Chapter 10 takes up the question of what these quantities are, and gets to the definition of the real number system. First, we look at what constituted the standards of rigour in mathematics from Greek times to the 19th century and give some of the reasons why people started to doubt some of the reasoning in the *Elements*. In particular, we show that there is a model of Euclidean plain geometry in which all the axioms of the plane are satisfied but Euclid I 1 is not. Then we show that the way quantities are used in the mathematics of the seventeenth and eighteenth centuries implies that they satisfy all the axioms of an ordered field and that the rational numbers, which are an ordered field, must all be quantities. But the model of plain geometry introduced at the beginning of the chapter shows that the rational numbers are not enough. Then the intuitively obvious theorem that a nondecreasing sequence of quantities that is bounded above has a limit, when compared with the $\varepsilon - N$ definition of the limit of a sequence, implies that the limit must be the least upper bound of the terms of the sequence. It is then shown that the set of all rational numbers whose squares are less than 2, which is clearly bounded above, has no least upper bound. This fact is used to introduce the Axiom of Completeness. We end the chapter by giving some consequences of the Axiom of Completeness and a modern resolution to each of the four paradoxes of Zeno discussed in Chapter 2.

- In Chapter 11 we introduce the continuity and derivatives concepts and show how continuity is preserved under different function operations and give a number of properties of derivatives such as the Quotient rule and the Chain rule.

- In Chapter 12 we will look at some results about sequences, limits, and continuity that depend on the Axiom of Completeness of the real numbers and demonstrate that the Axiom of Completeness has significant consequences for the notion of continuity. The chapter begins with a discussion of bounded sequences, defines Cauchy sequences, and proves the Bolzano-Weierstrass Theorem. Then, to make the definitions less formal, *tails* of sequences are defined, and lim inf and lim sup are defined in terms of them. Among the results proved from these definitions are the Intermediate Value Theorem, the continuity of the inverse of a continuous 1-1 function, and the Extreme Value Theorem. The chapter closes with a discussion of sequences of functions and uniform continuity.

- Chapter 13 takes up the Riemann integral and proves that the limiting process in its definition leads to proofs that continuous functions are Riemann integrable as well as other similar properties.

- Solutions to the exercises are available in a separate booklet.

- Throughout the book, we use "iff" to stand for "if and only if" and "IH" to stand for "the inductive hypothesis".

Acknowledgements

We would like to thank Martin Bunder, Mariangiola Dezani, Roger Hindley, Gérard Huet and Jan Willem Klop for their useful comments and suggestions.
As ever, Jane Spurr has been fantastic. Her knowledge, expertise and professionalism have been invaluable for achieving the beautiful look and feel of the book.

Fairouz D. Kamareddine and Jonathan P. Seldin
Edinburgh, UK and Lethbridge/Toronto Canada October 2023

Contents

Chapter 1

Introduction

Looking back at the stages in which you were taught mathematics at school, you see that first, you were taught instructions and little reasoning (as when asked to do arithmetical calculations) and then, you were taught more complex reasoning (e.g., when asked to solve equations or find unknown quantities). Soon you start to realise that apart from arithmetic, all of mathematics is concerned with reasoning and that some areas of mathematics (such as advanced calculus[1] and analysis[2]) are left to a late stage in the

---

[1]Calculus is the theory of change and differential calculus describes the way things change through differentiation. In calculus, problems are reduced to geometric curves which represent movement and where the tangent on the curve is used to give the direction of motion/change. Areas (quadratures) within curves were also studied in integration. Tangents and quadratures were studied well before Euclid's times but they were not formally linked until Newton's work on the calculus in the 17th century. Curve, tangent and quadrature were replaced by function, derivative and integral by the end of the 17th century.

As a theory of change, calculus and all pre-calculus developments relied on infinitesimals and it was always controversial how to accommodate infinitesimals without getting paradoxes. These controversies led in the nineteenth century to new developments in analysis and the theories of limits. They also led to the development of a more precise style of mathematics and proof and ultimately led to the development of an entirely new areas of foundational mathematics and the theory of computability (see [18, 24]).

[2]Analysis is the study of infinite processes. The word "Analysis" (or $\alpha\nu$ α $\lambda\upsilon\sigma\eta$ in Greek) means "loosening up" or "unravelling" which expresses the fact that analysis breaks down complex objects into (infinitely) smaller and simpler parts. Analysis studies real and complex-valued functions and is concerned with limits and integration. It was first developed as a branch in mathematics in the 18th century in the work of Euler who converted the calculus as studied by Newton and Leibniz from a geometrical field to a field where mathematical formulae are analysed. However, some traces of earlier work have been documented in the literature. E.g., in the 11th century, the mathematician Hasan Ibn al-Haytham (Alhazen) proved the earliest general formula for infinitesimal and

school/university curricula because they need a more developed reasoning process. In this chapter, we explain why mathematics is all about reasoning and why this reasoning requires much care as it is easy to make errors. We also briefly mention the fear of the infinite and the reasoning obstacle faced in the history of mathematics with respect to limits and infinitesimals (i.e., very very small values), and how this led to the development of calculus and analysis.

Since analysis is about the mathematical basis of the calculus and provides the calculus with a satisfactory foundation, in analysis, we answer to two questions:

1. Why does calculus work?

2. Why can we trust the results we obtain by using the calculus?

To answer these questions, we need to consider what it means to justify mathematical results. We will begin with this. After that, we will go on to consider why anybody would have reason to doubt the calculus.

1.1   Mathematics is about Reasoning

If you think back to your school days and to the difference between what you studied when you were doing arithmetic and what changed when you started studying algebra, you may recall that in every problem you were given in arithmetic, the problem told you what operations to perform, the numbers on which to perform them, and in what order to perform them. This changed when you started studying algebra.

Arithmetic is a Greek word which means "the skill of number". Arithmetic deals with basic computation on numbers using arithmetic operations like addition, multiplication, division, etc. On the other hand, algebra is the study of generalisations of arithmetic operations. The word algebra is derived from the Arabic word Al-Jabr which means "uniting broken parts" and deals with problem solving and with finding the unknown quantities.

Example 1.1.1. *Consider the following problem:*

> Find a number such that five more than three times the number is twenty.

---

integral calculus which he needed in his volume calculations.

This problem does not tell you directly what operations to perform on what numbers; it is necessary to reason this out indirectly. To do this, start with the sentence

> *Suppose five more than three times the number is twenty.*

This sentence tells us that we started with the number and performed two operations on it to get twenty. The last of those operations was adding five. So if we go back to the stage before that operation, we have to undo the operation of adding five to get twenty, and we undo that by subtracting five from twenty to get fifteen. This gives us

> *Hence three times the number is fifteen.*

Once again, we get back to the stage before the last operation by undoing it. The last operation here was to multiply by three, so we undo it by dividing fifteen by three:

> *Thus the number is five.*

This chain of reasoning proves the following:

> If five more than three times a number is twenty, then the number is five.

When we check this answer, we reason the other way: five more than three times five is five more than fifteen, which is twenty. This check proves

> If the number is five, then five more than three times the number is twenty.

This result is called the converse *of the preceding one.*   □

Note that this chain of reasoning is algebra, even though the usual notation of algebra is not used. *Algebra is not so much about using letters to stand for numbers as it is about reasoning.* And reasoning is used in algebra not only to solve particular problems as we just did, but it is also used to prove general properties of numbers, such as the property that the result of adding two numbers is not changed when the order of the numbers is changed.

> *All mathematics more advanced than arithmetic is mainly about reasoning.*

Of course, we can do without the standard algebraic notation only for simple problems. For complicated problems, the sentences we would need, if written out in English, would become so complicated that we would not be able to see the patterns necessary to use our algebraic reasoning to get to the solutions. In fact, the standard algebraic notation was originally developed as a shorthand to make it easier to see these patterns. We can see this in the above example if we rewrite the three key sentences of our solution in order as follows:

$$\text{Assume that } 5 + 3 \times \text{(the number)} = 20.$$
$$\text{Hence } 3 \times \text{(the number)} = 15.$$
$$\text{Thus, (the number)} = 5.$$

This represents what might have been an intermediate step in the development of algebraic notation. The final stage is as follows: let $n = $ the number. Then

$$\text{Assume that } 5 + 3n \ = \ 20.$$
$$\text{Hence } 3n \ = \ 15.$$
$$\text{Thus, } n \ = \ 5.$$

The corresponding use of the symbols for the check of the answer, the converse, would be

$$5 + (3 \times 5) = 5 + 15 = 20.$$

or

$$5 + 3 \cdot 5 = 5 + 15 = 20.$$

Note that the converse of an implication is not the same as the implication: it goes in the other direction.

1.1.1 Exercises

Exercise 1.1. Diophantus of Alexandria (see [21]) busied himself with Algebra to help him cope with the death of his son. Here is a puzzle about Diophantus:

> He was a boy for 1/6 of his life, after 1/12 more he acquired a beard, after another 1/7 he married, 5 years after marriage he had a son, the son died when he was half his father's life, four years later the father died.

Give the age to which Diophantus lived and say how old was his son when he died.

Exercise 1.2. Think of any natural number n. Take that number and multiply it by 5 and add 20 to the result. Then, divide that result by 5 and then subtract n. Show that no matter which n you chose, you always get the final answer 4.

Exercise 1.3. Think of any natural number n. Take that number and multiply it by 6 and add 48 to the result. Then, divide that result by 6 and then subtract n. Show that no matter which n you chose, you always get the final answer 8.

Exercise 1.4. Think of any natural numbers n and k and of a non zero natural number m. Take the number n and multiply it by m and add $m \times k$ to the result. Then, divide that result by m and then subtract n. Show that no matter which n, m and k as above you chose, you always get the final answer k.

Exercise 1.5. You have two identical cakes. There is the ballroom which has 6 people and the seminar room which has 8 people. You take one cake to each of the rooms and you divide it equally amongst the people in the room. Since $6 < 8$, Would a person in the ballroom get a smaller piece of cake than the person in the seminar room? Demonstrate this by drawing two circles and dividing them according to the numbers in each room. In general, if $m < n$ does it hold that $\frac{1}{m} < \frac{1}{n}$?

1.2 The Need for Care

The reasoning we do in mathematics must be done carefully. This is because mathematics is a subtle subject, and it is easy to make mistakes.

Consider the following example:[3]

Example 1.2.1. *Suppose you are offered the choice of two jobs with similar work, so that the difference is in the amount of pay. Suppose the pay for the two jobs is as follows:*

1. *Starting annual salary $20,000 per year with an increase of $4,000 per year.*

2. *Starting semiannual salary of $10,000 with an increase of $1,000 every six months.*

---

[3]This example comes from [29, pp. 17–18]. We have changed the numbers to reflect today's prices.

Which is the better offer (after the first year)? Perhaps you decided that Job 1 is better because it has an increase of $4,000, whereas Job 2 has an annual increase of twice $1,000, or $2,000, which is only half as much as the increase for Job 1.

Well, think again. Consider the following table of earnings for the first four years, where all values are in thousands of dollars:

	1st Year		2nd Year		3rd Year		4th Year	
	Job 1	Job 2	Job 1	Job 2	Job 1	Job 2	Job 1	Job 2
1st Half of Year	10	10	12	12	14	14	16	16
2nd Half of Year	10	11	12	13	14	15	16	17
Total for Year	20	21	24	25	28	29	32	33

Note that for every year of the first four years, the earnings from Job 2 are more than those for Job 1, and it is clear that this pattern continues indefinitely into the future.

The moral is that it is easy to make errors in reasoning about mathematics, and it is necessary to be very careful.   □

One place at which many students who are beginning their study of algebra make a mistake is to miss the difference between an implication and its converse. This is because most of the steps students carry out early in their study of algebra can be reversed. The first operation that most students learn which cannot be reversed is the squaring of both sides to eliminate radicals. Consider the following example:

Example 1.2.2. *Solve the equation*

$$\sqrt{x^2 - 2x + 1} = 2x.$$

The solution proceeds as follows:

$$\begin{aligned} x^2 - 2x + 1 &= 4x^2 \\ 3x^2 + 2x - 1 &= 0 \\ (3x - 1)(x + 1) &= 0 \end{aligned}$$

Thus,

$$3x - 1 = 0 \qquad or \qquad x + 1 = 0,$$

from which we get $x = \frac{1}{3}$ *or* $x = -1$. *If we check the answer* $\frac{1}{3}$, *we get*

$$\sqrt{\left(\frac{1}{3}\right)^2 - 2\left(\frac{1}{3}\right) + 1} \stackrel{?}{=} 2\left(\frac{1}{3}\right)$$

$$\sqrt{\frac{4}{9}} \stackrel{?}{=} \frac{2}{3}$$

$$\frac{2}{3} \stackrel{\checkmark}{=} \frac{2}{3}$$

but when we check the answer -1, *we get*

$$\sqrt{(-1)^2 - 2(-1) + 1} \stackrel{?}{=} 2(-1)$$

$$\sqrt{4} \stackrel{?}{=} -2$$

$$2 \stackrel{\times}{=} -2$$

So where does this extra root come from? It comes from the first step of the solution, which cannot be reversed. If we substitute -1 *for* x *into the step after we square to remove radicals, we get*

$$(-1)^2 - 2(-1) + 1 \stackrel{?}{=} 4(-1)^2$$

$$4 \stackrel{\checkmark}{=} 4$$

So with this substitution, the first step of the solution goes from $2 = -2$ *to* $4 = 4$, *which reflects the fact that* $2^2 = (-2)^2$. *This is why the step of squaring both sides of an equation cannot be reversed.* □

In this example, it is valid to say that

> *If two numbers are equal their squares are equal, but if the squares of two numbers are equal, it does not necessarily follow that the numbers are equal.*

Another example that shows the importance of careful reasoning is the following:[4]

Example 1.2.3. *Solve the equation*

$$2\sqrt{x+1} = \sqrt{4x+5}.$$

---

[4]This example comes from the preface of [31].

If this problem is solved mechanically, it will go as follows:

$$
\begin{aligned}
2\sqrt{x+1} &= \sqrt{4x+5} \\
4(x+1) &= 4x+5 \\
4x+4 &= 4x+5 \\
4 &= 5
\end{aligned}
$$

Students who do not understand that the important thing here is the reasoning may be lost at this point. However, reasoning can be emphasised by writing this as follows:

$$
\begin{aligned}
Assume\quad 2\sqrt{x+1} &= \sqrt{4x+5}. \\
Then\quad 4(x+1) &= 4x+5. \\
Hence\quad 4x+4 &= 4x+5. \\
Thus\quad 4 &= 5.
\end{aligned}
$$

Now it is clear that what we have is a proof of the following: if $2\sqrt{x+1} = \sqrt{4x+5}$, then $4 = 5$. Since $4 \neq 5$, it follows that there is no value of x which satisfies $2\sqrt{x+1} = \sqrt{4x+5}$.   □

The conclusion of this example is the result of a *proof by contradiction*: when a contradiction is deduced from an assumption, the conclusion is that the assumption is false. The idea is closely related to the debating tactic of *reductio ad absurdum*, or *reduction to an absurdity*. As you will see from the implication truth table on page 160, if A is true and B is false then deducing B from A is false. Hence, a contradiction B cannot be deduced from a true assumption A. Therefore, if we deduce a contradiction B from an assumption A, then the assumption A must be false.

1.2.1 Exercises

Exercise 1.6. To show that we need to be careful when calculating, here is one of the many false proofs that show $1 = 2$. Give the false steps in the proof.

$$
\begin{aligned}
\text{1. Let} \quad a &= b \\
\text{2. Then} \quad a^2 &= ab \\
\text{3. Then} \quad a^2 - b^2 &= ab - b^2 \\
\text{4. Then} \quad (a+b)(a-b) &= b(a-b) \\
\text{5. Then} \quad a+b &= b \\
\text{6. Then} \quad b+b &= b \qquad \text{since } a = b \\
\text{7. Hence} \quad 2b &= b \\
\text{8. Hence} \quad 2 &= 1
\end{aligned}
$$

Exercise 1.7. Here is a proof where we show that $1 = 0$. Give the false steps in the proof.

1.	-2	$=$	-2
2.	$4 - 6$	$=$	$1 - 3$
3.	$4 - 6 + (3/2)^2$	$=$	$1 - 3 + (3/2)^2$
4.	$2^2 - 2 \times 2 \times 3/2 + (3/2)^2$	$=$	$1^2 - 2 \times 1 \times 3/2 + (3/2)^2$
5.	$(2 - 3/2)^2$	$=$	$(1 - 3/2)^2$
6.	$2 - 3/2$	$=$	$1 - 3/2$
7.	2	$=$	1
8.	$2 - 1$	$=$	$1 - 1$
9.	1	$=$	0

Exercise 1.8. You have \$40,000 to invest. Suppose you are offered the choice of two interest rates as follows:

1. 20% of the total value is added to your sum at the end of every year.

2. 10% of the total value is added to your sum at the end of every six months.

Which is the better offer at the end of each of the first two years?

Exercise 1.9. Below is a proof that shows that $\sqrt{19} < \sqrt{3} + \sqrt{7} < \sqrt{20}$. Do you think this is a correct proof? Does the proof show that indeed $\sqrt{3} + \sqrt{7} < \sqrt{20}$ and $\sqrt{3} + \sqrt{7} > \sqrt{19}$. If it is not a correct proof give reasons why it is not correct and for each of the statements $\sqrt{3} + \sqrt{7} < \sqrt{20}$ and $\sqrt{3} + \sqrt{7} > \sqrt{19}$, say how you would either prove it or disprove it.

1.	$\sqrt{19}$	$<$	$\sqrt{3} + \sqrt{7}$	$<$	$\sqrt{20}$
2.	19	$<$	$10 + 2\sqrt{21}$	$<$	20
3.	9	$<$	$2\sqrt{21}$	$<$	10
4.	81	$<$	84	$<$	100

Exercise 1.10. Explain what is wrong in the following incorrect proof by induction which shows that for any number of people, all of them are of the same age. It is based on [35, 4]. (Note, you will study induction in Chapter 6, but for now, you can follow this exercise without much knowledge of induction.)

1. If we consider one person, then, that person is the same age as itself.

2. Assume the property holds for k. That is, for any k persons we consider, they all are of the same age. We call this the induction hypothesis IH.

3. Now we prove that for any $k + 1$ persons, they all are of the same age. Call the people $P_1, P_2, \cdots, P_k, P_{k+1}$. By IH, all people $P_1, P_2, \cdots, P_k$ are of the same age and also all of the people $P_2, \cdots, P_k, P_{k+1}$ are of the same age. Hence, all the people in $P_1, P_2, \cdots, P_k, P_{k+1}$ are of the same age.

Hence, no matter which number n of people we consider, all n persons are of the same age.

1.3 The fear of the infinite

Going back to your school days, after studying elementary algebra, you are introduced to geometry (the study of shapes and their properties) and trigonometry (the study of side lengths and angles of triangles) and then to more algebra. Only after this, you are introduced to a pre-calculus course which combines advanced algebra and geometry with trigonometry. After all this, you are introduced to calculus, which is considered to be the pinnacle of high-school mathematics.

The discovery of the *calculus* was the defining moment in the birth of modern mathematics in the 17th century. Calculus (originally called *infinitesimal calculus*) is the mathematical study of continuous change. An infinitesimal quantity is a very very small quantity but which is not 0. Even though infinitesimals were sometimes controversial and even faced a ban on religious grounds by clerics in Rome, their use became prominent in the 17th century when for example, it was postulated that:

- a curved line is made of infinitely small straight line segments, and that

- quantities that differ by an infinitely small quantity can be considered equal.

Although our ancient predecessors had done considerable work on arithmetic, algebra, geometry and trigonometry, they avoided the *limit* of infinite values as in the curved line above and the infinitesimal in general (as in the infinitely small quantity above that differentiates two quantities). The reasons for this are the problems that they faced when reasoning about the concepts of limits and infinitesimals.

> *Infinitesimals have a long and colourful history. They make an early appearance in the mathematics of the Greek atomist philosopher Democritus (c. 450 B.C.E.), only to be banished by the mathematician Eudoxus (c. 350 B.C.E.) in what was to become official "Euclidean" mathematics. Taking the somewhat obscure form of "indivisibles," they reappear in the mathematics of the late middle ages and later played an important role in the development of the calculus. Their doubtful logical status led in the nineteenth century to their abandonment and replacement by the limit concept. In recent years, however, the concept of infinitesimal has been refounded on a rigorous basis.*
>
> Bell [2]

Let us now see how limits and infinitesimals reflect in the calculus and why there might be a problem with the calculus. Consider the function $y = f(x) = x^2$ and the problem of finding its derivative at $x = 2$, $f'(2)$. The process for finding this derivative is to consider first the difference quotient

$$\frac{\Delta y}{\Delta x} = \frac{f(x) - f(2)}{x - 2} = \frac{x^2 - 2^2}{x - 2}.$$

We evaluate this quotient as follows:

$$\frac{x^2 - 4}{x - 2} = \frac{(x + 2)(x - 2)}{x - 2} = x + 2,$$

where the last step is valid only if $x \neq 2$. But then we "take the limit" at $x = 2$ by substituting 2 for x to get 4. Since we are only able to conclude that the difference quotient is equal to $x + 2$ on the assumption that $x \neq 2$, we appear to have taken an illegal step.

Calculus textbooks justify this last step by saying that we are not *evaluating* the difference quotient at $x = 2$, but taking its limit as $x \to 2$, and so we write

$$\frac{dy}{dx} = \lim_{x \to 2} \frac{\Delta y}{\Delta x},$$

and that all this really means is that if x is *near* 2, then the difference quotient is *near* 4. And elsewhere in these textbooks, we are given a number of rules for evaluating limits which can be used to make the difference quotient as close to 4 as we please by taking x close enough to 2. But what really justifies these rules about limits? Is this reasoning careful enough to satisfy the requirements of mathematics?

In the next chapter, we will see that it was arguments like these which involve limits and infinites that led the Greeks to avoid the use of the infinitesimal. The Greeks concentrated instead on geometric methods in order to reason about quadratures such as the process of constructing a square with an area equal to that of a given geometric figure. But, in the 17th century, despite the admiration for the Greek's geometric methods and their continuous application to reason about quadratures, fear of the infinite was abandoned. For example, Johannes Kepler[5] used infinitesimals to calculate the area of an ellipse and viewed the circumference of a circle as an infinite sided regular polygon. As can be seen in Figure 1.1, Kepler noted that for a circle of radius a and an ellipse of radiuses a and b:

- The ratio of each vertical line within the circle to the vertical line within the ellipse is a/b.

- The area of each of the circle/ellipse is the infinite sum of vertical lines contained in the circle/ellipse.

Hence the ratio of the area of the circle to that of the ellipse is also a/b. Since we know that the area of the circle is πa^2, we conclude that the area of the ellipse $= (\pi a^2) \times (b/a) = \pi ab$.

Further uses of the infinitesimal as in Kepler's method were later developed and used to calculate more advanced tangents and quadratures and to relate tangents with quadratures. But there was still a lack of a formal development of the concepts and rules in question. Part of this was filled independently by the work of Newton[6] and Leibniz[7] on the calculus. Infinite series and infinite sequences were crucial for their calculus methods and they

---

[5]Johannes Kepler was a German astronomer, mathematician and astrologer who discovered the elliptic movements of the planets around the sun and provided the foundations for Newton's theory of gravity. His contributions are numerous but even if you don't know of any, you may well have heard of his famous conjecture concerning the most dense packing of spheres which was again listed by Hilbert as part of his 18th problem in the famous list of his 23 open problems in mathematics.

[6]Isaac Newton, 1642-1727, was a "natural philosopher." He invented the calculus when he was a student. He also introduced the idea for universal gravity and successfully used it to predict the orbits of the planets of the solar system that were known at the time; see [32]. He also studied light, and was the first to use a prism to break white light down into its various colours; see [33].

[7]Gottfried Wilhelm Leibniz, 1646-1716, was a German philosopher. He invented the calculus independently of Newton and introduced the notation $\frac{dy}{dx}$, which we still use today.

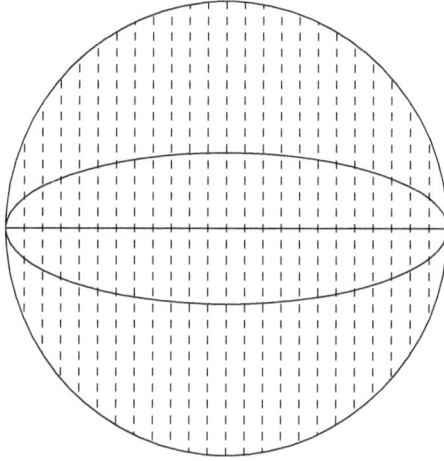

Figure 1.1: A circle of radius a and an ellipse of radiuses a and b. The area of each of the 2 shapes is the infinite sum of the dashed lines contained in it.

in this way legitimised the use of infinite processes. Although Newton and Leibniz's work had introduced differentiation and its reverse (integration), they were still short of formulating the definition of limit. This despite the superiority of the symbolic notation that Leibniz introduced. As a matter of fact, when Newton and Leibniz introduced the calculus in the late 17th century, they did not have a good explanation of what they were doing. Newton, for example, used explanations that fall short of the normal requirements of a mathematical definition when he described what we call

$$\lim_{x \to 2} \frac{\Delta y}{\Delta x}$$

as its "ultimate value," or its value at "the instant of its disappearance."

In fact, in the 1730s, the inadequacy of this definition was brought home by Berkeley[8] in [15]. The Anglican bishop Berkeley had become disturbed for the soul of Edmond Halley[9] who was proclaiming himself an atheist on the basis of Newton's physics. Berkeley happened to be an excellent satirist, and in [15], he jumped on Newton's explanation of the limit of the difference quotient as its value at "the instant of its disappearance" by asking why it should not be called "the ghost of a departed quantity." His idea was to

---

[8]George Berkeley, 1685-1753, was an Anglican bishop and philosopher.
[9]Edmond Halley, 1656-1742, the astronomer for whom Halley's Comet is named.

suggest that anybody who was prepared to accept Newton's calculus should also have no trouble accepting theology.

Despite all this criticism, infinite processes and calculus as developed by Newton and Leibniz continued to be very much in use. New developments by Euler on the generalisations of functions were followed by attempts at explaining the notion of limit. This followed in the 19th century by Cauchy's work that successfully combined the new ideas of function and limit in order to give a rigorous formulation of the calculus explaining convergence, divergence and continuous functions. More rigour was put into explaining the calculus, its notion of limit and continuous function and even the core on which it is based (the real numbers). This theoretical work is called *analysis*, or the *arithmetisation of analysis*. In fact, all analysis can be derived from a set of axioms about the real numbers.

Some of you may wonder why this theory is not taught right in your calculus courses. The reason it is not is that understanding this theory requires more mathematical sophistication than is required for the understanding and use of the tools of the calculus itself. This is why a mathematical analysis course comes after calculus courses. So, the order in which you are exposed to different branches of mathematics is as follows:

arithmetic → elementary algebra → geometry and trigonometry

→ pre-calculus → calculus → analysis

It is claimed sometimes that the arithmetisation of analysis is one of the greatest intellectual achievements of human history. However, it does not seem possible to appreciate its greatness just by looking at the latest version of the theory itself. It is necessary to consider the entire process by which this theory developed. This process involved major changes in the way mathematicians looked at their subject, at the sorts of things they studied, and even at how any mathematics can be justified. It really began over two thousand years ago in ancient Greece. In order to fully understand and appreciate analysis, we intend to begin with the ideas of the ancient Greeks. Some ideas developed then turned out to be important for the history of our subject.

1.3.1   Exercises

Exercise 1.11. Let the x-y axis given below.

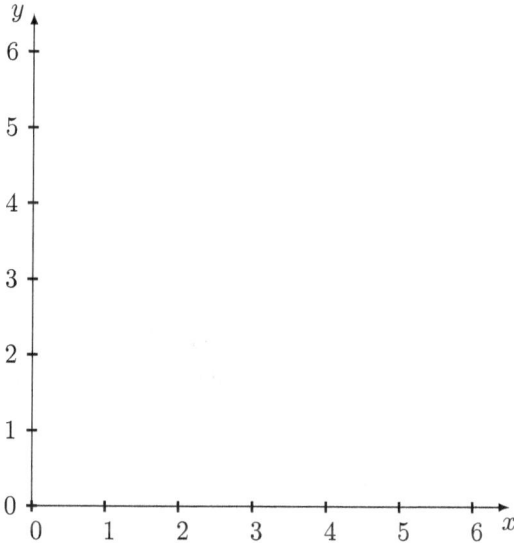

For each $1 \leq x \leq 6$ draw the line that connects the points $(0, x)$ and $(7 - x, 0)$. Now, draw new x-y axis and for each $1 \leq x \leq 6$ draw the line that connects the points $(0, x - 0.5)$ and $(7 - x, 0)$ and also the line that connects the points $(0, x)$ and $(6.5 - x, 0)$. What do you notice about the outer boundary of the shape between $(0, 6)$ and $(6, 0)$ and how does this boundary differ in each of the two figures you drew.

Exercise 1.12. Find the rectangle with maximum area whose perimeter is P.

Exercise 1.13. The Dido problem is one of the oldest problems in the calculus of variations. Dido (also known as Elissar or Elissa) was a Phoenician princess. Her brother did not want to share their father's kingdom with her, so he killed her husband and wanted to kill her too. To escape, she took her fortune and her fleet and went across the sea to modern day Tunisia. There, she built the city of Carthage. To get the land she needed to build Carthage, she persuaded the local king to sell her a piece of land whose area is that of a bull's hide. To get maximum land, Dido cut the Bull's hide into thin strips and joined them together. Amazingly, Dido-Elissar knew that the maximum area bounded by a curve is a circle and hence she gathered the bull's strips to be circular in shape. This result was only proven by Jakob Steiner in the nineteenth century. You are not asked to prove this result in this exercise. Instead, you are asked to show that the following two statements are equivalent:

1. The circle has the largest area of all planar shapes whose perimeter is P.

2. The circle has the lowest parameter of all the planar shapes whose area is A.

Exercise 1.14. In the Rhind mathematical papyrus, it seems that the ancient Egyptians knew that $1 + 7 + 7^2 + 7^3 + 7^4 = \frac{7^5 - 1}{7 - 1}$ and that $1 + 2 + 2^2 + 2^3 + \cdots + 2^{n-1} = 2^n - 1$. Can you prove these two formulas?

Exercise 1.15. In an Egyptian document written in 1850 B.C. (referred to as the Moscow Papyrus), we are told how to calculate the volume of a truncated pyramid:

> **Moscow Papyrus showing Ancient Egyptians calculations for the volume of a truncated pyramid:**
> If you are told: a truncated pyramid of 6 for the vertical height by 4 on the base by 2 on the top. You are to square this 4, result 16. You are to double 4, result 8. You are to square 2, result 4, You are to add the 16, the 8, and the 4, result 28. You are to take a third of 6, result 2. You are to take 28 twice, result 56. See, it is 56. You will find it right.

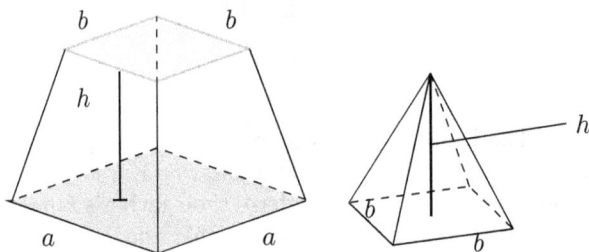

According to Devlin [12], it is clear that the particular numbers chosen were purely as examples, and could be replaced by other numbers. In modern notation, we would use an algebraic formula: the volume of the truncated pyramid with height h, base square of side a and top square of side b is given by:

$$V = \frac{1}{3}h(a^2 + ab + b^2).$$

Researchers have been guessing as to whether the Egyptians could have known how to calculate the volume of the truncated pyramid and whether they did discover the formula $V = 1/3h(a^2+ab+b^2)$. One such researcher [40] proposes that the Egyptians were aware of formulas and that they used formulas in their calculations. These include:

(a) $a^2 + ab + b^2 = \dfrac{a^3 - b^3}{a - b}$.

(b) $a^2 + ab + b^2 = a^3 - b^3$ when a and b are two consecutive integers.

1. Show that the numbers used in the Egyptian Moscow Papyrus satisfy the formula for the volume of the truncated pyramid with height h, base square of side a and top square of side b given above.

2. Show the formulas (a) and (b).

3. [40] also proposed that the Egyptians discovered from a mud model of a pyramid on a square base that they reached half the height of the pyramid when the base edge had shrunk to half its original size. Hence, in the figure above, assume the pyramid was the entire rightside put on top of the leftside, then the entire height 2h is halved at the point when the middle base edge b is half base edge a. The larger pyramid (rightside on top of leftside in above picture) is twice as high, twice as broad and twice as long than the smaller pyramid (that on the right-side of the picture above). Hence, [40] states that it appeared that the Egyptians knew that the volume resp. area of the larger pyramid was $2 \times 2 \times 2$ the volume resp. area of the smaller pyramid. In proportion times, the larger pyramid is 8, the smaller pyramid is 1 and the truncated pyramid (the rightside alone) is 7. Note that the truncated pyramid proportion $7 = 1 + 2 + 4 = 1 + 2 + 2^2 = 2^3 - 1$. Based the various discussions so far, look at the picture below which shows only one side of a pyramid with square bases of side lengths ranging from 1 to 6 cm and where the heights increase from one stage to another by 3 cm. Calculate the volumes of the various pyramids and truncated pyramids in the picture.

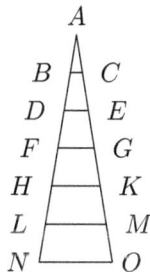

Chapter 2

Greek Background

In this chapter, we briefly discuss the development of proofs in mathematics. We start by emphasising the importance of the rhetorical style of arguments in Ancient Greece and move to discuss early Greek mathematics concentrating on the theory of odd and even numbers as developed by the Pythagorean school and documented in Euclid's *Elements*. On our way, we need to mention the motion paradoxes developed by Zeno which contributed to the Greeks fear of the infinite and their attempt to avoid infinite processes. Since Euclid's Elements developed mathematics in geometric terms and anything not expressible in such terms was excluded,[1] we look at proofs as diagrams. We follow Knorr's suggestion that the original proofs were proofs as diagrams using pebble diagrams and present the theory of even/odd numbers and pebble diagram proofs. We do this until we hit the point in history where proofs by diagrams or pebbles were replaced by proofs as sequences of statements. According to Knorr, this occurred at the same time as the incommensurability of the side and diagonal of a square was discovered which implied the existence of non rational numbers.

2.1   Style of Argument

What we accept today as mathematical proof is the result of a long line of development that goes back to ancient cultures including the ancient Greeks in Athens.

---

[1]The reason for this is that the Greeks used arithmetic and numerical forms to represent numbers (which they took as whole units), but they did not develop numerical formulations nor an arithmetic for any other quantities, and only used Geometry to describe these quantities.

Today, becoming a lawyer or advocate involves studying the law, including the statutes and legal precedents. In the ancient world of Greece and Rome, however, it did not mean this. Instead, it meant studying *rhetoric*, or how to argue effectively. In the fifth century B.C.E., there arose a class of men called *sophists*, who made their living teaching rhetoric to prospective advocates. Since ancient Athens was a very litigious society, a good teacher of rhetoric could make a good living at it. One of the most famous of these sophists was Protagoras.[2] Protagoras was, in fact, one of the first ancient Greeks who made his living by teaching rhetoric, and apparently he charged very high fees for his course. He became known for the view that there is no absolute truth, but that it is possible to make a reasonable case for any side of any issue. This characterisation of his views may not be completely accurate, since much of what we know of him comes from figures unfriendly to his position.

What is clear is that one had to be sharp to formulate an argument that was immune to an attack by a sophist. An example which illustrates the kind of arguments that occurred in court in Athens in the fifth century B.C.E. is given in Figure 2.1.

2.1.1   Topics for Discussion

Topic 2.1. In 399 B.C., the famous philosopher and mathematician Socrates who was 70 years old was accused of impiety and corrupting the youth by a certain unknown Meletus and taken to face trial. At the end of the trial, Socrates lost and was put to death. Read the Euthyphro dialogue [34] which describes Socrates dialogue with Euthyphro on the definition of piety that Socrates hoped to find in order to defend himself. Give a number of definitions of piety provided by Euthyphro and describe how Socrates disqualifies each of them. Say whether in the end, Socrates obtains the desired definition of piety.

Topic 2.2. In the apology dialogue of Plato [34], Socrates is defending himself against the accusations of Meletus. Socrates, an honest direct man who describes himself to "have a benevolent habit of pouring out myself to everybody, and would even pay for a listener" defends himself in an emotional and provocative manner. Here, Socrates, the "father of the dialogue" is in top form and his defence is a must read. In this apology dialogue, Socrates makes interesting observations that include him classifying himself as wiser

---

[2]Protagoras of Abdera, c. 590–c. 520 B.C.E., ancient Greek sophist.

[3]De Long's added a footnote that stated: "*I have altered this story in inessential ways in order to bring out its logical form. For those who are interested in looking up the original story, see [37, pp. 404ff].*"

... We do know that there existed a class of teachers who came to be known as sophists. These sophists would travel, much like wandering minstrels, and for a fee would teach their students how to speak persuasively on many different kinds of topics. Sophists were also prepared to defeat any opponent in a public argument. The competitiveness of such a spectacle must have been very keen and the arguments often dramatic, so that we can understand why the arrival of an important sophist in town was the occasion of much excitement and why sophists were often able to command large fees.

Protagoras, often considered to be the greatest of the sophists, would no doubt be thought a great thinker if his works had survived. He is best known for his saying that "man is the measure of all things" and his humanism probably exhibited itself in ways we consider uniquely modern. The following ancient story about him, although probably apocryphal, indicates the kind of verbal pyrotechnics of which the sophists were capable. Protagoras had contracted to teach Euathlus rhetoric so that he could become a lawyer. Euathlus initially paid only half of the large fee, and they agreed that the second instalment should be paid after Euathlus had won his first case in court. Euathlus, however, delayed going into practice for quite some time. Protagoras, worrying about his reputation as well as wanting the money, decided to sue. In court Protagoras argued to the jury:

> Euathlus maintains he should not pay me but this is absurd. For suppose he wins this case. Since this is his maiden appearance in court he then ought to pay me because he won his first case. On the other hand, suppose he loses the case. Then he ought to pay me by the judgement of the court. Since he must either win or lose the case he must pay me.

Euathlus had been a good student and was able to answer Protagoras' argument with a similar one of his own:

> Protagoras maintains that I should pay him but it is this which is absurd. For suppose he wins this case. Since I will not have won my first case I do not need to pay him according to our agreement. On the other hand, suppose he loses the case. Then I do not have to pay him by judgement of the court. Since he must either win or lose the case I do not have to pay him.[3]

Delong [11, p. 10]

Figure 2.1: Example of arguments that occurred in court in Athens

than other men only in that "he knows that he knows nothing" and questioning whether Meletus had devised an intelligence test to identify logical contradictions. Despite his extreme wit, Socrates was still condemned to death. Read the apology (defence) dialogue in [34] and give the arguments that may have led to Socrates losing the case.

Topic 2.3. In Plato's dialogue Meno (see [23]), we learn that in order to teach Meno about virtue, Socrates is trying to convince him of something about the nature of mathematical knowledge, and for that purpose, Socrates questions one of Meno's slave boys about some geometric problems. Read the dialogue given in Appendix A.1 and explain why you think Socrates is questioning the boy on geometric problems. Then, give these geometric problems and their solutions.

2.2   Early Greek Mathematics

By the fifth century B.C.E., there was already a tradition of philosophy in the Greek world, and some of this philosophy dealt with what we now call mathematics. This tradition started in Greek colonies in Asia Minor (now Turkey). One of the first was Thales of Miletus,[4] who is credited with the first proofs in geometry. However, his proofs have been lost, and all we have left are descriptions of them which do not really tell us what they were like.

More famous today was Pythagoras of Samos,[5] who founded a school which survived for over a century. The school was actually a kind of brotherhood whose members were required to keep many of their results secret, and for this reason it is difficult to separate results of Pythagoras himself from those of his followers. One important thing that Pythagoras taught is that *number is everything*. Members of this brotherhood were probably the developers of the oldest deductive theory of mathematics which still exists, the theory of *odd and even numbers*, which can be found in Propositions 21–34 of Book IX of Euclid's *Elements* [20].[6]

In the sixth century B.C.E., Zeno of Elea,[7] who was not a mathematician, devised some paradoxical arguments against the possibility of motion. Since the calculus was developed partly to deal with motion, these paradoxical arguments are important for the foundations of analysis. Four of the most important of these are preserved by Aristotle in his *Physics* [1]:

---

[4]C. 624–546 B.C.E.

[5]Flourished c. 570 B.C.E.

[6]The definitions are at the beginning of Book VII. Euclid, 325–265 B.C.E., was a mathematician who spent his career in Alexandria, Egypt.

[7]C. 490–430 B.C.E.

1. *The Dichotomy Paradox.* Anything in motion, must get halfway first to its destination. For example, to leave the room, you first have to get halfway to the door, then you have to get halfway from that point to the door, etc. No matter how close you are to the door, you have to go half the remaining distance before proceeding. Hence, there is no finite motion because the above process of always going half way while in motion is infinite.

> There is no motion, because what moves must arrive at the middle of its course before it reaches the end.

2. *The Achilles Paradox.*

> The slower in a race will never be overtaken by the quicker; because the pursuer must first reach the starting point of the pursued, so that the slower must always be some distance ahead.

If the fastest runner in the world Achilles is at A and the tortoise is at B (which is ahead of A) then Achilles can never reach the tortoise since when Achilles reaches B, the tortoise will have moved to C and when Achilles moves to C, the tortoise will have moved to D, etc., ad infinitum. So Achilles has to complete an infinite series of discrete separate tasks before overtaking the tortoise which is impossible.[8]

A		B	C	D	E

3. *The Arrow Paradox.* A thing is at rest when occupying its own space at a given time, as the arrow does at every instant of its alleged flight.

> The flying arrow is at rest

Unlike the Achilles and Tortoise paradox, this paradox assumes that space and time are not infinitely divisible and hence there is no smallest point of space or smallest instant of time.

---

[8]The solution here is to note that although time and space are infinitely divisible, they are not necessarily infinitely divided in practice. Achilles can take a finite number of steps to reach the tortoise.

4. *The Stadium Paradox.* This argument concerns 3 rows of bodies equal in number, two of which pass one another on a race course as they move in opposite directions. This is illustrated in the passage of the bodies $B_1 \cdots B_6$ and $C_1 \cdots C_6$ from

$$
\begin{array}{cccccc}
A & A & A & A & A & A \\
\end{array}
$$
$$
\begin{array}{cccccc}
B_6 & B_5 & B_4 & B_3 & B_2 & B_1 \\
\end{array}
$$
$$
\begin{array}{cccccc}
C_1 & C_2 & C_3 & C_4 & C_5 & C_6 \\
\end{array}
$$

to

$$
\begin{array}{cccccc}
A & A & A & A & A & A \\
B_6 & B_5 & B_4 & B_3 & B_2 & B_1 \\
C_1 & C_2 & C_3 & C_4 & C_5 & C_6 \\
\end{array}
$$

The A's are stationary. As we see, B_1 will have passed alongside 3 A's and 6 C's during the same time and without changing. The same would be true of C_1 which will have passed alongside 3 A's and 6 B's during the same time and without changing. Zeno's idea was to show that a time is equal to its double.

> Half a given time is equal to double that time.

We will need to refer back to these arguments as we proceed. For now, note that these paradoxes were faced even when opposite hypothesis were assumed (as in the time and space being/or not being infinitely divisible). This was behind the fact that the Greeks tried to avoid infinite processes as is the case in Euclid's *Elements*.

> Impossibility arguments were behind the fact that the Greeks tried to avoid infinite processes.

We will revisit Zeno's paradoxes again in Section 10.5.

2.2.1   Exercises

Exercise 2.1. Apply the Dichotomy paradox to a runner who has to run 100 km stating the middle distances that the runner has to reach. According to Dichotomy, will the runner reach its destination?

Exercise 2.2. Apply the Dichotomy paradox to the frog who has to reach the pond. According to Dichotomy, will the frog reach the pond?

Exercise 2.3. Apply the Achilles paradox to Achilles and the tortoise assuming that to start with, Achilles is at point 0 whereas the tortoise is 100 meters ahead of Achilles and that Achilles is ten times faster than the tortoise.

2.3 Proofs as Diagrams

Many early ancient Greek proofs are based on looking at diagrams, seeing the result proved from the diagrams, and seeing that the diagrams properly represent all the possibilities. The following examples show how this can work in some results from elementary geometry.

Example 2.3.1 (The sum of the angles of a triangle). *The diagram below shows, if looked at the right way, that the angles of a triangle add up to two right angles.*

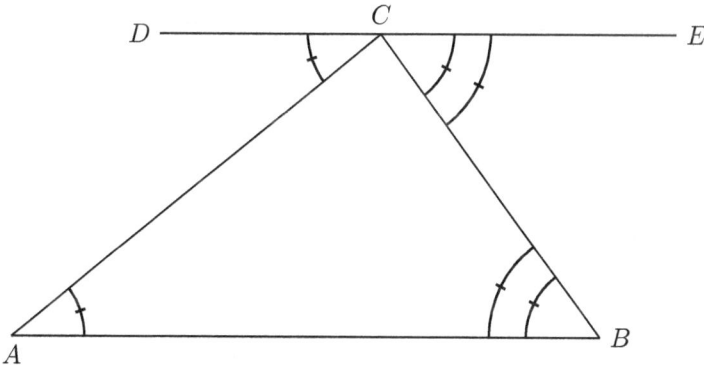

The key idea is that DE is constructed to be parallel to AB (and to pass through C) and this means that $\angle DCA = \angle CAB$ and $\angle ABC = \angle BCE$. Since $\angle DCA + \angle ACB + \angle BCE = 180°$, we get $\angle CAB + \angle ACB + \angle ABC = 180° = 90° \times 2$. □

Example 2.3.2 (The area of a parallelogram). *Similarly, the diagram below can be used to "see" the formula for the area of a parallelogram with sides of length a and c.*

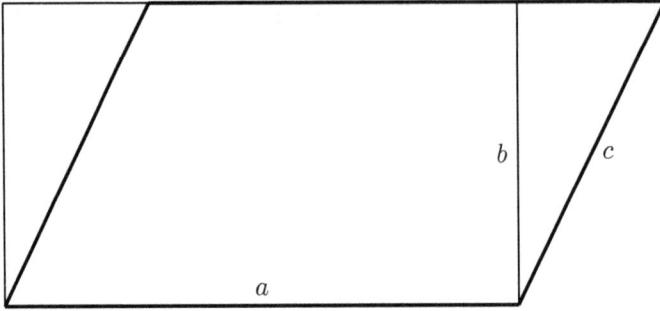

It is clear from this diagram that the area of the parallelogram is the same as that of the rectangle whose sides are of length b and a (where b is the length of the orthogonal to the side of length a).

Example 2.3.3 (The area of a triangle is one-half that of the parallelogram). *The following diagram shows that the area of a triangle is one-half that of the parallelogram.*

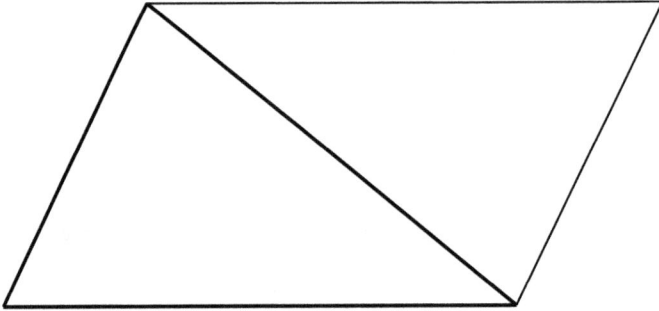

Example 2.3.4 (A kind of proof of the Pythagorean Theorem). *The following diagram is a kind of proof of the Pythagorean Theorem. For example, on the left hand side of this diagram, if you add the areas of the various parts you get* $2ab+c^2 = (a+b)^2$, *and since by the righthand side* $(a+b)^2 = a^2+b^2+2ab$, *we get* $c^2 = a^2 + b^2$.

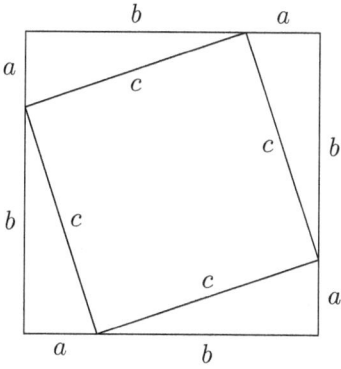

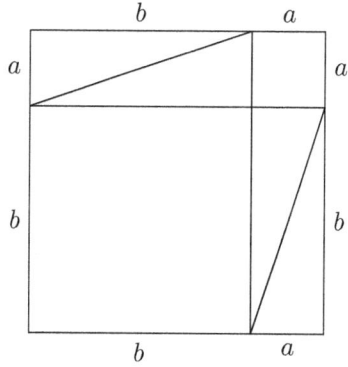

2.3.1 Exercises

Exercise 2.4. Find all the angles of the triangle in the following diagram:

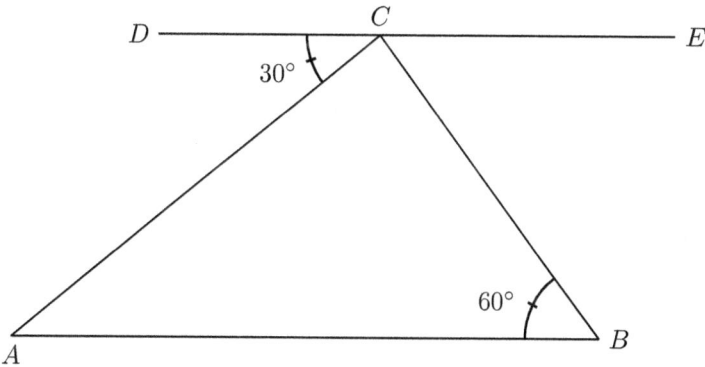

Exercise 2.5. In the following diagram, show that $\angle ACB = \angle ADB = 90°$ (this theorem is attributed to the Greek Mathematician Thales of of Miletus c. 624 – c. 547 BC).

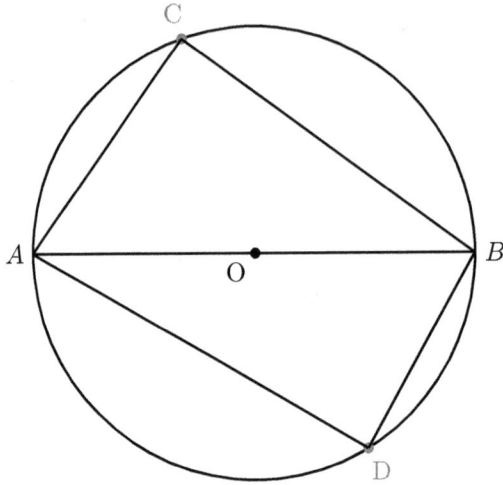

Exercise 2.6. In the following diagram, show that $\angle AOB = 2\angle ACB$.

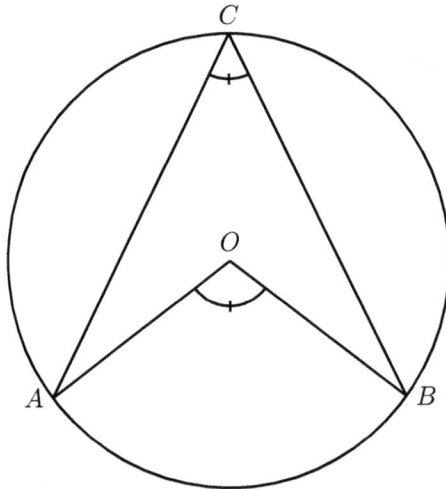

Exercise 2.7. Let square $ABCD$ inscribed in the circle $ABCD$. Show that the area of the square is greater than half the area of the circle.

[Hint: draw the tangents to the circle at each of the points A, B, C and D, and let $EFGH$ be the square formed by the intersection points of these tangents. Then, show that the area of $EFGH$ is twice that of $ABCD$ and note that the circle $ABCD$ encloses the square $ABCD$ and is enclosed in the square $EFGH$.]

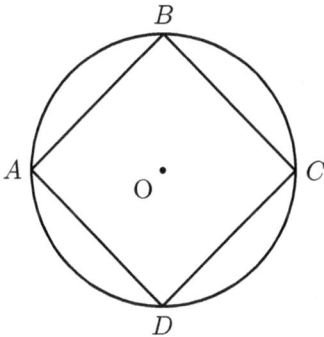

Exercise 2.8. Let square $ABCD$ inscribed in the circle $ABCD$ and let I be the mid points of the arc AB. Show that the area of the triangle IAB is greater than half the area of the part of the circle that encloses it.

[Hint: draw the tangent to the circle at point I and let the tangent meet BC and AD at L and K respectively. Then, show that the area of the parallelogram $ABLK$ is twice that of the triangle IAB and note that the part of the circle in question encloses triangle IAB and is enclosed in the parallelogram $ABLK$.]

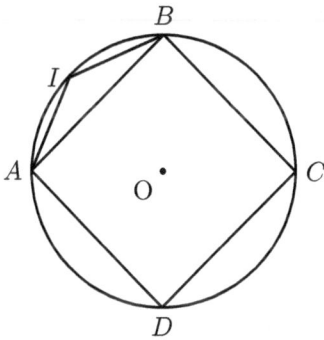

2.4 The Theory of Even/Odd Numbers and Pebble Diagram Proofs

As remarked in § 2.2, the theory of even and odd numbers is thought to be the oldest deductive theory of which we have a record. The record we have today is found in Euclid's *Elements* [20]. Euclid's *Elements* developed mathematics in geometric terms and anything not expressible in such terms was excluded. Geometry could accommodate the whole numbers and their

ratios as well as irrational magnitudes. As an example, take the spiral of
Theodorus of Cyrene shown in Figure 2.2 which established that the square
roots of non square integers from 3 to 17 are irrationals. In that figure, we
see that there are 16 right triangles, each of which has a side of length 1. The
smallest triangle has a second side of length 1 and the length of the third
side is $\sqrt{2}$ which was known to be irrational before Theodorus. Theodorus
however is assumed to have proven that the remaining square roots of non
square integers from 3 to 17 are irrationals. We will see in the next section
how we can show that $\sqrt{2}$ is irrational. For now, remember that in a right
triangle of sides of lengths a resp. b which surround the right angle, the
length of the third side is $\sqrt{a^2 + b^2}$.

In a right triangle of sides of lengths a resp. b which surround
the right angle, the length of the third side is $c = \sqrt{a^2 + b^2}$.

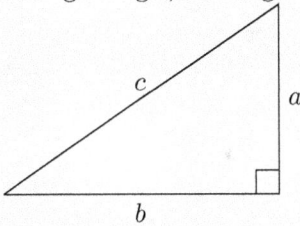

In the manuscripts of the *Elements* which have survived, the diagrams
there are all line segments. But Knorr [26, Chapter V, Section III] suggests
that the original proofs were proofs as diagrams using *pebble diagrams*. It
is known that the ancient Greeks did arithmetic by counting with pebbles,
and pebble diagrams give these calculations by representing the pebbles by
using small circles.

Let us see how pebble diagrams can be used to illustrate proofs. Consider
the key definitions (6. and 7.) from Book VII of the *Elements* and the key
Propositions 21–34 of Book IX of the *Elements*. The definitions (6. and 7.)
and Proposition 21 are stated as follows:

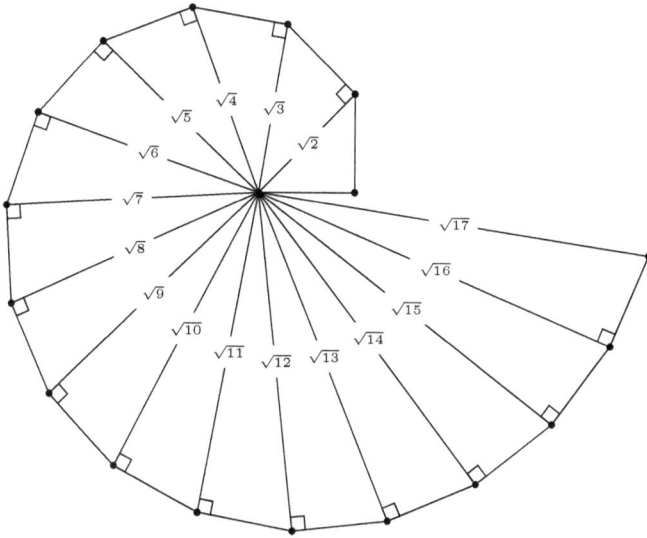

Figure 2.2: Spiral of Theodorus of Cyrene

DEFINITIONS (6. AND 7.) FROM BOOK VII OF THE *Elements*
6. An *even number* is that which is divisible into two equal parts.

7. An *odd number* is that which is not divisible into two equal parts, or that which differs by a unit from an even number.

PROPOSITION 21 OF BOOK IX OF THE *Elements*.
If as many even numbers as we please be added together, the sum is even.
For let as many even numbers as we please, AB, BC, CD, DE, be added together; I say that the sum AE is even.
For, since each of the numbers AB, BC, CD, DE is even, therefore each has a half part;   [VII. Def. 6]
so that the sum AE also has a half part. But an even number is that which is divisible into two equal parts;   [id.]
therefore AE is even.   Q. E. D.

The diagram that goes with Proposition 21 in the *Elements* is as follows:

A   B   C   D   E

Now consider the "pebble" diagram of Figure 2.3 that Knorr gives to illustrate this proof.

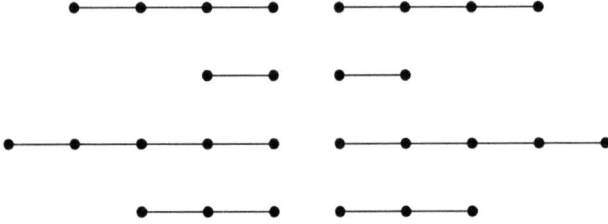

Figure 2.3: Diagram for Euclid IX 21 by Knorr

Note that this diagram makes it possible to visualise the proof stated in words. Of course, the notations "AB", "BC', etc. are no longer relevant, and we should re-read the proof without these names, reading "the sum" for "AE". Instead, we have an example consisting of four even numbers, namely 8, 4, 10, and 6. The key to reasoning about diagrams this way is that which particular numbers are given in the diagram and how many even numbers there are is not the important thing: it is enough to see that a similar diagram could (at least in principle) be created for any number of even numbers.

Now consider Proposition 22:

> PROPOSITION 22 OF BOOK IX OF THE *Elements*.
> *If as many odd numbers as we please be added together, and their multitude be even, then the sum is even.*
> For let as many odd numbers as we please, AB, BC, CD, DE, even in multitude, be added together;
> I say that the whole AE is even.
> For, since each of the numbers AB, BC, CD, DE is odd, if a unit is subtracted from each, each of the remainders is even,
>   [VII. Def. 7]
> so that the sum of them will be even.   [IX. 21]
> But the multitude of the units is also even. Therefore the sum AE is even.   [IX. 21]
>   Q. E. D.

The diagram that goes with Euclid's original is as follows:

Knorr illustrates this theorem with a pair of diagrams: see Figure 2.4

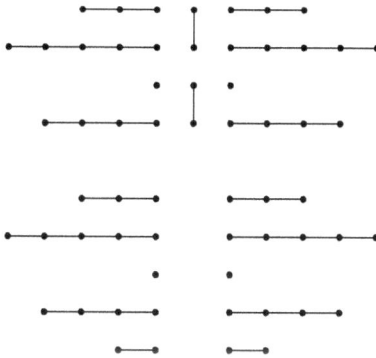

Figure 2.4: Diagram for Euclid IX 22 by Knorr

The first four lines represent the first of the two diagrams, and the next five lines represent the second. In the first diagram, again, we have one example, consisting of the four numbers 7, 11, 3, and 9. The numbers are displayed so that the unit to be subtracted from each of them is in the middle, and the vertical lines here show that the number of numbers is even. In the next five lines, the parts of each of the four original numbers which is left after the unit is subtracted are shown in the first four lines, and the subtracted units are displayed horizontally (instead of vertically) in the last line. This leads to five numbers each of which is even, so the total sum is even.

Below, we give a couple more theorems and their proofs by pebble diagrams. These theorems will be needed in the proof of incommensurability that will follow.

Theorem 2.4.1. *The square of every even number is a multiple of 4.*

Proof. The proof can be seen from Figure 2.5. □

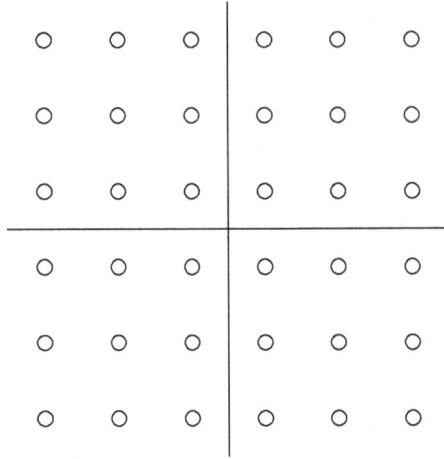

Figure 2.5: Diagram proof of Theorem 2.4.1

It should be clear that a diagram like this can be created for any even square.

In modern notation, we see this result from $(2n)^2 = 4n^2$.

Theorem 2.4.2. *The square of every odd number is one more than a multiple of 4.*

Proof. The proof can be seen in the same way from Figure 2.6.   □

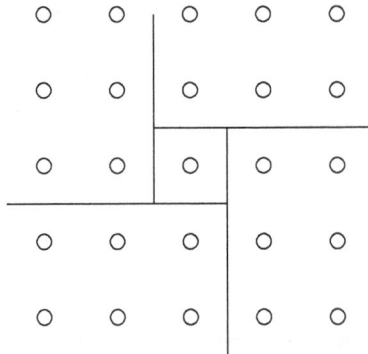

Figure 2.6: Diagram proof of Theorem 2.4.2

Note that this actually proves more. Each of the four rectangular arrays has one side one unit more than the other, and so one of them must be even. This results in Corollary 2.4.3.

Corollary 2.4.3. *The square of every odd number is one more than a multiple of 8.*

In modern notation, we see this from $(2n+1)^2 = 4n^2 + 4n + 1 = 4n(n+1) + 1$. Since at least one of n and $n+1$ must be even, $4n(n+1)$ is a multiple of 8.

From this, we get the following:

Corollary 2.4.4. *It is impossible for the square of an even number to be the sum of the squares of two odd numbers.*

Proof. By Corollary 2.4.3, the sum of the squares of two odd numbers is two more than a multiple of 8, which is never a multiple of 4. ☐

2.4.1   Exercises

Exercise 2.9. In Figure 2.2, we have assumed that for each of the 16 triangles, the outerside adjacent to the right angle is of length 1. We also assumed that for the smaller triangle, the second side adjacent to the right angle is also of length 1. Explain how we get the length of all remaining sides.

Exercise 2.10. Figure 2.6 illustrated Theorem 2.4.2 for the number 5 (i.e., by showing that $5^2 = 4 \times 6 + 1$). Can you similarly illustrate Theorem 2.4.2 for the number 7.

Exercise 2.11. Give a pebble proof of Corollary 2.4.3.

Exercise 2.12. Look up the proofs of Book IX Propositions 23–29 as stated below and construct pebble diagrams to illustrate the proofs in the style of Knorr.

23. If as many odd numbers as we please are added together, and their multitude is odd, then the sum is also odd.

24. If an even number is subtracted from an even number, then the remainder is even.

25. If an odd number is subtracted from an even number, then the remainder is odd.

26. If an odd number is subtracted from an odd number, then the remainder is even.

27. If an even number is subtracted from an odd number, then the remainder is odd.

28. If an odd number is multiplied by an even number, then the product is even.

29. If an odd number is multiplied by an odd number, then the product is odd.

Exercise 2.13. Look up the remaining key Propositions 30–34 of Book IX and check their proofs.

2.5 Proofs as Sequences of Statements

Most proofs today are not proofs by diagrams or using pebbles, and neither are those of later Greek mathematics. Rather, they are sequences of statements, each one justified by a reason. How did this change come about? One opinion, that of Wilbur R. Knorr in [26], is that the change occurred at the same time as the *incommensurability* of the side and diagonal of a square was discovered.

According to Webster's dictionary, *commensurability* is "divisibility without remainder by a common unit." Hence 6 and 9 are commensurable (since they are both divisible by 3). Saying that a and b are commensurable if and only if they are divisible without remainder by a common unit, is equivalent to saying that there is a common measure c that exactly measures both a and b. Note that there is a common unit 3 which measures both 6 and 9. In fact, 6 is twice 3 and 9 is three times 3. That is:

Definition 2.5.1. *We say that a and b are* commensurable *if and only if there is a common measure c that exactly measures both a and b. I.e., there there are a common measure c and integer numbers m and n such that $a = mc$ and $b = nc$ and hence $\frac{a}{b} = \frac{m}{n}$.*
We say that a and b are incommensurable *if and only if they are not commensurable.*

Let us see how these definitions are given in Euclid's Elements..

PROPOSITION 5. OF BOOK X OF THE *Elements*.
Commensurable magnitudes have to one another the ratio which a number has to a number.

PROPOSITION 6. OF BOOK X OF THE *Elements*.
If two magnitudes have to one another the ratio which a number has to a number, then the magnitudes are commensurable.

PROPOSITION 7. OF BOOK X OF THE *Elements*.
Incommensurable magnitudes do not have to one another the ratio which a number has to a number.

PROPOSITION 8. OF BOOK X OF THE *Elements*.
If two magnitudes do not have to one another the ratio which a number has to a number, then the magnitudes are incommensurable.

PROPOSITION 9. OF BOOK X OF THE *Elements*.
The squares on straight lines commensurable in length have to one another the ratio which a square number has to a square number; and squares which have to one another the ratio which a square number has to a square number also have their sides commensurable in length. But the squares on straight lines incommensurable in length do not have to one another the ratio which a square number has to a square number; and squares which do not have to one another the ratio which a square number has to a square number also do not have their sides commensurable in length either.

According to Knorr [26], attempts to find the unit which measures exactly the side and diagonal of a square led to the proof of the *incommensurability* of the side and diagonal of a square and it was discovered then that there is no unit which measures both the side and diagonal of a square. This explanation can be found in [38, pp. 163–168], and is based on the theory of even and odd numbers.

The historic change from proofs using diagrams/pebbles to proofs as sequences of statements occurred at the same time as the incommensurability of the side and diagonal of a square was discovered.

The key results needed for the proof of this incommensurability of the side and diagonal of a square (see Theorem 2.5.11 below) can be proved from diagrams. Those key results consist of Theorems 2.4.1, 2.4.2, 2.5.4, 2.5.5, 2.5.8, 2.5.9, 2.5.10 and Corollaries 2.4.3, 2.4.4, 2.5.6, 2.5.7. We already saw Theorems 2.4.1, 2.4.2 and Corollaries 2.4.3, 2.4.4 in the previous section. Here, we start briefly with these remaining theorems and lemmas before giving the incommensurability theorem.

Definition 2.5.2 (Pythagorean Triples). *Pythagorean triples are triples of positive (whole) numbers which can represent the lengths of two legs and the hypotenuse of a right triangle. In other words, a Pythagorean triple is a triple of positive integers (a, b, c) if and only if $a^2 + b^2 = c^2$. When we write Pythagorean triples, we will always write the largest, the length of the hypotenuse, in the last position.*

Example 2.5.3. *The following are all Pythagorean triples:*
$(3, 4, 5)$, $(6, 8, 10)$, $(5, 12, 13)$, $(9, 12, 15)$, $(8, 15, 17)$, $(12, 16, 20)$, $(15, 20, 25)$, $(7, 24, 25)$, $(10, 24, 26)$, $(20, 21, 29)$, $(18, 24, 30)$, $(16, 30, 34)$, $(21, 28, 35)$.

Theorem 2.5.4. *There is no pythagorean triple (a, b, c) in which c is even and both a and b are odd.*

Proof. By Corollary 2.4.4, an even square cannot be the sum of two odd squares. $\qquad\square$

In fact, we can tell more:

Theorem 2.5.5. *In a Pythagorean triple (a, b, c), if c is even, then both a and b are even.*

Proof. By Corollary 2.4.4, an even square cannot be the sum of two odd squares. If one is even (say a) and one is odd (b), we get that a multiple of 4 (c^2, by Theorem 2.4.1) is equal to one more than a multiple of 4 ($a^2 + b^2$, by Theorems 2.4.1 and 2.4.2). Absurd. $\qquad\square$

Note that this means the following:

Corollary 2.5.6. *In a Pythagorean triple (a, b, c), if c is even, then $(\frac{a}{2}, \frac{b}{2}, \frac{c}{2})$ is also a Pythagorean triple.*

For example, $(6, 8, 10)$ is a Pythagorean triple since

$$6^2 + 8^2 = 36 + 64 = 100 = 10^2.$$

By the corollary, we would expect $(3, 4, 5)$ to be a Pythagorean triple, and it is, since

$$3^2 + 4^2 = 9 + 16 = 25 = 5^2.$$

In fact, this is one of the best known Pythagorean triples.

As an application to a particular right triangle, we can regard this as doubling the unit, since this will halve the numerical length of each side.

Corollary 2.5.7. *In a Pythagorean triple (a, b, c), if c is a multiple of four, so are a and b.*

Proof. If c is a multiple of four, then it is even, so by Theorem 2.5.5 so are a and b. But c is a multiple of four, so $\frac{c}{2}$ is even. By Corollary 2.5.6, $(\frac{a}{2}, \frac{b}{2}, \frac{c}{2})$ is also a Pythagorean triple. Hence, by Theorem 2.5.5, $\frac{a}{2}$ and $\frac{b}{2}$ are even, and it follows that a and b are multiples of four. □

Now since every odd square is one more than a multiple of 8, the sum of two odd squares is not a square since it can neither be a multiple of 4 nor one more than a multiple of 8. From this follows:

Theorem 2.5.8. *In a Pythagorean triple (a, b, c), if c is odd, then one of a and b is odd and the other is even.*

Proof. • If a and b are both even then by Theorem 2.4.1, $a^2 + b^2$ is a multiple of 4 and hence even, contradicting the implication of Theorem 2.4.2 that c^2 is not even.

• If a and b are both odd then by Theorem 2.4.2, $a^2 + b^2$ is two more than a multiple of 8 and hence cannot be a square since it can neither be a multiple of 4 nor one more than a multiple of 8.

Hence one of a, b is odd and the other is even. □

Theorem 2.5.9. *In a Pythagorean triple (a, b, c), if any two of the numbers is even, the third is also even.*

Proof. The square of the third is either the sum or difference of the squares of the other two, and so is even. Hence by Theorem 2.4.2, the third number itself cannot be odd and is even. □

Theorem 2.5.10. *In a Pythagorean triple (a, b, c), if one of the numbers is odd, then two of them are odd and one is even.*

Proof. • If the odd number is c then by Theorem 2.5.8, one of a, b is even and the other is odd.
• If the odd number is a then by Theorem 2.5.5, c cannot be even. Hence c is odd and by Theorem 2.5.8, b is even.
• The case where the odd number is b is similar to the above case. □

We can now see how the incommensurability of the side and diagonal of a square can be proved. Knorr [26] believes that this proof arose from an attempt to find the unit involved.

Theorem 2.5.11 (Incommensurability). *There is no unit which measures exactly the side and diagonal of a square.*

Proof. Suppose there is such a unit, and suppose in terms of that unit the side of the square is a and the diagonal is c. Then, we have a right triangle whose legs are both of length a and whose hypotenuse is of length c, and so (a, a, c) is a Pythagorean triple.

Now c must either be even or odd. Suppose it is even. Then, by Theorem 2.5.5, a is even. So by Corollary 2.5.6, we can double the unit and halve all the dimensions. Clearly, we cannot do this indefinitely, since otherwise the unit will grow larger than a. So we must have a Pythagorean triple of the form (a, a, c) in which c is odd. But then, by Theorem 2.5.8, a is both even and odd, a contradiction.[9]   □

This proof differs from the previous proofs in at least two important ways.

- First, it is a *proof by contradiction*. That is, the proof begins with the assumption of the negation of the conclusion and then proves that from this assumption a contradiction follows. This method of proof is similar to the debating strategy of *reductio ad absurdum*, or reduction to an absurdity, which often occurs in formal debates. Knorr [26] expresses the view that this is the first proof by contradiction in the history of proof.

- Second, the proof cannot be "seen" by looking at a diagram: it is necessary to follow a sequence of sentences with reasons.

> The proof of the incommensurability of the side and diagonal of a square is the first known proof by contradiction in the history of proof, it cannot be done by diagrams and follows instead a sequence of sentences with reasons.

---

[9]Another way of doing this proof is to assume a, b to be commensurable and to let $a = um$ and $c = un$ where u, m, n are chosen so that um/un is in lowest terms (i.e., there is no common factor between m and n). Note that m and n cannot be both even for otherwise, um/un is not in its lowest terms. By the Pythagorean triples, $2a^2 = c^2$ and hence $2u^2m^2 = u^2n^2$. Since $u \neq 0$, we have $2m^2 = n^2$ and hence by Theorem 2.4.2, n is even (say $n = 2p$). Hence $m^2 = 2p^2$. Again, by Theorem 2.4.2, m is even. This contradicts that m and n are not both even.

It is important to note the idealisation involved here: the ancient Greeks were looking for a unit which divides evenly and *exactly* into both the side and diagonal of a square. An approximation will not do. This is in line with the Pythagorean idea that number is everything. On the other hand, it does not coincide with our idea of measurement. For us, every measurement is an approximation. But what the ancient Greeks mathematicians were trying to do about 430 B.C.E. when this took place was to find a unit that divides evenly and completely exactly into the side and diagonal of a square. The attempt to find such a unit led to a proof by contradiction that it is impossible for there to be such a unit. This means that the side and diagonal of a square are *incommensurable*.

Remark 2.5.12. *This result on incommensurability[10] implies that $\sqrt{2}$ is not a rational number, that is, cannot be represented as the quotient of two integers. To see this, assume $\sqrt{2} = \frac{p}{q}$, then $2q^2 = p^2$ and hence (q, q, p) forms a Pythagorean triple and there is a unit which measures exactly the side and diagonal of a square contradicting the theorem proven above. For a geometric construction, see Page 181.*

Of course this means that you cannot draw right isoceles triangles whose side and hypotenuse are commensurable with the unit. In all the triangles below, we see that the side n is not commensurable with the hypotenuse $n\sqrt{2}$.

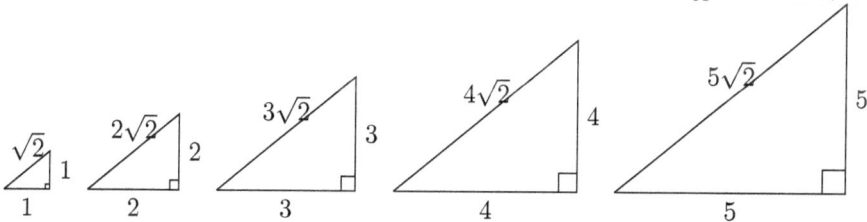

2.5.1 Exercises

Exercise 2.14. We say that a Pythagorean triple (a, b, c) is primitive if a, b and c have no common divisors (factors) larger than 1.

1. Give all the primitive Pythagorean triples of Example 2.5.3.

2. Give a primitive Pythagorean triple (a, b, c) such that c is the smallest integer where (a, b, c) a Pythagorean triple.

3. Use the primitive Pythagorean triple you found in the above item to give 100 more Pythagorean triples.

---

[10] The discovery of the incommensurability of the side and diagonal of a square showed that the Pythagorean idea that number is everything would not work.

4. Assume (a, b, c) is a primitive Pythagorean triple. Show that a and b cannot both be even.

5. Assume (a, b, c) is a primitive Pythagorean triple. Show that a and b cannot both be odd.

6. Assume (a, b, c) is a primitive Pythagorean triple and a is odd. Show that $c + b$ and $c - b$ do not have any common factors larger than 1.

7. Assume (a, b, c) is a primitive Pythagorean triple. Show that c is odd.

Exercise 2.15. Let m, n be integers such that $m > n > 0$. Show that $(m^2 - n^2, 2mn, m^2 + n^2)$ is a Pythagorean triple.

Chapter 3

Elementary Proofs and Theories

Reflecting a bit on the history of numbers, the nonzero natural numbers were always first. Counting as in 1, 2, 3, $\cdots$ was made in different ways (using pebbles/stones, strokes, etc.). See for example the Hieroglyphic Numbers depicted in Table 3.1.[1] There we see that 1 is represented by a single stroke, 100 is represented by a coil of rope, 1,000 by a lotus plant, 10,000 by a finger, 100,000 by frog, etc. Work out how 3,244 and 21,237 are represented in that system.

The number 0 was not always present in ancient times and this explains why Roman numerals (which did not have 0) were replaced by the Indian-Arabic numerals that we use today. As to the rationals, when you think of a rational, do you know what it is? Do you think of the connection between $\frac{2}{3}$ and $\frac{4}{6}$? Would $\frac{2}{3}$ and $\frac{4}{6}$ represent the same rational? What about the integers? When did our ancestors start working with negative numbers? You may be surprised to learn that in fourteenth century Italy, negative numbers were not known and so a double entry bookkeeping system (which is widely used throughout the world) was originally invented in Italy during the fourteenth century to compensate for the absence of negative numbers. The intention was to make it possible, without using negative integers, to compare accounts in which the debits may be greater than the credits.

In this chapter we will take up proofs as statements as applied to some elementary algebra. We will do this in the context of defining the positive rationals $\mathbb{Q}^+$ and the integers $\mathbb{Z}$ from nonzero natural numbers $\mathbb{N}^+$.

---

[1] See https://discoveringegypt.com/egyptian-hieroglyphic-writing/egyptian-mathcm atics-numbers-hieroglyphs/

> Nonzero natural numbers $\mathbb{N}^+$: $1, 2, 3, \cdots$.
> Positive rationals $\mathbb{Q}^+$: rationals representing $\frac{m}{n}$ for m, n in $\mathbb{N}^+$.
> Integers $\mathbb{Z}$: all elements of $\mathbb{N}^+$, their negatives and 0.
> That is $\cdots, -3, -2, -1, 0, 1, 2, 3, \cdots$.

We start by assuming the nonzero natural numbers $\mathbb{N}^+$ and the usual notations for the standard operations of addition and multiplication on $\mathbb{N}^+$ and we review the basic rules for these operations on $\mathbb{N}^+$. Then we show how we can build the positive rationals $\mathbb{Q}^+$ as values of fractions (formal quotients) of nonzero natural numbers and we introduce the addition and multiplication operations on these newly built positive rationals and establish the properties of these operations. Although in $\mathbb{N}^+$, inverses[2] did not exist (since we have no fractions and no negatives), we see that in $\mathbb{Q}^+$, multiplicative inverses do exist and that even if we start with a natural number system without a multiplicative identity (which is 1 in $\mathbb{N}^+$), the rational number system we get adds one.

Afterwards, we move to build the integers as equivalence classes of accounts of credits and debits and define the addition and multiplication operations on these integers. For these operations, we will see that identity for both addition and multiplication exists and that additive inverses exist but not multiplicative ones. Comparing the positive rationals and the integers, we finally build starting from a set with an operation satisfying certain "axioms", a unified theory for adding inverses and, if none is present identity elements. We will do this so that if we identify the starting set with the nonzero naturals and the operation with multiplication we will have the theory of fractions and if we identify the set with the nonzero naturals and the operation with addition we will have the theory of accounts.

3.1   The Nonzero Natural Numbers

We will begin by assuming that the reader knows about the nonzero natural numbers: 1, 2, 3, ... We denote the collection of nonzero natural numbers by $\mathbb{N}^+$. Note that 0 is not in $\mathbb{N}^+$.

We write $=$ for the usual equality relation on $\mathbb{N}^+$ (where $n = m$ stands for n and m are the same number). We will use the usual notations for the standard operations on the nonzero natural numbers, so that $m + n$ and mn

---

[2]We are speaking of additive inverses (such as -3 being the additive inverse of 3 since $3 + (-3) = 0$) and multiplicative inverses (such as $\frac{2}{3}$ being the multiplicative inverse of $\frac{3}{2}$ since $\frac{2}{3} \cdot \frac{3}{2} = 1$.

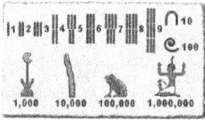

Table 3.1: Hieroglyphic Numbers

are the notations for the sum and product respectively of m and n. To avoid ambiguity, sometimes mn will be written as $m \cdot n$.

We will also assume that the reader is familiar with the basic rules for these operations. In particular, we will assume that the reader knows about the properties of the standard operations given in Figure 3.1.

Since 0 is not a nonzero natural number, the nonzero natural numbers do not have an identity for addition.

Note that if we took the natural numbers instead of the nonzero natural numbers so that we would have 0, we would have to specify in the Cancellation Law for Multiplication that $m \neq 0$.

Much in this chapter will be defined through the so-called *equivalence relations*. Basically, we will define some operations/relations on our various collections and we will show that these relations are equivalence relations.

Definition 3.1.1 (Equivalence relation). • *A relation R on a collection C is* reflexive *iff for any a in C, aRa.*

- *A relation R on a collection C is* symmetric *iff for any a, b in C, if aRb then bRa.*

- *A relation R on a collection C is* transitive *iff for any a, b and c in C, if aRb and bRc then aRc.*

- *A relation R on a collection C is an* equivalence relation *iff it is reflexive, symmetric, and transitive.*

Example 3.1.2. *If we take the equality relation $=$ on $\mathbb{N}^+$, then purely on logical grounds, we see that $=$ is:*

1. *Closure Law of Addition.* Given any two nonzero natural numbers m and n, there is a unique nonzero natural number $m + n$ called the *sum* of m and n.

2. *Closure Law of Multiplication.* Given any two nonzero natural numbers m and n, there is a unique nonzero natural number mn called the *product* of m and n.

3. *Commutative Law of Addition.* For any two nonzero natural number m and n, $m + n = n + m$.

4. *Commutative Law of Multiplication.* For any two nonzero natural numbers m and n, $mn = nm$.

5. *Associative Law of Addition.* For any three nonzero natural numbers m, n, and p, $(m + n) + p = m + (n + p)$.

6. *Associative Law of Multiplication.* For any three nonzero natural numbers m, n, and p, $(mn)p = m(np)$.

7. *Distributive Law of Multiplication over Addition.* For any three nonzero natural numbers m, n, and p, $m(n + p) = mn + mp$.

8. *Cancellation Law for Addition.* For any three nonzero natural numbers m, n, and p, if $m + n = m + p$, then $n = p$.

9. *Cancellation Law for Multiplication.* For any three nonzero natural numbers m, n, and p, if $mn = mp$, then $n = p$.

10. *Identity for Multiplication.* $1 \cdot m = m \cdot 1 = m$.

Figure 3.1: Properties of $(\mathbb{N}^+, +, \cdot, 1)$

- *Reflexive: For any m in $\mathbb{N}^+$, $m = m$.*

- *Symmetric: For any m, n in $\mathbb{N}^+$, if $m = n$ then $n = m$.*

- *Transitive: For any m, n and p in $\mathbb{N}^+$, if $m = n$ and $n = p$ then $m = p$.*

- *Equivalence relation: since it is reflexive, symmetric, and transitive.*

3.1.1   Exercises

Exercise 3.1. For each of the following relations on $\mathbb{N}^+$, establish whether they satisfy the reflexivity, symmetry, transitivity or equivalence properties (in which case give proofs) or not (in which case, give a counter example).

1. mR_1n iff $m \cdot m = n \cdot n$.

2. mR_2n iff $m + m = n + n$.

3. mR_3n iff there is a p in $\mathbb{N}^+$ such that $m \cdot p = n + p$.

4. mR_4n iff there is a p in $\mathbb{N}^+$ such that $m \cdot n = n + p$.

3.2   Fractions and the Positive Rationals

In school, when you learned about fractions, you learned about adding and multiplying fractions:

$$\frac{1}{2} + \frac{3}{5} = \frac{1 \cdot 5 + 3 \cdot 2}{2 \cdot 5} = \frac{11}{10}$$

and

$$\frac{1}{2} \cdot \frac{3}{5} = \frac{1 \cdot 3}{2 \cdot 5} = \frac{3}{10}.$$

The general formulas for adding and multiplying fractions are

$$\frac{m}{n} + \frac{p}{q} = \frac{mq + np}{nq}$$

and

$$\frac{m}{n} \cdot \frac{p}{q} = \frac{mp}{nq}.$$

Note that above, we have used the same symbols $+$ and $\cdot$ for fractions and for naturals. It should be clear that this is an abuse of notation because it seems that we are defining the operations on fractions in terms of the operations on the naturals. For example in $\frac{m}{n} + \frac{p}{q} = \frac{mq + np}{nq}$, the leftmost $+$ is between fractions whereas the rightmost $+$ and all the multiplications are between naturals. Similarly, in $\frac{m}{n} \cdot \frac{p}{q} = \frac{mp}{nq}$, the leftmost $\cdot$ is between fractions whereas the rightmost multiplications are between naturals. In order to make this clear, let us use $+_f$ and $\cdot_f$ for addition and multiplication on fractions.

You also learned that to determine whether or not two fractions are equal, you cross-multiply:

$$\frac{1}{2} = \frac{3}{6} \text{ if and only if } 1 \cdot 6 = 2 \cdot 3,$$

which is true in this case. In general, this is

$$\frac{m}{n} = \frac{p}{q} \text{ if and only if } mq = np.$$

But what do we really mean by writing the equal sign between fractions? Clearly, $\frac{1}{2}$ and $\frac{3}{6}$ are not identical as fractions. What we really mean to say here is that these two fractions *have the same value* (as rational numbers). If we want to distinguish the fraction from the rational number which is its value, we should probably use a different symbol.

For this reason, from now on, we will say that two fractions are equal if and only if they are identical. Thus, we will have[3]

$$\frac{m}{n} = \frac{p}{q} \text{ if and only if } m \text{ is identical to } p \text{ and } n \text{ is identical to } q.$$

To say that two fractions have the same value, we will write

$$\frac{m}{n} \asymp \frac{p}{q} \text{ if and only if } mq = np,$$

where the equal sign between mq and np will be the usual equality of nonzero natural numbers. Thus,

$\asymp$ will mean "has the same value as".

Obviously, if $\frac{m}{n} = \frac{p}{q}$ then $\frac{m}{n} \asymp \frac{p}{q}$ but the other way does not necessarily hold.

---

[3]Here we are again using $=$ for both the natural numbers and the fractions. We will not bother with this since $=$ stands for "identical" and we will use it freely to express this.

Definition 3.2.1 (fraction, =, $\asymp$, $+_f$, $\cdot_f$). *We define the following:*

- *A fraction $\frac{m}{n}$ is a formal quotient of two nonzero natural numbers m and n.*

- *Let two fractions be $\frac{m}{n}$ and $\frac{p}{q}$.*

  - *We define* equality *on fractions (written =) as follows:*

  $$\frac{m}{n} = \frac{p}{q} \text{ if and only if}$$
  $$m \text{ is identical to } p \text{ and } n \text{ is identical to } q$$

  - *We define* have the same value *(written $\asymp$) on fractions as follows:*

  $$\frac{m}{n} \asymp \frac{p}{q} \text{ if and only if } mq = np$$

  - *We define addition $+_f$ on fractions as follows:*

  $$\frac{m}{n} +_f \frac{p}{q} = \frac{mq + np}{nq}$$

  - *We define multiplication $\cdot_f$ on fractions as follows:*

  $$\frac{m}{n} \cdot_f \frac{p}{q} = \frac{mp}{nq}$$

Example 3.2.2.   *1. $\frac{1}{3} = \frac{1}{3}$ but $\frac{1}{3} \neq \frac{2}{6}$.*

2. $\frac{1}{2} \asymp \frac{3}{6}$ since $1 \cdot 6 = 3 \cdot 2$.

3. $\frac{1}{3} \asymp \frac{2}{6}$ since $1 \cdot 6 = 2 \cdot 3$.

4. $\frac{4}{10} \asymp \frac{2}{5}$ since $4 \cdot 5 = 2 \cdot 10$.

5. $\frac{7}{6} \asymp \frac{28}{24}$ since $7 \cdot 24 = 6 \cdot 28$.

6. $\frac{1}{3} +_f \frac{4}{10} = \frac{22}{30}$.

7. $\frac{1}{3} \cdot_f \frac{4}{10} = \frac{4}{30}$.   □

The relation $\asymp$ has many of the same properties as equality:

Lemma 3.2.3 ($\asymp$ is reflexive). *For any fraction $\frac{m}{n}$, we have $\frac{m}{n} \asymp \frac{m}{n}$.*

This is immediate, since $mn = nm$.

Lemma 3.2.4 ($\asymp$ is symmetric). *For any fractions $\frac{m}{n}$ and $\frac{p}{q}$, if $\frac{m}{n} \asymp \frac{p}{q}$, then $\frac{p}{q} \asymp \frac{m}{n}$.*

This follows immediately from the fact that if $mq = np$ then $pn = qm$.

Lemma 3.2.5 ($\asymp$ is *transitive*). *For any fractions $\frac{m}{n}$, $\frac{p}{q}$, and $\frac{r}{s}$, if $\frac{m}{n} \asymp \frac{p}{q}$ and $\frac{p}{q} \asymp \frac{r}{s}$, then $\frac{m}{n} \asymp \frac{r}{s}$.*

Proof. The hypothesis implies that $mq = np$ and $ps = qr$, and we want to prove that $ms = nr$. From the first of these, we get $mqps = npps$, and then, using the second $mqps = npqr$. Hence, by dividing out qp (which is the same as pq), we get $ms = nr$, as desired. $\qquad\qquad\square$

Remark 3.2.6. *Note that we are assuming that the nonzero natural numbers we are using are all nonzero, so we do not have to worry about pq being 0. If we were using the entire natural numbers, we would have to worry about this.*

By Lemmas 3.2.3, 3.2.4, and 3.2.5, we have the following:

Theorem 3.2.7. *The relation $\asymp$ is an equivalence relation.*

We usually think of a rational number as being the quotient of two integers. But clearly the fractions, or formal quotients, are not the rational numbers, two fractions that have the same value are in some sense "the same" rational number. Thus, $\frac{1}{2}$ and $\frac{3}{6}$ both represent the same rational number, since $\frac{1}{2} \asymp \frac{3}{6}$. In general, we should have that

$$\frac{m}{n} \text{ and } \frac{p}{q} \text{ have the same value if } mq = np.$$

But what are these values that we are talking about? What is this *same value* of $\frac{m}{n}$ and $\frac{p}{q}$? We have defined fractions having the same value without saying what those values are.

Let us start by defining the *equivalence class* of a fraction under the equivalence relation $\asymp$:

Definition 3.2.8. *The* equivalence class *of a fraction $\frac{m}{n}$ under the equivalence relation $\asymp$, $[\frac{m}{n}]$, is defined to be the set of fractions related to $\frac{m}{n}$ by $\asymp$. In the usual symbolism for sets (see Chapter 4), we can write this as follows:*

$$\left[\frac{m}{n}\right] = \left\{\frac{p}{q} : \frac{p}{q} \asymp \frac{m}{n}\right\}.$$

This means that

$$\frac{p}{q} \in \left[\frac{m}{n}\right] \text{ if and only if } \frac{p}{q} \asymp \frac{m}{n},$$

where $\in$ is the standard symbol for membership in a set.[4]

Note that $\frac{m}{n} \in \left[\frac{m}{n}\right]$ and that two equivalence classes $\left[\frac{m}{n}\right]$ and $\left[\frac{p}{q}\right]$ are equal if and only if $\frac{m}{n}$ and $\frac{p}{q}$ are related by $\asymp$:

Lemma 3.2.9. *The following hold:*

1. $\frac{m}{n} \in \left[\frac{m}{n}\right]$

2. $\left[\frac{m}{n}\right] = \left[\frac{p}{q}\right]$ *if and only if* $\frac{m}{n} \asymp \frac{p}{q}$.[5]

Proof. 1. Since by Lemma 3.2.3, $\asymp$ is reflexive then $\frac{m}{n} \asymp \frac{m}{n}$ and hence $\frac{m}{n} \in \left[\frac{m}{n}\right]$.

2. Assume $\left[\frac{m}{n}\right] = \left[\frac{p}{q}\right]$. By 1. above, $\frac{m}{n} \in \left[\frac{m}{n}\right]$ and hence $\frac{m}{n} \in \left[\frac{p}{q}\right]$ and so, $\frac{m}{n} \asymp \frac{p}{q}$.
On the other hand, assume $\frac{m}{n} \asymp \frac{p}{q}$. Then, $\frac{r}{s} \in \left[\frac{m}{n}\right]$ iff $\frac{r}{s} \asymp \frac{m}{n}$ iff $\frac{r}{s} \asymp \frac{p}{q}$ iff $\frac{r}{s} \in \left[\frac{p}{q}\right]$. Hence, $\left[\frac{m}{n}\right] = \left[\frac{p}{q}\right]$.

□

We also have some results relating the equivalence classes to the operations on fractions.

Lemma 3.2.10. *Suppose*

$$\left[\frac{m}{n}\right] = \left[\frac{m'}{n'}\right],$$

---

[4]We will have more to say about set theory later in this book. For now, it is enough to think about a set as being a collection of things, called its "members" or its "elements", and also to think of sets as being specified by listing the elements between curly braces or by specifying a property $P(x)$ that elements of the set satisfy by writing $\{x : P(x)\}$ for the set of those things x for which $P(x)$ is true. See Chapter 4.

[5]Note that the equality used here is set equality where two sets are equal iff they have the same elements. See Chapter 4.

and

$$\left[\frac{p}{q}\right] = \left[\frac{p'}{q'}\right].$$

Then

1. $\left[\frac{m}{n} +_f \frac{p}{q}\right] = \left[\frac{m'}{n'} +_f \frac{p'}{q'}\right]$

2. $\left[\frac{m}{n} \cdot_f \frac{p}{q}\right] = \left[\frac{m'}{n'} \cdot_f \frac{p'}{q'}\right].$

Proof. Since by definition of $+_f$, we have $\frac{m}{n} +_f \frac{p}{q} = \frac{mq + np}{nq}$ and $\frac{m'}{n'} +_f \frac{p'}{q'} = \frac{m'q' + n'p'}{n'q'}$, then:

$$\left[\frac{m}{n} +_f \frac{p}{q}\right] = \left[\frac{mq + np}{nq}\right]$$

and

$$\left[\frac{m'}{n'} +_f \frac{p'}{q'}\right] = \left[\frac{m'q' + n'p'}{n'q'}\right].$$

Similarly, by definition of $\cdot_f$, we have:

$$\left[\frac{m}{n} \cdot_f \frac{p}{q}\right] = \left[\frac{mp}{nq}\right]$$

and

$$\left[\frac{m'}{n'} \cdot_f \frac{p'}{q'}\right] = \left[\frac{m'p'}{n'q'}\right].$$

1. By Lemma 3.2.9.2, the premises and the definition of $\asymp$, we have:

$$mn' = nm' \text{ and } pq' = qp'$$

and so

$$(mn')(qq') = (nm')(qq') \text{ and } (pq')(nn') = (qp')(nn')$$

and hence

$$(mq)(n'q') = (m'q')(nq) \text{ and } (pn)(n'q') = (p'n')(nq),$$

and so

$$(mq)(n'q') + (np)(n'q') = (m'q')(nq) + (p'n')(nq)$$

and thus
$$(mq + np)(n'q') = (m'q' + n'p')(nq),$$

and from this follows:
$$\frac{m}{n} +_f \frac{p}{q} = \frac{mq + np}{nq} \asymp \frac{m'q' + n'p'}{n'q'} = \frac{m'}{n'} +_f \frac{p'}{q'},$$

from which it follows by Lemma 3.2.9.2, that
$$\left[\frac{m}{n} +_f \frac{p}{q}\right] = \left[\frac{m'}{n'} +_f \frac{p'}{q'}\right].$$

2. By Lemma 3.2.9.2, the premises and the definition of $\asymp$, we have:
$$mn' = nm' \text{ and } pq' = qp'$$

and so,

$$
\begin{aligned}
mpn'q' &= mn'pq' \\
&= nm'qp' \\
&= nqm'p',
\end{aligned}
$$

from which it follows that
$$\frac{m}{n} \cdot_f \frac{p}{q} = \frac{mp}{nq} \asymp \frac{m'p'}{n'q'} = \frac{m'}{n'} \cdot_f \frac{p'}{q'}$$

and hence by Lemma 3.2.9.2,
$$\left[\frac{m}{n} \cdot_f \frac{p}{q}\right] = \left[\frac{m'}{n'} \cdot_f \frac{p'}{q'}\right].$$

$\square$

We also have that a fraction can belong to one and only one equivalence class:

Theorem 3.2.11. *If $\frac{m}{n}$ is in both $[\frac{p}{q}]$ and $[\frac{r}{s}]$ then $[\frac{p}{q}] = [\frac{r}{s}] = [\frac{m}{n}]$.*

Proof. By Definition 3.2.8, $\frac{m}{n} \asymp \frac{p}{q}$ and $\frac{m}{n} \asymp \frac{r}{s}$. By Lemma 3.2.9.2, $[\frac{m}{n}] = [\frac{p}{q}]$ and $[\frac{m}{n}] = [\frac{r}{s}]$. Hence, $[\frac{m}{n}] = [\frac{p}{q}] = [\frac{r}{s}]$. $\square$

This theorem tells us that the value of a fraction must have some relation to its equivalence class. In fact, we can *define* the value of a fraction to be its equivalence class:

Definition 3.2.12 (Positive Rational Numbers, $\mathbb{Q}^+$). *The value of a fraction is its equivalence class under the relation $\asymp$. Each value is also called a* positive rational number. *Hence,*

> The value of the fraction $\frac{m}{n}$ is the equivalence class $\left[\frac{m}{n}\right]$.

> We call $\left[\frac{m}{n}\right]$, a positive rational number and denote the set of positive rational numbers by $\mathbb{Q}^+$.

Note that a fraction is a member of its value.

> In the rest of this chapter, the positive rational numbers will be denoted by *boldface* characters, such as $\mathbf{a}, \mathbf{b}$ etc. If $\mathbf{a}$ is the value of a fraction $\frac{m}{n}$, we will write
> $$\frac{m}{n} \in \mathbf{a}.$$

The notations $m+n$ and $m \cdot n$ or mn will continue to be used for operations on nonzero natural numbers.

The operations addition and multiplication on positive rational numbers $\mathbb{Q}^+$ are defined as follows:

Definition 3.2.13 ($+_r$ and $\cdot_r$ on $\mathbb{Q}^+$). *If $\mathbf{a}, \mathbf{b}$ are two positive rational numbers, and if fractions with these positive rational numbers as values are respectively $\frac{m}{n}$ and $\frac{p}{q}$, then the* sum *and* product *of these positive rational numbers are given by*

> $$\mathbf{a} +_r \mathbf{b} = \left[\frac{m}{n} +_f \frac{p}{q}\right] = \left[\frac{mq + np}{nq}\right]$$

and

> $$\mathbf{a} \cdot_r \mathbf{b} = \left[\frac{m}{n} \cdot_f \frac{p}{q}\right] = \left[\frac{mp}{nq}\right].$$

When no confusion is likely to result, we may write **ab** *for* **a** $\cdot_r$ **b**.

These operations have many important properties that the operations on nonzero natural numbers have.

Theorem 3.2.14 (Closure Law for $+_r$ **and** $\cdot_r$**).** *For all positive rational numbers* **a** *and* **b**, *the positive rational numbers* **a** $+_r$ **b** *and* **a** $\cdot_r$ **b** *are positive rational numbers uniquely determined by* **a** *and* **b**.

Proof. By Definition 3.2.13 and Lemma 3.2.10. $\square$

Theorem 3.2.15 (Commutative Law for $+_r$ **and** $\cdot_r$**).** *For any two positive rational numbers* **a** *and* **b**, *we have*

1. **a** $+_r$ **b** = **b** $+_r$ **a**; *and*

2. **a** $\cdot_r$ **b** = **b** $\cdot_r$ **a**

Proof. Suppose that $\mathbf{a} = \left[\frac{m}{n}\right]$ and $\mathbf{b} = \left[\frac{p}{q}\right]$. Then

1.

$$\mathbf{a} +_r \mathbf{b} = \left[\frac{mq + np}{nq}\right] = \left[\frac{pn + qm}{qn}\right] = \left[\frac{p}{q}\right] +_r \left[\frac{m}{n}\right] = \mathbf{b} +_r \mathbf{a}.$$

2.

$$\mathbf{a} \cdot_r \mathbf{b} = \left[\frac{mp}{nq}\right] = \left[\frac{pm}{qn}\right] = \mathbf{b} \cdot_r \mathbf{a}.$$

$\square$

Theorem 3.2.16 (Associative Law for $+_r$ **and** $\cdot_r$**).** *For any positive rational numbers* **a**, **b** *and* **c**, *we have:*

1. $(\mathbf{a} +_r \mathbf{b}) +_r \mathbf{c} = \mathbf{a} +_r (\mathbf{b} +_r \mathbf{c})$.

2. $(\mathbf{a} \cdot_r \mathbf{b}) \cdot_r \mathbf{c} = \mathbf{a} \cdot_r (\mathbf{b} \cdot_r \mathbf{c})$

Proof. Suppose that $\mathbf{a} = \left[\frac{m}{n}\right]$, $\mathbf{b} = \left[\frac{p}{q}\right]$ and $\mathbf{c} = \left[\frac{r}{s}\right]$. Then

1.

$$\left(\left[\frac{m}{n}\right] +_r \left[\frac{p}{q}\right]\right) +_r \left[\frac{r}{s}\right] = \left[\frac{mq + np}{nq}\right] +_r \left[\frac{r}{s}\right]$$

$$= \left[\frac{(mq + np)s + (nq)r}{(nq)s}\right]$$

$$= \left[\frac{m(qs) + n(ps + qr)}{n(qs)}\right]$$

$$= \left[\frac{m}{n}\right] +_r \left[\frac{ps + qr}{qs}\right]$$

$$= \left[\frac{m}{n}\right] +_r \left(\left[\frac{p}{q}\right] +_r \left[\frac{r}{s}\right]\right).$$

2.

$$\left(\left[\frac{m}{n}\right] \left[\frac{p}{q}\right]\right) \left[\frac{r}{s}\right] = \left[\frac{mp}{nq}\right] \left[\frac{r}{s}\right]$$

$$= \left[\frac{(mp)r}{(nq)s}\right]$$

$$= \left[\frac{m(pr)}{n(qs)}\right]$$

$$= \left[\frac{m}{n}\right] \left[\frac{pr}{qs}\right]$$

$$= \left[\frac{m}{n}\right] \left(\left[\frac{p}{n}\right] \left[\frac{r}{s}\right]\right).$$

□

Theorem 3.2.17 (Distributive Law for $+_r$ and $\cdot_r$). *If* **a, b,** *and* **c** *are positive rational numbers, then*

$$\mathbf{a} \cdot_r (\mathbf{b} +_r \mathbf{c}) = \mathbf{a} \cdot_r \mathbf{b} +_r \mathbf{a} \cdot_r \mathbf{c}.$$

The proof of this theorem is left as an exercise (see Exercise 3.2).

The positive rational numbers can be viewed as an extension of the nonzero natural numbers.

Definition 3.2.18 ($\mathbb{N}^+ \subseteq \mathbb{Q}^+$). *The nonzero natural rational n_r, which corresponds to the nonzero natural number n is defined by*

$$n_r = \left[\frac{n}{1}\right].$$

Note that for every nonzero natural numbers n and m we also have

$$n_r = \left[\frac{mn}{m}\right].$$

Theorem 3.2.19 (Connection of $+$ and $\cdot$ to $+_r$, $\cdot_r$). *For nonzero natural numbers m and n,*

1. $m_r +_r n_r = (m+n)_r.$

2. $m_r \cdot_r n_r = (mn)_r.$

The proof of this theorem is left as an easy exercise (see Exercise 3.3).

The nonzero natural rational 1_r, which corresponds to the nonzero natural number 1, shares an important property with it.

Theorem 3.2.20 (Identity for Multiplication $\cdot_r$). *The nonzero natural rational $1_r = \left[\frac{1}{1}\right]$ has the property for the positive rationals that the nonzero natural number 1 has for the nonzero natural numbers: it is an identity for multiplication in the sense that*

$$1_r \cdot_r \mathbf{a} = \mathbf{a} \cdot_r 1_r = \mathbf{a}.$$

Proof. This follows since if $\mathbf{a} = \left[\frac{m}{n}\right]$, then

$$1_r \cdot_r \mathbf{a} = \left[\frac{1}{1}\frac{m}{n}\right] = \left[\frac{m}{n}\right] = \mathbf{a} = \left[\frac{m}{n}\frac{1}{1}\right] = \mathbf{a} \cdot_r 1_r.$$

$\square$

Exercise 3.6 illustrates the fact that even if we start with a natural number system without a multiplicative identity, the rational number system we get adds one.

The positive rational numbers have an important property that the nonzero natural numbers do not have:

Theorem 3.2.21 (Inverse for Multiplication $\cdot_r$). *For every positive rational number $\mathbf{a}$, there is another positive rational number $\mathbf{a}^{-1}$ with the property that*

$$\mathbf{a}\mathbf{a}^{-1} = 1_r = \mathbf{a}^{-1}\mathbf{a}, \text{ moreover, if } \mathbf{a} = \left[\frac{m}{n}\right] \text{ then } \mathbf{a}^{-1} = \left[\frac{n}{m}\right].$$

Proof. Let $\mathbf{a} = \left[\frac{m}{n}\right]$ and $\mathbf{a}^{-1} = \left[\frac{n}{m}\right]$. Then

$$\mathbf{a}\mathbf{a}^{-1} = \left[\frac{m}{n}\right]\left[\frac{n}{m}\right] = \left[\frac{m}{n}\frac{n}{m}\right] = \left[\frac{mn}{nm}\right] = \left[\frac{1}{1}\right] = 1_r,$$

1. **(fractions, $\asymp, +_f, \cdot_f$)**.
   - A *fraction* $\frac{m}{n}$ is a *formal quotient* of m, n for $m, n \in \mathbb{N}^+$.
   - $\frac{m}{n} = \frac{p}{q}$ iff m is identical to p and n is identical to q.
   - $\frac{m}{n} +_f \frac{p}{q} = \frac{mq + np}{nq}$ and $\frac{m}{n} \cdot_f \frac{p}{q} = \frac{mp}{nq}$.
   - Let $\frac{m}{n} \asymp \frac{p}{q}$ iff $mq = np$. Then $\asymp$ is an equivalence relation.

2. The **value of the fraction** $\frac{m}{n}$ is the equivalence class
   $$\left[\frac{m}{n}\right] = \left\{ \frac{p}{q} : \frac{p}{q} \asymp \frac{m}{n} \right\}.$$

3. $\mathbb{Q}^+ = \left\{ \left[\frac{m}{n}\right] : m, n \in \mathbb{N}^+ \right\}$. We call $\left[\frac{m}{n}\right]$ a positive rational number and use $\mathbb{Q}^+$ for the set of positive rational numbers.

4. For any two positive rational numbers $\mathbf{a} = \left[\frac{m}{n}\right]$ and $\mathbf{b} = \left[\frac{p}{q}\right]$,
   let $\mathbf{a} +_r \mathbf{b} = \left[\frac{m}{n} +_f \frac{p}{q}\right]$ and $\mathbf{a} \cdot_r \mathbf{b} = \left[\frac{m}{n} \cdot_f \frac{p}{q}\right]$.

5. **Closure Law for $+_r$ and $\cdot_r$.** The positive rational numbers $\mathbf{a} +_r \mathbf{b}$ and $\mathbf{a} \cdot_r \mathbf{b}$ are uniquely determined by $\mathbf{a}, \mathbf{b} \in \mathbb{Q}^+$.

6. **Commutative Law for $+_r$ and $\cdot_r$.** If $\mathbf{a}, \mathbf{b} \in \mathbb{Q}^+$, then:
   $\mathbf{a} +_r \mathbf{b} = \mathbf{b} +_r \mathbf{a}$ and $\mathbf{a} \cdot_r \mathbf{b} = \mathbf{b} \cdot_r \mathbf{a}$.

7. **Associative Law for $+_r$ and $\cdot_r$.** If $\mathbf{a}, \mathbf{b}, \mathbf{c} \in \mathbb{Q}^+$, then:
   $(\mathbf{a} +_r \mathbf{b}) +_r \mathbf{c} = \mathbf{a} +_r (\mathbf{b} +_r \mathbf{c})$.   and $(\mathbf{a} \cdot_r \mathbf{b}) \cdot_r \mathbf{c} = \mathbf{a} \cdot_r (\mathbf{b} \cdot_r \mathbf{c})$.

8. **Distributive Law for $+_r$ and $\cdot_r$.** If $\mathbf{a}, \mathbf{b}, \mathbf{c} \in \mathbb{Q}^+$, then:
   $\mathbf{a} \cdot_r (\mathbf{b} +_r \mathbf{c}) = \mathbf{a} \cdot_r \mathbf{b} +_r \mathbf{a} \cdot_r \mathbf{c}$.

9. **Cancellation Law for $+_r$ and $\cdot_r$.** If $\mathbf{a}, \mathbf{b}, \mathbf{c} \in \mathbb{Q}^+$, then:
   (a) If $\mathbf{a} +_r \mathbf{b} = \mathbf{a} +_r \mathbf{c}$ then $\mathbf{b} = \mathbf{c}$.
   (b) If $\mathbf{a} \cdot_r \mathbf{b} = \mathbf{a} \cdot_r \mathbf{c}$ then $\mathbf{b} = \mathbf{c}$.

10. $\mathbb{N}^+ \subseteq \mathbb{Q}^+$. The **nonzero natural rational** n_r, which corresponds to the nonzero natural number n is $\left[\frac{n}{1}\right]$.

11. **Identity fo** $\cdot_r$. $1_r \cdot_r \mathbf{a} = \mathbf{a} \cdot_r 1_r = \mathbf{a}$ where $1_r = \left[\frac{1}{1}\right]$.

12. **Inverse for** $\cdot_r$. If $\mathbf{a} \in \mathbb{Q}^+$, then there is $\mathbf{a}^{-1} \in \mathbb{Q}^+$ such that $\mathbf{a}\mathbf{a}^{-1} = 1_r = \mathbf{a}^{-1}\mathbf{a}$, moreover, if $\mathbf{a} = \left[\frac{m}{n}\right]$ then $\mathbf{a}^{-1} = \left[\frac{n}{m}\right]$.

Figure 3.2: Properties of $(\mathbb{Q}^+, (\text{fraction}, \asymp, +_f, \cdot_f), +_r, \cdot_r, 1_r, \mathbf{a}^{-1})$

and

$$\mathbf{a}^{-1}\mathbf{a} = \begin{bmatrix} n \\ m \end{bmatrix} \begin{bmatrix} m \\ n \end{bmatrix} = \begin{bmatrix} n & m \\ m & n \end{bmatrix} = \begin{bmatrix} nm \\ mn \end{bmatrix} = \begin{bmatrix} 1 \\ 1 \end{bmatrix} = 1_r.$$

□

It is common to write $\mathbf{a}^{-1}$ as

$$\frac{1_r}{\mathbf{a}},$$

and to call it the *reciprocal* of $\mathbf{a}$.

Note that this allows us to define the operation of *division* on positive rational numbers.

> The quotient of $\mathbf{a}$ divided by $\mathbf{b}$ is given by
>
> $$\frac{\mathbf{a}}{\mathbf{b}} = \mathbf{a} \cdot_r \frac{1_r}{\mathbf{b}} \text{ or simply } \frac{\mathbf{a}}{\mathbf{b}} = \mathbf{a} \cdot_r \mathbf{b}^{-1}$$

For the properties of $(\mathbb{Q}^+, (\text{fractions}, \asymp, +_f, \cdot_f), +_r, \cdot_r, 1_r, \mathbf{a}^{-1})$, see Figure 3.2.

3.2.1 Exercises

Exercise 3.2. Prove Theorem 3.2.17.

Exercise 3.3. Prove Theorem 3.2.19.

Exercise 3.4. Show that

1. $(\mathbf{a}^{-1})^{-1} = \mathbf{a}$.

2. $(\mathbf{ab})^{-1} = \mathbf{a}^{-1}\mathbf{b}^{-1}$.

3. $\left(\frac{\mathbf{a}}{\mathbf{b}}\right)^{-1} = \frac{\mathbf{b}}{\mathbf{a}}$.

4. $\left(\frac{\mathbf{a}}{\mathbf{b}}\right)^{-1} \cdot_r \mathbf{a} = \mathbf{b}$.

5. $(\mathbf{ab})^{-1} \cdot_r \mathbf{a} = \mathbf{b}^{-1}$.

Exercise 3.5. Show the cancellation Law for Addition $+_r$ and multiplication $\cdot_r$ for positive rationals. That is, show that for any three positive rationals $\mathbf{a}$, $\mathbf{b}$, and $\mathbf{c}$:

1. If $a +_r b = a +_r c$ then $b = c$.

2. If $a \cdot_r b = a \cdot_r c$ then $b = c$.

Exercise 3.6. Suppose instead of starting with the nonzero natural numbers, we started forming fractions with the nonzero *even* natural numbers. How would the set of "positive rational numbers" be different from the set we have in the text?

Exercise 3.7. Let a nonzero natural number $k \geq 3$. Repeat Exercise 3.6 but instead of starting with the nonzero natural numbers or the nonzero *even* natural numbers, we started forming fractions with the nonzero k multiples of natural numbers. How would the set of "positive rational numbers" be different from the set we have in the text?

3.3   Formal Differences and the Integers

Double entry bookkeeping is a system of bookkeeping widely used throughout the world that was originally invented in Italy during the fourteenth century. In this system, every account has both a *credit* and a *debit*. The amount of money in an account should be calculated by subtracting the debit from the credit. But in fourteenth century Italy, negative numbers were not known, so the value of an account could not be calculated this way if the debit exceeded the credit. Hence, a different method of comparing accounts was needed. This method allows us to compare, without using negative integers, accounts in which the debits may be greater than the credits.

Definition 3.3.1 (accounts, $\ominus$, $\cong$, $+_c$, $\cdot_c$). *We define the following:*

- *Let c and d be nonzero natural numbers. We write an* account *with credit c and debit d as*
$$c \ominus d.$$

- *Let two accounts, $m \ominus n$ and $p \ominus q$.*

  - *We define* equality *on accounts (written* $=$*) as follows:*

    > $m \ominus n = p \ominus q$ *if and only if*
    > m *is identical to* p *and* n *is identical to* q

  - *We define* have the same value *(written $m \ominus n \cong p \ominus q$) on accounts as follows:*

$$m \ominus n \cong p \ominus q \text{ if and only if } m + q = n + p.$$

– *We define addition $+_c$ on accounts as follows:*

$$(m \ominus n) +_c (p \ominus q) = (m + p) \ominus (n + q),$$

– *We define multiplication $\cdot_c$ on accounts as follows:*

$$(m \ominus n) \cdot_c (p \ominus q) = (mp + nq) \ominus (mq + np).$$

Example 3.3.2. *1. $5 \ominus 3 \cong 8 \ominus 6$ since $5 + 6 = 3 + 8 = 11$.*

2. $3 \ominus 5 \cong 6 \ominus 8$ since $3 + 8 = 5 + 6 = 11$.

3. $(5 \ominus 3) +_c (3 \ominus 5) = 8 \ominus 8$ and $(5 \ominus 3) +_c (9 \ominus 4) = 14 \ominus 7$.

4. $(5 \ominus 3) \cdot_c (3 \ominus 5) = 30 \ominus 34$ and $(5 \ominus 3) \cdot_c (9 \ominus 4) = 57 \ominus 47$. □

The relation $\cong$ has many of the same properties of equality:

Lemma 3.3.3 ($\cong$ is reflexive). *The relation $\cong$ is reflexive; i.e., for any account $m \ominus n$, $m \ominus n \cong m \ominus n$.*

This is immediate since $m + n = n + m$.

Lemma 3.3.4 ($\cong$ is symmetric). *The relation $\cong$ is symmetric; i.e., for any accounts $m \ominus n$ and $p \ominus q$, if $m \ominus n \cong p \ominus q$, then $p \ominus q \cong m \ominus n$.*

This follows immediately from the fact that if $m + q = n + p$ then $p + n = q + m$.

Lemma 3.3.5 ($\cong$ is transitive). *The relation $\cong$ is transitive; i.e., for any accounts $m \ominus n$, $p \ominus q$, and $r \ominus s$, if $m \ominus n \cong p \ominus q$ and $p \ominus q \cong r \ominus s$, then $m \ominus n \cong r \ominus s$.*

Proof. The hypothesis implies that $m + q = n + p$ and $p + s = q + r$, and we want to prove that $m + s = n + r$. From the first of these we get $m + q + p + s = n + p + p + s$, and then, using the second, $m + q + p + s = n + p + q + r$. Hence, by canceling $q + p$ (which is the same as $p + q$), we get $m + s = n + r$, as desired. □

By Definition 3.1.1, and Lemmas 3.3.3, 3.3.4, and 3.3.5, we have the following:

Theorem 3.3.6. *The relation* $\cong$ *is an equivalence relation.*

Since we clearly want two accounts to have the same value if and only if they are related by $\cong$, we can define the equivalence classes of accounts under $\cong$, and then we can take these equivalence classes as the values:

Definition 3.3.7. *The equivalence class* $[m \ominus n]$, *of an account* $m \ominus n$ *under the equivalence relation* $\cong$, *is defined to be the set of accounts related to* $m \ominus n$ *by* $\cong$. *In the usual symbolism for sets, we can write this as follows:*

$$[m \ominus n] = \{p \ominus q : p \ominus q \cong m \ominus n\}.$$

This means that

$$p \ominus q \in [m \ominus n] \text{ if and only if } p \ominus q \cong m \ominus n.$$

Note that $m \ominus n \in [m \ominus n]$ and that two equivalence classes $[m \ominus n]$ and $[p \ominus q]$ are equal if and only if $m \ominus n$ and $p \ominus q$ are related by $\cong$:

Lemma 3.3.8. *The following hold:*

1. $m \ominus n \in [m \ominus n]$.

2. $[m \ominus n] = [p \ominus q]$ *if and only if* $m \ominus n \cong p \ominus q$.

Proof.   1. Since by Lemma 3.3.3, $\cong$ is reflexive then $m \ominus n \cong m \ominus n$ and hence $m \ominus n \in [m \ominus n]$.

2. Assume $[m \ominus n] = [p \ominus q]$. By 1. above, $m \ominus n \in [m \ominus n]$ and hence $m \ominus n \in [p \ominus q]$ and so, $m \ominus n \cong p \ominus q$.
   On the other hand, assume $m \ominus n \cong p \ominus q$. Then, $r \ominus s \in [m \ominus n]$ iff $r \ominus s \cong m \ominus n$ iff $r \ominus s \cong p \ominus q$ iff $r \ominus s \in [p \ominus q]$. Hence, $[m \ominus n] = [p \ominus q]$. $\square$

As with fractions, we have some results relating the equivalence classes to the operations on accounts.

Lemma 3.3.9. *Suppose*

$$[m \ominus n] = [m' \ominus n']$$

and

$$[p \ominus q] = [p' \ominus q'].$$

Then

1. $[(m \ominus n) +_c (p \ominus q)] = [(m' \ominus n') +_c (p' \ominus q')]$

2. $[(m \ominus n) \cdot_c (p \ominus q)] = [(m' \ominus n') \cdot_c (p' \ominus q')]$.

Proof. By definition,

$$[(m \ominus n) +_c (p \ominus q)] = [(m + p) \ominus (n + q)]$$

and

$$[(m' \ominus n') +_c (p' \ominus q')] = [(m' + p') \ominus (n' + q')],$$

and

$$[(m \ominus n) \cdot_c (p \ominus q)] = [(mp + nq) \ominus (mq + np)]$$

and

$$[(m' \ominus n') \cdot_c (p' \ominus q')] = [(m'p' + n'q') \ominus (m'q' + n'p')],$$

By the premises, $m + n' = n + m'$ and $p + q' = q + p'$.

1. We have

$$\begin{aligned} m + p + n' + q' &= m + n' + p + q' \\ &= m' + n + p' + q \\ &= m' + p' + n + q, \end{aligned}$$

from which it follows that

$(m \ominus n) +_c (p \ominus q) =$
$(m + p) \ominus (n + q) \cong$
$(m' + p') \ominus (n' + q') =$
$(m' \ominus n') +_c (p' \ominus q')$

and hence

$$[(m \ominus n) +_c (p \ominus q)] = [(m' \ominus n') +_c (p' \ominus q')].$$

2. By tricotomy, $m' = n'$, or $m' < n'$, or $n' < m'$.

Case 1. If $m' = n'$, then

$$m + n' = n + m' = n + n',$$

and so $m = n$. Hence

$$\begin{aligned} (mp + nq) + (m'q' + n'p') &= \\ (mp + mq) + (m'q' + m'p') &= \\ (mq + mp) + (m'p' + m'q') &= \\ (mq + np) + (m'p' + n'q'). \end{aligned}$$

It follows that

$$(mp + nq) \ominus (mq + np) \cong (m'p' + n'q') \ominus (m'q' + n'p'),$$

and hence,

$$(m \ominus n) \cdot_c (p \ominus q) \cong (m' \ominus n') \cdot_c (p' \ominus q'),$$

from which we get easily that

$$[(m \ominus n) \cdot_c (p \ominus q)] = [(m' \ominus n') \cdot_c (p' \ominus q')].$$

Case 2. If $m' > n'$, then $m' = u + n'$ for some u. Then we have

$$
\begin{aligned}
m + n' &= n + m' \\
&= n + (u + n') \\
&= (n + u) + n'.
\end{aligned}
$$

Hence, $m = n + u$. Now

$$
\begin{aligned}
up + uq' &= u(p + q') \\
&= u(q + p') \\
&= uq + up'.
\end{aligned}
$$

Hence,

$$
\begin{aligned}
(mp + nq) + (m'q' + n'p') &= \\
((n + u)p + nq) + ((u + n')q' + n'p') &= \\
((np + up) + nq) + (uq' + n'q') + (n'p') &= \\
(up + uq') + (np + nq) + (n'q' + n'p') &= \\
(uq + up') + (np + nq) + (n'q' + n'p') &= \\
((nq + uq) + np) + ((up' + n'p') + n'q') &= \\
((n + u)q + np) + ((u + n')p' + n'q') &= \\
(mq + np) + (m'p' + n'q').
\end{aligned}
$$

As in Case 1, it follows that

$$[(m \ominus n) \cdot_c (p \ominus q)] = [(m' \ominus n') \cdot_c (p' \ominus q')].$$

Case 3. Similar to Case 2, except that instead of $m' = u + n'$ for some u, we have $n' = v + m'$ for some v.

□

As with rational numbers, an account can belong to one and only one equivalence class:

Theorem 3.3.10. *If $m \ominus n$ is in both $[p \ominus q]$ and $[r \ominus s]$, then $[m \ominus n] = [p \ominus q] = [r \ominus s]$.*

Proof. By Definition 3.3.7, $m \ominus n \cong p \ominus q$ and $m \ominus n \cong r \ominus s$. By Lemma 3.3.8, $[m \ominus n] = [p \ominus q]$ and $[m \ominus n] = [r \ominus s]$. Hence, $[m \ominus n] = [p \ominus q] = [r \ominus s]$. □

Definition 3.3.11 (Integers $\mathbb{Z}$). *The value of an account $m \ominus n$ is its equivalence class $[m \ominus n]$ under the relation $\cong$. Each value is also called an integer. We denote the set of integers by $\mathbb{Z}$.*

Note that by Lemma 3.3.8.1, each account is a member of its value.

In the rest of this chapter, the integers will be denoted by *lower case Greek* characters, such as α, β, etc. If α is the value of an account $m \ominus n$, we will write

$$m \ominus n \in \alpha.$$

The operations of addition and multiplication on integers are defined as follows:

Definition 3.3.12 ($+_i$ and $\cdot_i$ on $\mathbb{Z}$). *If α and β are two integers, and if accounts with these integers as values are respectively $m \ominus n$ and $p \ominus q$, then the sum and product of these integer numbers are given by*

$$\alpha +_i \beta = [(m \ominus n) +_c (p \ominus q)] = [(m + p) \ominus (n + q)]$$

and

$$\alpha \cdot_i \beta = [(m \ominus n) \cdot_c (p \ominus q)] = [(mp + nq) \ominus (mq + np)].$$

When no confusion is likely to result, we may write $\alpha\beta$ for $\alpha \cdot_i \beta$.

These operations have many important properties that the operations on positive rational numbers have.

Theorem 3.3.13 (Closure Law for $+_i$ and $\cdot_i$). *For all integers α and β, the integers $\alpha +_i \beta$ and $\alpha \cdot_i \beta$ are integers uniquely determined by α and β.*

Proof. By Definition 3.3.12 and Lemma 3.3.9. □

Theorem 3.3.14 (Commutative Law for $+_i$ and $\cdot_i$). *For any integers α and β, we have*

1. $\alpha +_i \beta = \beta +_i \alpha$; and

2. $\alpha \cdot_i \beta = \beta \cdot_i \alpha$.

Proof. Assume $m \ominus n \in \alpha$ and $p \ominus q \in \beta$. Then

1. $\alpha +_i \beta = [(m+p) \ominus (n+q)] = [(p+m) \ominus (q+n)] = \beta +_i \alpha$.

2. $\alpha \cdot_i \beta = [(mp+nq) \ominus (mq+np)] = [(pm+qn) \ominus (pn+qm)] = \beta \cdot_i \alpha$.

$\qquad\qquad\qquad\qquad\qquad\qquad\qquad\qquad\qquad\qquad\qquad\qquad\qquad$ $\square$

Theorem 3.3.15 (Associative Law for $+_i$ and $\cdot_i$). *Let integers α, β, and γ. Then*

1. $(\alpha +_i \beta) +_i \gamma = \alpha +_i (\beta +_i \gamma)$.

2. $(\alpha \cdot_i \beta) +_i \gamma = \alpha \cdot_i (\beta \cdot_i \gamma)$.

Proof. Assume $m \ominus n \in \alpha$, $p \ominus q \in \beta$ and $r \ominus s \in \gamma$. Then
1.

$$
\begin{aligned}
(\alpha +_i \beta) +_i \gamma &= [((m \ominus n) + (p \ominus q)) + (r \ominus s)] \\
&= [((m+p) \ominus (n+q)) + (r \ominus s)] \\
&= [((m+p) + r) \ominus ((n+q) + s)] \\
&= [(m + (p+r)) \ominus (n + (q+s))] \\
&= [(m \ominus n) + ((p+r) \ominus (q+s))] \\
&= [(m \ominus n) + ((p \ominus q) + (r \ominus s))] \\
&= \alpha +_i (\beta +_i \gamma).
\end{aligned}
$$

2.

$$
\begin{aligned}
(\alpha \cdot_i \beta) \cdot_i \gamma &= [((m \ominus n)(p \ominus q))(r \ominus s)] \\
&= [((mp+nq) \ominus (mq+np))(r \ominus s)] \\
&= [((mp+nq)r + (mq+np)s) \ominus ((mp+nq)s + (mq+np)r)] \\
&= [(mpr + nqr + mqs + nps) \ominus (mps + nqs + mqr + npr)] \\
&= [(m(pr+qs) + n(ps+qr)) \ominus (m(ps+qr) + n(pr+qs))] \\
&= [(m \ominus n)((pr+qs) \ominus (ps+qr))] \\
&= [(m \ominus n)((p \ominus q)(r \ominus s))] \\
&= \alpha \cdot_i (\beta \cdot_i \gamma)
\end{aligned}
$$

$\qquad\qquad\qquad\qquad\qquad\qquad\qquad\qquad\qquad\qquad\qquad\qquad\qquad$ $\square$

Theorem 3.3.16 (Distributive Law of $\cdot_i$ over $+_i$). *If α, β, and γ are integers, then*

$$\alpha \cdot_i (\beta +_i \gamma) = \alpha \cdot_i \beta +_i \alpha \cdot_i \gamma.$$

The proof of this theorem is left as an exercise (see Exercise 3.8).

The following lemma is needed to define the integers as an extension of the nonzero natural numbers (Definition 3.3.18).

Lemma 3.3.17. *For any nonzero natural numbers m, n, and p, we have*

$$[(m + n) \ominus m] = [(p + n) \ominus p].$$

Proof. $(m + n) + p = m + (p + n)$. □

Like the positive rational numbers (see Definition 3.2.18), the integers can be viewed as an extension of the nonzero natural numbers.

Definition 3.3.18 ($\mathbb{N}^+ \subseteq \mathbb{Z}$). *The positive integer n_i, which corresponds to the nonzero natural number n, is defined by*

$$n_i = [(m + n) \ominus m]$$

for some nonzero natural number m.

This n_i is well defined by Lemma 3.3.17.

The next lemma prepares us to define the integer 0.

Lemma 3.3.19. *For any nonzero natural numbers m and n, $[m \ominus m] = [n \ominus n]$.*

Proof. This is true because $m + n = n + m$. □

Definition 3.3.20 (Integer 0). *The integer 0 is defined to be $[m \ominus m]$ for some m.*

This is well defined by Lemma 3.3.19.

Theorem 3.3.21 (Identity for Addition $+_i$). *The integer 0 is an identity for addition. That is: for any integer α,*

$$0 +_i \alpha = \alpha +_i 0 = \alpha.$$

Proof. If $\alpha = [m \ominus n]$, then for any nonzero natural number p,

$$
\begin{aligned}
0 +_i \alpha &= [p \ominus p] +_i [m \ominus n] \\
&= [(p + m) \ominus (p + n)] \\
&= [m \ominus n] \\
&= \alpha \\
&= [(m + p) \ominus (n + p)] \\
&= \alpha +_i 0.
\end{aligned}
$$

$\square$

In Exercise 3.11, we will also define an identity 1_i for multiplication.

Note that we have added an identity element for addition to the nonzero natural numbers. We can also define an inverse under addition, or, a negative.

Definition 3.3.22 (The Negative of an Integer). *The* negative *of an integer α, denoted $-\alpha$ is defined as follows: if $\alpha = [m \ominus n]$, then $-\alpha = [n \ominus m]$.*

Theorem 3.3.23 (Inverses for Addition $+_i$). *The negative of an integer α is its inverse under addition: i.e.,*

$$
\alpha +_i -\alpha = 0 = -\alpha +_i \alpha.
$$

Proof. If $\alpha = [m \ominus n]$, then we have

$$
\alpha +_i -\alpha = [m \ominus n] +_i [n \ominus m] = [(m + n) \ominus (n + m)] = 0
$$

and

$$
-\alpha +_i \alpha = [n \ominus m] +_i [m \ominus n] = [(n + m) \ominus (m + n)] = 0.
$$

$\square$

Note that we can now define subtraction by

$$
\alpha - \beta = \alpha +_i (-\beta).
$$

Finally, for the properties of $(\mathbb{Z}, (\text{accounts}, \cong, +_c, \cdot_c), +_i, \cdot_i, 0, 1_i, -\alpha)$, see Figure 3.3.

1. **(accounts, $\cong, +_c, \cdot_c$)**.
   - For $c, d \in \mathbb{N}^+$, $c \ominus d$ is an *account* with credit c and debit d.
   - $m \ominus n = p \ominus q$ iff m is identical to p and n is identical to q.
   - $(m \ominus n) +_c (p \ominus q) = (m + p) \ominus (n + q)$ and
     $(m \ominus n) \cdot_c (p \ominus q) = (mp + nq) \ominus (mq + np)$.
   - $m \ominus n \cong p \ominus q$ iff $m + q = n + p$ is an equivalence relation.

2. $[m \ominus n] = \{p \ominus q : p \ominus q \cong m \ominus n\}$. The **value** $[m \ominus n]$ of an account $m \ominus n$ is its equivalence class $[m \ominus n]$ under $\cong$.

3. $\mathbb{Z} = \{m \ominus n : m, n \in \mathbb{N}^+\}$. We call $[m \ominus n]$ an **integer** and denote the set of integers by $\mathbb{Z}$. Use α, β, etc. to range over $\mathbb{Z}$.

4. **Addition $+_i$/Multiplication $\cdot_i$.** For $\alpha = m \ominus n$ and $\beta = p \ominus q$, $\alpha +_i \beta = [(m \ominus n) +_c (p \ominus q)]$ and $\alpha \cdot_i \beta = [(m \ominus n) \cdot_c (p \ominus q)]$.

5. **Closure Law for $+_i$ and $\cdot_i$.** If $\alpha, \beta \in \mathbb{Z}$, then $\alpha +_i \beta$ and $\alpha \cdot_i \beta$ are integers uniquely determined by α and β.

6. **Commutativity.** $\alpha +_i \beta = \beta +_i \alpha$ and $\alpha \cdot_i \beta = \beta \cdot_i \alpha$.

7. **Associative Law for $+_i$ and $\cdot_i$.** $(\alpha +_i \beta) +_i \gamma = \alpha +_i (\beta +_i \gamma)$ and $(\alpha \cdot_i \beta) +_i \gamma = \alpha \cdot_i (\beta \cdot_i \gamma)$.

8. **Distributivity.** $\alpha \cdot_i (\beta +_i \gamma) = \alpha \cdot_i \beta +_i \alpha \cdot_i \gamma$.

9. $\mathbb{N}^+ \subseteq \mathbb{Z}$. The **positive integer** n_i, which corresponds to the nonzero natural number n, is given by $[(m+n) \ominus m]$ for $m \in \mathbb{N}^+$.

10. **Integer 0.** The integer 0 is defined as $[m \ominus m]$ for some $m \in \mathbb{N}^+$.

11. **Cancellation Law for $+_i$ and $\cdot_i$.** If $\alpha, \beta, \gamma \in \mathbb{Z}$, then:
    (a) If $\alpha +_i \beta = \alpha +_i \gamma$ then $\beta = \gamma$.
    (b) If α is not 0 and $\alpha \cdot_i \beta = \alpha \cdot_i \gamma$ then $\beta = \gamma$.

12. **Identity for $\cdot_i$.** $1_i \cdot_i \alpha = \alpha \cdot 1_i = \alpha$ where $1_i = (p+1) \ominus p$.

13. **Identity for Addition $+_i$.** $0 +_i \alpha = \alpha +_i 0 = \alpha$.

14. **The Negative of an Integer.** $-[m \ominus n] = [n \ominus m]$.

15. **Inverse for Addition $+_i$.** $\alpha +_i -\alpha = 0 = -\alpha +_i \alpha$.

Figure 3.3: Properties of $(\mathbb{Z}, (\text{accounts}, \cong, +_c, \cdot_c), +_i, \cdot_i, 0, 1_i, -\alpha)$

3.3.1   Exercises

Exercise 3.8. Prove Theorem 3.3.16.

Exercise 3.9. Show the cancellation law for addition and multiplication on the integers. That is, show that for any three integers α, β, and γ:

1. If $\alpha +_i \beta = \alpha +_i \gamma$ then $\beta = \gamma$.

2. If α is not 0 and $\alpha \cdot_i \beta = \alpha \cdot_i \gamma$ then $\beta = \gamma$.

Exercise 3.10. Show the equivalent of Theorem 3.2.19 for $+_i$ and $\cdot_i$. That is: show that for nonzero natural numbers m and n,

1. $m_i +_i n_i = (m + n)_i$.

2. $m_i \cdot_i n_i = (mn)_i$.

Exercise 3.11. Show the equivalent of Theorem 3.2.20. That is, show that the positive integer 1_i which is an identity for multiplication $\cdot_i$ in the sense that:
$$1_i \cdot_i \alpha = \alpha \cdot 1_i = \alpha.$$

3.4   Adding Identity Elements and Inverses

The comparison between the theory of fractions and the theory of accounts given in Table 3.2 suggests that we can define a unified theory for adding inverses and, if none is present, identity elements. We need to start with a set with an operation satisfying certain "axioms" as we did with the set the nonzero naturals $\mathbb{N}^+$ for example. We will do this so that

- if we identify the starting set with the nonzero naturals and the operation with multiplication we will have the theory of fractions; and

- if we identify the set with the nonzero naturals and the operation with addition we will have the theory of accounts.

Definition 3.4.1 (Commutative Cancellation Semigroup $(S, \circ)$). *A commutative cancellation semigroup consists of a set S together with a binary operation $\circ$, which satisfies the following statements (or axioms):*

1. Closure *holds for the operation $\circ$ on S: for any x and y in S, $x \circ y$ is an element of S uniquely determined by x and y.*

2. *The operation $\circ$ is* commutative *on S: for any x and y in S, $x \circ y = y \circ x$.*

	$\mathbb{N}^+$	$\mathbb{Q}^+$	$\mathbb{Z}$
Equivalence relations		$\asymp$, Def. 3.2.1 The. 3.2.7	$\cong$, Def. 3.3.1 The. 3.3.6
Sets		$\mathbb{Q}^+$, Def. 3.2.12	$\mathbb{Z}$, Def. 3.3.11
Operations: addition, multiplication	$+$, $\cdot$	$+_r$, $\cdot_r$, Def. 3.2.13	$+_i$, $\cdot_i$, Def. 3.3.12
Closure	$\checkmark$	$\checkmark$ The. 3.2.14	$\checkmark$ The. 3.3.13
Commutativity	$\checkmark$	$\checkmark$ The. 3.2.15	$\checkmark$ The. 3.3.14
Associativity	$\checkmark$	$\checkmark$ The. 3.2.16	$\checkmark$ The. 3.3.15
Distributivity of multiplication over addition	$\checkmark$	$\checkmark$ The. 3.2.17	$\checkmark$ The. 3.3.16
$\mathbb{N}^+ \subseteq$ set		$\checkmark$ Def. 3.2.18	$\checkmark$ Def. 3.3.18
Connection		$\checkmark$ The. 3.2.19	$\checkmark$ Exer. 3.10
Cancellation	$\checkmark$	$\checkmark$ Exer. 3.5	$\checkmark$ Exer. 3.9
Identity for addition	$\times$	$\times$	$\checkmark$ 0, The. 3.3.21
Identity for multiplication	$\checkmark$ 1	$\checkmark$ 1_r, The. 3.2.20	$\checkmark$ 1_i, Exer. 3.11
Inverse for addition	$\times$	$\times$	$\checkmark$ The. 3.3.23
Inverse for multiplication	$\times$	$\checkmark$ The. 3.2.21	$\times$

Table 3.2: Comparison of Fractions and Accounts

3. *The operation $\circ$ is* associative *on S: for any x, y, and z in S, $(x \circ y) \circ z = x \circ (y \circ z)$.*

4. *The* cancellation law *holds for $\circ$ on S: for any x, y, and z in S, if $x \circ z = y \circ z$, then $x = y$.*

We are not assuming anything about the existence of inverse elements in S, and there may or may not be any identity elements in S.

Note that the nonzero naturals with addition and the nonzero naturals with multiplication are both commutative cancellation semigroups.

Lemma 3.4.2. $(\mathbb{N}^+, +)$ *and* $(\mathbb{N}^+, \cdot)$ *are commutative cancellation semigroups.*

The proof is left as an exercise (see Exercise 3.12).

We will show in this section how to add an identity (if there is none) and an inverse for each element to this kind of system.

Definition 3.4.3 (Cartesian Product and Ordered Pairs). *If A and B are sets, we define $A \times B = \{(a, b) : a \in A \text{ and } b \in B\}$. We call $A \times B$*

the cartesian product of A and B and we call the elements of $A \times B$, ordered pairs.

Definition 3.4.4 ($\approx$ **on** $S \times S$ **based on** $(S, \circ)$)**.** *Let $(S, \circ)$ be a commutative cancellation semigroup. Two elements (x, y) and (u, v) of $S \times S$ are said to be* congruent, *notation $(x, y) \approx (u, v)$, if and only if $x \circ v = y \circ u$. We call $\approx$ a congruence based on $(S, \circ)$.*

Example 3.4.5. *By Lemma 3.4.2, $(\mathbb{N}^+, +)$ and $(\mathbb{N}^+, \cdot)$ are commutative cancellation semigroups. Note that:*

- $\approx$ *of Definition 3.4.4 based on $(\mathbb{N}^+, \cdot)$ gives $\asymp$ for fractions as in Definition 3.2.1.*

- $\approx$ *of Definition 3.4.4 based on $(\mathbb{N}^+, +)$ gives and $\cong$ for accounts as in Definition 3.3.1.*

Lemma 3.4.6. *Let $\approx$ be a congruence based on a commutative cancellation semigroup $(S, \circ)$. The relation $\approx$ is* reflexive; *i.e., for any ordered pair (x, y) of elements of S, $(x, y) \approx (x, y)$.*

This is immediate since $x \circ y = y \circ x$.

Lemma 3.4.7. *Let $\approx$ be a congruence based on a commutative cancellation semigroup $(S, \circ)$. The relation $\approx$ is* symmetric; *i.e., for any ordered pairs (x, y) and (u, v) of elements of S, if $(x, y) \approx (u, v)$, then $(u, v) \approx (x, y)$.*

This follows immediately from the facts that if $x \circ v = y \circ u$, then $y \circ u = x \circ v$ and by commutativity of $\circ$, $u \circ y = v \circ x$.

Lemma 3.4.8. *Let $\approx$ be a congruence based on a commutative cancellation semigroup $(S, \circ)$. The relation $\approx$ is* transitive; *i.e., for any ordered pairs (x, y), (u, v), and (w, z) of elements of S, if $(x, y) \approx (u, v)$, and $(u, v) \approx (w, z)$, then $(x, y) \approx (w, z)$.*

Proof. The hypothesis implies that $x \circ v = y \circ u$ and $u \circ z = v \circ w$. From the first of these, we get $x \circ v \circ u \circ z = y \circ u \circ u \circ z$, and then, using the second, $x \circ v \circ u \circ z = y \circ u \circ v \circ w$. If we rearrange the last one (by commutativity and associativity of $\circ$), we get $x \circ z \circ v \circ u = y \circ w \circ v \circ u$, and then by the Cancellation law for $\circ$, we get $x \circ z = y \circ w$, as desired.   □

By Lemmas 3.4.6, 3.4.7, and 3.4.8, we have the following:

Theorem 3.4.9 ($\approx$ **is an equivalence relation**)**.** *Let $\approx$ be a congruence based on a commutative cancellation semigroup $(S, \circ)$. The relation $\approx$ is an equivalence relation.*

Definition 3.4.10 (Operation on $S \times S$ inherited from $\circ$). *Let $\approx$ be a congruence based on a commutative cancellation semigroup $(S, \circ)$. The natural operation on ordered pairs is the one defined by*

$$(x, y) * (u, v) = (x \circ u, y \circ v).$$

We call $*$ the operation on $S \times S$ inherited from $\circ$.

Example 3.4.11. *Consider the commutative cancellation semigroups $(\mathbb{N}^+, \cdot)$ and $(\mathbb{N}^+, +)$.*

- *In Definition 3.2.1 we wrote elements (m, n) of $\mathbb{N}^+ \times \mathbb{N}^+$ as fractions $\frac{m}{n}$ and defined the operation $\cdot_f$ on fractions by $\frac{m}{n} \cdot_f \frac{p}{q} = \frac{mp}{nq}$ where mp stands for $m \cdot p$ and the $\cdot$ is that of $\mathbb{N}^+$ (the same holds for nq). In our notation of Definition 3.4.4, $*$ is the $\cdot_f$ of $\mathbb{N}^+ \times \mathbb{N}^+$ and $\circ$ is the $\cdot$ of $\mathbb{N}^+$.*

- *In Definition 3.3.1 we wrote elements (m, n) of $\mathbb{N}^+ \times \mathbb{N}^+$ as accounts $m \ominus n$ and defined the operation $+_c$ on accounts by $(m \ominus n) +_c (p \ominus q) = (m + p) \ominus (n + q)$. In our notation of Definition 3.4.4, $*$ is the $+_c$ of $\mathbb{N}^+ \times \mathbb{N}^+$ and $\circ$ is the $+$ of $\mathbb{N}^+$.*

Note that historically, the same symbol $\cdot$ is used for the operations on $\mathbb{N}^+ \times \mathbb{N}^+$ and $\mathbb{N}^+$. In Section 3.2, we decided to use $\cdot_f$ instead of $\cdot$ for $\mathbb{N}^+ \times \mathbb{N}^+$. Of course $\cdot$ could also have worked well and even makes sense from the historical point of view, however, it makes more sense to use a different symbol when the operation on the individuals of S is not really known to us. So, we decided to be consistent.

Definition 3.4.12. *Let $\approx$ be a congruence based on a commutative cancellation semigroup $(S, \circ)$. The equivalence class of a pair (x, y) under the equivalence relation $\approx$ is defined to be the set of pairs related to (x, y) by $\approx$. In the usual symbolism for sets we can write this as follows:*

$$[(x, y)] = \{(u, v) : (u, v) \approx (x, y)\}.$$

This means that

$$(x, y) \in [(u, v)] \text{ if and only if } (x, y) \approx (u, v).$$

Note also that

$$[(x, y)] = [(u, v)] \text{ if and only if } (x, y) \approx (u, v).$$

The next lemma shows that the operation on pairs is compatible with the relation $\approx$.

Lemma 3.4.13. *Let $\approx$ be a congruence based on a commutative cancellation semigroup $(S, \circ)$ and let $*$ be the operation on $S \times S$ inherited from $\circ$. Suppose*

$$[(x, y)] = [(x', y')]$$

and

$$[(u, v)] = [(u', v')].$$

Then

$$[(x, y) * (u, v)] = [(x', y') * (u', v')].$$

Proof. By the premises, $x \circ y' = y \circ x'$ and $u \circ v' = v \circ u'$. We also have that

$$[(x, y) * (u, v)] = [(x \circ u, y \circ v)]$$

and

$$[(x', y') * (u', v')] = [(x' \circ u', y' \circ v')].$$

Now we have

$$
\begin{aligned}
x \circ u \circ y' \circ v' &= x \circ y' \circ u \circ v' \\
&= x' \circ y \circ u' \circ v \\
&= x' \circ u' \circ y \circ v,
\end{aligned}
$$

from which it follows that

$$(x, y) * (u, v) = (x \circ u, y \circ v) \approx (x' \circ u', y' \circ v') = (x', y') * (u', v')$$

and hence

$$[(x, y) * (u, v)] = [(x', y') * (u', v')].$$

$\square$

We also have that a pair belongs to one and only one equivalence class:

Theorem 3.4.14. *Let $\approx$ be a congruence based on a commutative cancellation semigroup $(S, \circ)$.*

1. *For every pair (x, y), we have $(x, y) \in [(x, y)]$.*

2. *If (x, y) is in both $[(u, v)]$ and $[(z, w)]$, then $[(x, y)] = [(u, v)] = [(z, w)]$.*

Proof. 1. follows by Lemma 3.4.6.
2. follows by Lemmas 3.4.7 and 3.4.8. $\square$

Definition 3.4.15 (Dyad). *Let $\approx$ be a congruence based on a commutative cancellation semigroup $(S, \circ)$. The value of a pair is its equivalence class under the relation $\approx$. Each value is also called a dyad.*

Note that by Theorem 3.4.14, each pair is a member of its value.

Example 3.4.16. *In Definition 3.2.12 we defined the positive rational number dyad and in Definition 3.3.11 we defined the integer dyad.*

In the rest of this chapter, dyads will be denoted by lower case German letters, such as $\mathfrak{a}$, $\mathfrak{b}$, etc. If $\mathfrak{a}$ is the value of a pair (x, y), we will write

$$(x, y) \in \mathfrak{a}.$$

The operation corresponding to the operation $\circ$ on elements of S is defined as follows:

Definition 3.4.17 (Operation $\circ_d$ corresponding to $\circ$). *Let $\approx$ be a congruence based on a commutative cancellation semigroup $(S, \circ)$ and let $*$ be the operation on $S \times S$ inherited from $\circ$. If $\mathfrak{a} = [(x, y)]$ and $\mathfrak{b} = [(u, v)]$ are two dyads, the composition of these dyads is given by*

$$\mathfrak{a} \circ_d \mathfrak{b} = [(x, y) * (u, v)] = [(x \circ u, y \circ v)].$$

Theorem 3.4.18 (Closure Law). *For all dyads $\mathfrak{a}$ and $\mathfrak{b}$, the dyad $\mathfrak{a} \circ_d \mathfrak{b}$ is a dyad uniquely determined by $\mathfrak{a}$ and $\mathfrak{b}$.*

Proof. By Definition 3.4.17 and Lemma 3.4.13. □

Theorem 3.4.19 (Commutative Law). *For any dyads $\mathfrak{a} = [(x, y)]$ and $\mathfrak{b} = [(u, v)]$, we have $\mathfrak{a} \circ_d \mathfrak{b} = \mathfrak{b} \circ_d \mathfrak{a}$.*

Proof. We have

$$
\begin{aligned}
\mathfrak{a} \circ_d \mathfrak{b} &= [(x, y) * (u, v)] \\
&= [(x \circ u, y \circ v)] \\
&= [(u \circ x, v \circ y)] \\
&= [(u, v) * (x, y)] \\
&= \mathfrak{b} \circ_d \mathfrak{a}.
\end{aligned}
$$

□

Theorem 3.4.20 (Associative Law). *For dyads $\mathfrak{a} = [(x, y)]$, $\mathfrak{b} = [(u, v)]$, and $\mathfrak{c} = [(z, w)]$, we have*

$$(\mathfrak{a} \circ_d \mathfrak{b}) \circ_d \mathfrak{c} = \mathfrak{a} \circ_d (\mathfrak{b} \circ_d \mathfrak{c}).$$

Proof. We have

$$
\begin{aligned}
(\mathfrak{a} \circ_d \mathfrak{b}) \circ_d \mathfrak{c} &= [((x, y) * (u, v)) * (z, w)] \\
&= [(x \circ u, y \circ v) * (z, w)] \\
&= [((x \circ u) \circ z), ((y \circ v) \circ w)] \\
&= [(x \circ (u \circ z), y \circ (v \circ w))] \\
&= [(x, y) * (u \circ z, v \circ w)] \\
&= [(x, y) * ((u, v) * (z, w))] \\
&= \mathfrak{a} \circ_d (\mathfrak{b} \circ_d \mathfrak{c}).
\end{aligned}
$$

$\square$

Lemma 3.4.21. *Let $\approx$ be a congruence based on a commutative cancellation semigroup $(S, \circ)$. For any elements x, y, and z of S,*

$$
[(x \circ y, x)] = [(z \circ y, z)]
$$

Proof. $(x \circ y) \circ z = (z \circ y) \circ x.$ $\qquad\square$

Definition 3.4.22 (S is a subset of Dyads). *Let $\approx$ be a congruence based on a commutative cancellation semigroup $(S, \circ)$. The object which corresponds to the element x of S is given by*

$$
x_d = [(y \circ x, y)]
$$

for some y in S.

This x_d is well defined by Lemma 3.4.21.

Example 3.4.23. *Consider the commutative cancellation semigroups $(\mathbb{N}^+, \cdot)$ and $(\mathbb{N}^+, +)$. Recall Example 3.4.11.*

- *In Definition 3.2.18, we gave $n_r = \left[\frac{n}{1}\right]$ for the positive rational that corresponds to a nonzero natural. Considering the notation here that $\frac{m}{n}$ is (m, n) and $\circ = \cdot$, then $n_r = \left[\frac{n}{1}\right]$ is the same as $n_r = [(1 \cdot n, 1)]$.*

- *In Definition 3.3.18, we gave $n_i = [(m + n) \ominus m]$ for the integer that corresponds to a nonzero natural. Considering the notation here that $r \ominus s$ is (r, s) and $\circ = +$, then $n_i = [(m + n) \ominus m]$ is the same as $n_i = [(m \circ n, m)]$.*

We have not specified whether or not S with the operation $\circ$ has an identity element; i.e., an element e such that $e \circ x = x \circ e = x$ for every x in S. The set of nonzero naturals on which we based both fractions and formal differences has an identity for multiplication but not for addition.

If S has an identity e, then we have

$$e_d = [(x \circ e, x)]$$

for any x in S. Note that this e_d equals $[(x, x)]$ (since $x \circ e = x$).

Lemma 3.4.24. *Let $\approx$ be a congruence based on a commutative cancellation semigroup $(S, \circ)$. For every elements x, y of S, we have $[(x, x)] = [(y, y)]$.*

Proof. This is immediate since $x \circ y = y \circ x$. $\qquad\square$

If there is no such identity element of S, we still have an identity among the dyads:

Definition 3.4.25. *Let $\approx$ be a congruence based on a commutative cancellation semigroup $(S, \circ)$. The dyad e_d is defined to be $[(x, x)]$ for some x in S.*

Theorem 3.4.26 (Identity for Dyads). *Let $\approx$ be a congruence based on a commutative cancellation semigroup $(S, \circ)$. The dyad e_d is an identity for the operation $\circ_d$. That is: for all dyads $\mathfrak{a}$,*

$$e_d \circ_d \mathfrak{a} = \mathfrak{a} \circ_d e_d = \mathfrak{a}.$$

Proof. If $\mathfrak{a} = [(x, y)]$, then for any z in S,

$$
\begin{aligned}
e_d \circ_d \mathfrak{a} &= [(z, z)] \circ_d [(x, y)] \\
&= [(z \circ x, z \circ y)] \\
&= [(x, y)] \\
&= \mathfrak{a} \\
&= [(x \circ z, y \circ z)] \\
&= \mathfrak{a} \circ_d e_d.
\end{aligned}
$$

$\qquad\square$

Thus, if S has no identity for $\circ$, then we have added one for the dyads.

Example 3.4.27. *Consider the commutative cancellation semigroup $(\mathbb{N}^+, +)$. Although $(\mathbb{N}^+, +)$ does not have an identity element, we have shown in Theorem 3.3.21 that the set of Dyads $(\mathbb{Z}, +_i)$ based on $(\mathbb{N}^+, +)$ has the identity element $0 = [m \ominus m]$ for some m in $\mathbb{N}^+$.*

commutative cancellation semigroup	$(\mathbb{N}^+, +)$	$(\mathbb{N}^+, \cdot)$
inverses	$\times$	$\times$
Identity element	$\times$	$\checkmark$
commutative cancellation semigroup with identity and inverses	$(\mathbb{Z}, +_i)$ $\checkmark$	$(\mathbb{Q}^+, \cdot_r)$ $\checkmark$

Table 3.3: Extending commutative cancellation semigroups

We can also define an inverse for the operation $\circ_d$:

Definition 3.4.28. *For any dyad* $\mathfrak{a} = [(x, y)]$, *its inverse,* $\mathfrak{a}^{-1}$ *is defined to be* $[(y, x)]$.

Theorem 3.4.29 (Inverses for $\circ_d$**).** *Let* $\approx$ *be a congruence based on a commutative cancellation semigroup* $(S, \circ)$. *For any dyad* $\mathfrak{a}$, *its inverse* $\mathfrak{a}^{-1}$ *exists and we have*

$$\mathfrak{a} \circ_d \mathfrak{a}^{-1} = e_d = \mathfrak{a}^{-1} \circ_d \mathfrak{a} \text{ where } e_d \text{ is given in Definition 3.4.25.}$$

Proof.
$$\mathfrak{a} \circ_d \mathfrak{a}^{-1} = [(x, y)] \circ_d [(y, x)] = [(x + y, y + x)] = e_d$$

and

$$\mathfrak{a}^{-1} \circ_d \mathfrak{a} = [(y, x)] \circ_d [(x, y)] = [(y + x, x + y)] = e_d.$$

$\square$

Example 3.4.30. *Consider the commutative cancellation semigroup* $(\mathbb{N}^+, +)$. *Although elements in* $(\mathbb{N}^+, +)$ *do not have inverses, we have shown in Theorem 3.3.23 that each element* $\alpha = [m \ominus n]$ *in the set of Dyads* $(\mathbb{Z}, +_i)$ *based on* $(\mathbb{N}^+, +)$ *has an inverse element* $-\alpha = [n \ominus m]$.

Note that we can now define the inverse operation to $\circ_d$ by

$$\mathfrak{a} \circ_d^{-1} \mathfrak{b} = \mathfrak{a} \circ_d \mathfrak{b}^{-1}.$$

Example 3.4.31. *Consider the commutative cancellation semigroups* $(\mathbb{N}^+, +)$ *and the set of Dyads* $(\mathbb{Z}, +_i)$ *based on it. The inverse of* $+_i$ *is the negation operator. That is:* $\alpha +_i^{-1} \beta = \alpha +_i (-\beta)$.

Let $(S, \circ)$ **be a Commutative Cancellation Semigroup** (i.e., $\circ$ satisfies closure, commutativity, associativity and cancellation Law on S).

1. **A congruence** $\approx$ **on** $S \times S$ **based on** $(S, \circ)$ is defined by: $(x, y) \approx (u, v)$ iff $x \circ v = y \circ u$. Then $\approx$ is an equivalence relation.

2. Let $\approx$ be a congruence on $S \times S$ based on $(S, \circ)$. **The operation** $*$ **on** $S \times S$ **inherited from** $\circ$ is defined by $(x, y) * (u, v) = (x \circ u, y \circ v)$.

3. Let $\approx$ be a congruence based on $(S, \circ)$ and let $*$ on $S \times S$ be inherited from $\circ$.

   - The **value** of (x, y) (also called a **dyad**) is $[(x, y)] = \{(u, v) : (u, v) \approx (x, y)\}$. The set of dyads $S_d = \{[(x, y)] : x, y \in S\}$.

   - **Operation** $\circ_d$ **corresponding to** $\circ$. If $\mathfrak{a} = [(x, y)]$ and $\mathfrak{b} = [(u, v)]$, define $\mathfrak{a} \circ_d \mathfrak{b} = [(x, y) * (u, v)] = [(x \circ u, y \circ v)]$.

   - **Closure Law.** For all dyads $\mathfrak{a}$ and $\mathfrak{b}$, the dyad $\mathfrak{a} \circ_d \mathfrak{b}$ is a dyad uniquely determined by $\mathfrak{a}$ and $\mathfrak{b}$.

   - **Commutative Law.** For any dyads $\mathfrak{a} = [(x, y)]$ and $\mathfrak{b} = [(u, v)]$, we have $\mathfrak{a} \circ_d \mathfrak{b} = \mathfrak{b} \circ_d \mathfrak{a}$.

   - **Associative Law.** For dyads $\mathfrak{a} = [(x, y)]$, $\mathfrak{b} = [(u, v)]$, and $\mathfrak{c} = [(z, w)]$, we have $(\mathfrak{a} \circ_d \mathfrak{b}) \circ_d \mathfrak{c} = \mathfrak{a} \circ_d (\mathfrak{b} \circ_d \mathfrak{c})$.

   - **Cancellation law for** $\circ_d$ **on** S_d. If $\mathfrak{a} \circ_d \mathfrak{c} = \mathfrak{b} \circ_d \mathfrak{c}$, then $\mathfrak{a} = \mathfrak{b}$.

   - $(S_d, \circ_d)$ **is a commutative cancellation semigroup.**

   - S **is a subset of** S_d. For each x in S, we have a dyad $x_d = [(y \circ x, y)]$.

   - **Identity for Dyads.** Define e_d to be $[(x, x)]$ for some x in S. For all dyads $\mathfrak{a}$, we have $e_d \circ_d \mathfrak{a} = \mathfrak{a} \circ_d e_d = \mathfrak{a}$.

   - **Inverses for Dyads.** If $\mathfrak{a} = [(x, y)]$, define $\mathfrak{a}^{-1}$ to be $[(y, x)]$. We have $\mathfrak{a} \circ_d \mathfrak{a}^{-1} = e_d = \mathfrak{a}^{-1} \circ_d \mathfrak{a}$.

   - Define the inverse operation to $\circ_d$ by $\mathfrak{a} \circ_d^{-1} \mathfrak{b} = \mathfrak{a} \circ_d \mathfrak{b}^{-1}$.

Figure 3.4: Building commutative cancellation semigroup $(S_d, \circ_d)$ with Identity Element and Inverses from commutative cancellation semigroup $(S, \circ)$

3.4.1   Exercises

Exercise 3.12. Prove Lemma 3.4.2.

Exercise 3.13. Let $(S, \circ)$ be a commutative cancellation semigroup. Let x, y, z, w be elements of S such that $x \circ z = y \circ w$ and $y \circ z = w \circ y$. Which of the elements x, y, z, w are equal and which do not need to be equal? For those not needing to be equal, give an example in the commutative cancellation semigroup $(\mathbb{N}^+, \cdot)$.

Exercise 3.14. Let $(S, \circ)$ be a commutative cancellation semigroup and construct $(S_d, \circ_d)$ the set of Dyads and the operation on it as in Definitions 3.4.15 and 3.4.17. Show that $(S_d, \circ_d)$ is a commutative cancellation semigroup.

Exercise 3.15. Show that $(\mathbb{Q}^+, \cdot_r)$ and $(\mathbb{Z}, +_i)$ as given in Definitions 3.2.12 and 3.2.13 (resp. Definitions 3.3.11 and 3.3.12) are commutative cancellation semigroups.
Repeat the question for $(\mathbb{Q}^+, +_r)$ and $(\mathbb{Z}, \cdot_i)$.

Exercise 3.16. Follow the steps described in Figure 3.4 to build respectively from the commutative cancellation semigroups $(\mathbb{N}^+, \cdot)$ and $(\mathbb{N}^+, +)$, the commutative cancellation semigroups $(\mathbb{Q}^+, \cdot_r)$ and $(\mathbb{Z}, +_i)$ stating what their Identity Elements and Inverses are.
Choose a commutative cancellation semigroup and use it to build a commutative cancellation semigroup $(\mathbb{Q}, +_r)$ which contains all the rational numbers, not just the positive rationals, showing all the steps described in Figure 3.4.

Chapter 4

Set Theory

We have already had occasion to mention sets, but we did not say much about them. In this chapter we will say a lot about them.

Recall that in Chapter 3, we spoke about the sets:

- Nonzero natural numbers $\mathbb{N}^+$: $1, 2, 3, \cdots$.

- Positive rationals $\mathbb{Q}^+$: rationals representing $\frac{m}{n}$ for m, n in $\mathbb{N}^+$.

- Integers $\mathbb{Z}$: integers representing elements of $\mathbb{N}^+$, their negation and 0.

- Rationals $\mathbb{Q}$: rationals representing elements elements of $\mathbb{Q}^+$, their negation and 0.

There, we showed how we could build the sets $\mathbb{Q}^+$ and $\mathbb{Z}$ from $\mathbb{N}^+$. Just like we built $\mathbb{Z}$ from $\mathbb{N}^+$, we can build $\mathbb{Q}$ from $\mathbb{Q}^+$ (see Exercise 3.16).

In this chapter, we will talk about the size of these sets as well as the size of a number of other sets including:

- $\mathbb{N}$: $0, 1, 2, 3, \cdots$.

- $\mathbb{R}$ the real numbers (the algebraic and transcendental numbers).

Sets became important in mathematics in connection with *infinite sets*, about which some problems arose in thinking about their sizes (also known as their number of elements). In the first section we give first thoughts on sizes of sets, we define the size of a finite set, and introduce the notions *(infinitely) countable* and *uncountable*, but fall short of defining the size of infinite sets in this chapter. Instead, we focus on the notions of countable and uncountable and show that subsets of countable sets are also countable, that $\mathbb{N}$, $\mathbb{Z}$, $\mathbb{Q}$ and the set of algebraic numbers are countable whereas the interval $(0,1]$ of $\mathbb{R}$ is uncountable. In this first section, we keep everything at the informal level and don't go into the formal details of one-to-one correspondences and functions. In the second section we introduce basic operations on sets such as union, intersection, powerset, and then formally define functions and their properties (including injectivity, surjectivity and bijectivity) which will help us establish further properties of countable sets. We also establish the Cantor-Schröder-Bernstein theorem which states that there is a bijection between two sets iff there is an injection from each of the sets to the other set. In the third section we reflect further on finite and infinite sets and the sizes of finite sets as well as countability and uncountability.

4.1   First Thoughts on Sizes of Sets

Calculating the size of an infinite set brought in many puzzling questions that were not settled before the late 19th century and even then, the given answers were controversial. In this section, we look at some puzzles that arise from thinking about the size of infinite sets and explain how this led to a more precise definition of set and how Cantor laid the way to the correct thinking about sets and their sizes.

Think of the sets of nonzero natural numbers $\mathbb{N}^+$, even nonzero natural numbers E and odd nonzero natural numbers O which contain the following elements:

$$\mathbb{N}^+ \quad : \quad 1, 2, 3, \cdots$$
$$E \quad : \quad 2, 4, 6, \cdots$$
$$O \quad : \quad 1, 3, 5, \cdots$$

Do you think they are finite or infinite? Do they have the same size? It is clear that $\mathbb{N}^+$ contains all the elements of E and all the elements of O, so is $\mathbb{N}^+$ twice as large as E or O? On the other hand, we can find a matching operation that matches each element of $\mathbb{N}^+$ with a unique element of E such that no element is left unmatched. This surely means that there are at least as many elements in E as there are in $\mathbb{N}^+$ for otherwise, some elements of $\mathbb{N}^+$ would not have been paired with their unique match in E. If you need more persuading, look at this matching:

$\mathbb{N}^+$	$\leftrightarrow$	E
1	$\leftrightarrow$	2
2	$\leftrightarrow$	4
3	$\leftrightarrow$	6
4	$\leftrightarrow$	8
$\vdots$		
n	$\leftrightarrow$	$2n$
$\vdots$		

We could do something similar for O. So, how can $\mathbb{N}^+$ appear twice as large as O and E and yet again, have its elements paired uniquely one by one with elements of either E or O which makes it likely that the size of $\mathbb{N}^+$ is the same as the size of E and also the same as the size of O. It doesn't seem to make sense. What is going on?

Confused? You are not the only one. Even Galileo asked these questions. Read on.

4.1.1 Galileo and the Size of Infinite Sets

One of the earliest examples of problems faced when dealing with sizes of infinite sets was first mentioned by Galileo[1]:

> *Simplicio*: Here a difficulty presents itself which appears to me insoluble. Since it is clear that we may have one line greater than another, each containing an infinite number of points, we are forced to admit that, within one and the same class, we may have something greater than infinity, because the infinity of points in the long line is greater than the infinity of points in the short line. This assigning to an infinite quantity a value greater than infinity is quite beyond my comprehension.

> *Salviati*: This is one of the difficulties which arise when we attempt, with our finite minds, to discuss the infinite, assigning to it those properties which we give to the finite and limited; but this I think is wrong, for we cannot speak of infinite quantities as being the one greater or less than or equal to another. To prove this I have in mind an argument which, for the sake of clearness, I shall put in the form of questions to Simplicio who raised this difficulty.

---

[1]See [16, pp. 31–33]. The book is written as a dialogue between three characters: Simplicio, Salviati, and Sagredo.

I take it for granted that you know which of the numbers are squares and which are not.

Simplicio: I am quite aware that a squared number is one which results from the multiplication of another number by itself; thus 4, 9, etc., are squared numbers which come from multiplying 2, 3, etc., by themselves.

Salviati: Very well; and you also know that just as the products are called squares so the factors are called sides or roots; while on the other hand those numbers which do not consist of two equal factors are not squares. Therefore if I assert that all numbers, including both squares and non-squares, are more than the squares alone, I shall speak the truth, shall I not?

Simplicio: Most certainly.

Salviati: If I should ask further how many squares there are one might reply truly that there are as many as the corresponding number of roots, since every square has its own root and every root its own square, while no square has more than one root and no root more than one square.

Simplicio: Precisely so.

Salviati: But if I inquire how many roots there are, it cannot be denied that there are as many as the numbers because every number is the root of some square. This being granted, we must say that there are as many squares as there are numbers because they are just as numerous as their roots, and all the numbers are roots. Yet at the outset we said that there are many more numbers than squares, since the larger portion of them are not squares. Not only so, but the proportionate number of squares diminishes as we pass to larger numbers, Thus up to 100 we have 10 squares, that is, the squares constitute 1/10 part of all the numbers; up to 10000, we find only 1/100 part to be squares; and up to a million only 1/1000 part; on the other hand in an infinite number, if one could conceive of such a thing, he would be forced to admit that there are as many squares as there are numbers taken all together.

Sagredo: What then must one conclude under these circumstances?

Salviati: So far as I see we can only infer that the totality of all numbers is infinite, that the number of squares is infinite, and that the number of their roots is infinite; neither is the number

> of squares less than the totality of all the numbers, nor the latter greater than the former; and finally the attributes "equal," "greater," and "less," are not applicable to infinite, but only to finite, quantities. When therefore Simplicio introduces several lines of different lengths and asks me how it is possible that the longer ones do not contain more points than the shorter, I answer him that one line does not contain more or less or just as many points as another, but that each line contains an infinite number.

Here, Galileo is implicitly referring to Euclid's Common Notion 5, "The whole is greater than the part."[2] He is also implying that infinite sets cannot be differentiated by their size; i.e., he is implying that all infinite sets have to be treated as having the same size.

This was a very common attitude in Galileo's day, and represented a reluctance of mathematicians to accept infinite sets as objects. As Gauss once put it[3]

> I protest ... against the use of infinite magnitude as if it were something finished; this use is not admissible in mathematics. The infinite is only a *façon de parler*: one has in mind limits approached by certain ratios as closely as desirable while other ratios may increase indefintely.

However, in the late 19th century, Georg Cantor, a German mathematician took up the subject with a different idea.

4.1.2 Basic Notions of Sets

Cantor considered infinite sets as mathematical objects that are complete and can thus be treated mathematically. This acceptance of infinite sets as mathematical objects required some assumptions about them, assumptions, in fact, about sets in general. Among those assumptions are that there is always a method for deciding, given a set and an object, whether or not the object is in the set.

> If the object is in the set, we say that the object is a *member* or *element* of the set. If a is an element of a set S, we write $a \in S$. If a is not an element of S, we write $a \notin S$.

---

[2]See [20, vol. I, p. 155].

[3]From *Briefwechsel Gauss-Schumacher*, vol. II (1860), p. 269; *Gauss' Werke*, vol. VIII (1900), p. 216, as quoted in [14, p. 1].

We also count as a set one which has no elements. This is called the *empty set*, and is denoted by $\emptyset$. This set is to other sets something like what 0 is to other numbers.

> For all objects a, $a \notin \emptyset$.

We also assume that sets are determined by their elements: two sets are considered *equal* just when they have exactly the same elements. We call this property of sets the principle of *extensionality*. This principle can be written as follows:

> **Principle of Extensionality.** If S and T are any sets, then $S = T$ if and only if for every x, $x \in S$ if and only if $x \in T$.

Note that this implies that the only set equal to $\emptyset$ is $\emptyset$.

Lemma 4.1.1. *If S is a set such that for all objects a, $a \notin S$, then $S = \emptyset$.*

Proof. We know that for all objects a, $a \notin \emptyset$. Hence, for all objects a, $a \notin \emptyset$ iff $a \notin S$. This means, for all objects a, $a \in \emptyset$ iff $a \in S$. By extensionality, $S = \emptyset$. $\qquad\qquad\qquad\qquad\qquad\qquad\qquad\qquad\qquad\qquad\qquad\qquad$ $\square$

It is also possible for one set to be a *part*, or a *subset* of another set.

> - A set S is a *subset* of a set T, notation $S \subseteq T$, if and only if every element $x \in S$ is also in T. That is, for every x, if $x \in S$, then $x \in T$.
> - If $S \subseteq T$ and there is $x \in T$ such that $x \notin S$, we write $S \subset T$ and say that S is a strict subset of T.
> - A set S is a *proper subset* of T, notation $S \subset T$, if and only if $S \subseteq T$ and $S \neq T$.

Lemma 4.1.2. *The following hold:*

1. *$\emptyset \subseteq S$ for every set S.*

2. *For every set S, if $S \neq \emptyset$ then $\emptyset \subset S$.*

3. *For any sets S and T, we have $S = T$ if and only if $S \subseteq T$ and $T \subseteq S$.*

4. For any set S, we have $S \subseteq S$.

5. For any sets S, T and V, if $S \subseteq T$ and $T \subseteq V$ then $S \subseteq V$.

The proof is left as an exercise (see Exercise 4.1).

Note that we need to distinguish between $\emptyset$ and $\{\emptyset\}$. The latter is a set with one element, that element being $\emptyset$.

There are two methods of writing sets that are in common use. One, that usually works only for finite sets, is to list the elements between curly braces. So, for example, to give the set whose elements are the first three odd numbers, 1, 3, and 5, we can write this as

$$\{1, 3, 5\}.$$

In writing sets in this way, the order in which the elements is written does not matter, nor does it matter whether or not an element is repeated in the list. Thus

$$\{1, 3, 5\} = \{5, 1, 3\} = \{1, 1, 3, 3, 3, 5\}.$$

Usually, writing sets by listing their elements in curly braces is only used for small finite sets, but not always; the set $\mathbb{N}^+$ of all nonzero natural numbers, can be written as follows:

$$\mathbb{N}^+ = \{1, 2, 3, \ldots\}.$$

We can also write a set by stating a property that is satisfied by elements of the set and only elements of the set. For this we need to list the set in connection with a *universe of discourse*, which, in general, we will call U. For the above set $\{1, 3, 5\}$, we can use $\mathbb{N}^+$ as the universe of discourse. With this, we can write the above set as follows:

$$\{x \in \mathbb{N}^+ : x \text{ is one of the first three odd numbers}\}$$

or

$$\{x \in \mathbb{N}^+ : x = 1 \text{ or } x = 3 \text{ or } x = 5\}.$$

Until further notice, we will not say much about the universe of discourse U, but will assume that we have a suitable set for U.

Some sets can be written by listing their elements, for example:

$$\{1,3,5\} \qquad \mathbb{N}^+ = \{1,2,3,\ldots\}$$

Any set can be written by stating a property that is satisfied by elements of the set and only elements of the set. For this we need to list it in connection with a *universe of discourse*, say U. For example:

$$\{x \in \mathbb{N}^+ : x \text{ is one of the first three odd numbers}\}$$

When the universe is clear, we may not write the universe of discourse, but it is always assumed.

Examples of other sets we will use in this chapter include the sets

$$\mathbb{N}_{\geq m} = \{n \in \mathbb{N} : n \geq m\} \text{ where } m \in \mathbb{N}.$$

Note that $\mathbb{N}_{\geq 0} = \mathbb{N}$, $\qquad \mathbb{N}_{\geq 1} = \mathbb{N}^+$, $\qquad \mathbb{N}_{\geq 2} = \{2,3,4,\cdots\}$, and so on.

You see that we build these sets $\mathbb{N}_{\geq m}$ where $m \in \mathbb{N}$ by continuously removing the smallest element of each set. So, it is clear that each new set has 1 less element than the previous one.

BUT.... In later sections we will show that all these sets $\mathbb{N}_{\geq m}$ where $m \in \mathbb{N}$ have the same size as $\mathbb{N}$.

In the rest of this section we will complete our basic information about sets and especially sets connected to the natural numbers. Recall that we have already used in earlier chapters, and are familiar with the laws of addition, multiplication and order on the sets $\mathbb{N}$ and $\mathbb{N}^+$. We will also assume the induction axiom for $\mathbb{N}$ which is crucial for establishing properties about natural numbers.

Induction axiom for $\mathbb{N}$: If $S \subseteq \mathbb{N}$, $0 \in S$, and whenever $n \in S$ we also have $n + 1 \in S$, then $S = \mathbb{N}$.

We will also apply this axiom to all sets $\mathbb{N}_{\geq m} = \{n \in \mathbb{N} : n \geq m\}$ where $m \in \mathbb{N}$ as follows:

> **Induction axiom for** $\mathbb{N}_{\geq m}$**:** If $S \subseteq \mathbb{N}_{\geq m}$, $m \in S$, and whenever $n \in S$ we also have $n + 1 \in S$, then $S = \mathbb{N}_{\geq m}$.

We also assume the following least number principle which states that even if the set grows infinitely with larger and larger numbers, it will always have a smallest element.

> **Least number principle:** If $S \subseteq \mathbb{N}$ and $S \neq \emptyset$, then S has a least element m where $m \leq n$ for every $n \in S$.

4.1.3 Cantor's One-to-one Correspondences as a Measure for Sizes of Sets

Once infinite sets are accepted as mathematical objects, it is possible to compare them by looking at sets which can be put into *one-to-one correspondence* (also called bijection, see Definition 4.2.20). This is exactly what Galileo did in the third and fourth statements by Salviati in the above quote. For he shows that the set of perfect squares is in one-to-one correspondence with the nonzero natural numbers. This may be shown graphically as follows:

$\mathbb{N}^+$	$\leftrightarrow$	perfect squares
1	$\leftrightarrow$	1
2	$\leftrightarrow$	4
3	$\leftrightarrow$	9
4	$\leftrightarrow$	16
$\vdots$		
n	$\leftrightarrow$	n^2
$\vdots$		

Cantor took this to mean that the set of nonzero natural numbers has the **same size** as the set of perfect squares.

As we said above, we will give a precise definition of a one-to-one correspondence (also called bijection) in Definition 4.2.20. For now, it is enough to remember that:

> A set A can be put in one-to-one correspondence with a set
> B iff every element of A corresponds uniquely to one element
> of B and every element of B has a corresponding element in
> A.

Lemma 4.1.3. $\mathbb{N}$ *can be put in one-to-one correspondence with* $\mathbb{N}_{\geq m}$ *for every* $m \in \mathbb{N}$.

Proof. We show the one-to-one correspondence between $\mathbb{N}$ and $\mathbb{N}_{\geq m}$ as follows:

$\mathbb{N}$	$\leftrightarrow$	$\mathbb{N}_{\geq m}$
0	$\leftrightarrow$	$0 + m$
1	$\leftrightarrow$	$1 + m$
2	$\leftrightarrow$	$2 + m$
3	$\leftrightarrow$	$3 + m$
$\vdots$		
n	$\leftrightarrow$	$n + m$
$\vdots$		

$\square$

Lemma 4.1.4. *The following hold:*

1. *Any set A can be put in one-to-one correspondence with itself.*

2. *If A can be put in one-to-one correspondence with B then B can be put in one-to-one correspondence with A.*

3. *If A can be put in one-to-one correspondence with B and B can be put in one-to-one correspondence with C, then A can be put in one-to-one correspondence with C.*

Proof.   1. Here is the one-to-one correspondence from A with itself:

A	$\leftrightarrow$	A
a	$\leftrightarrow$	a

2. Assume A can be put in one-to-one correspondence with B as on the left, then B can be put in one-to-one correspondence with A as on the right.

A	$\leftrightarrow$	B
a	$\leftrightarrow$	b

B	$\leftrightarrow$	A
b	$\leftrightarrow$	a

3. Assume we have the one-to-one correspondences on the left, then we can create the one-to-one correspondence on the right.

A	$\leftrightarrow$	B	B	$\leftrightarrow$	C
a	$\leftrightarrow$	b	b	$\leftrightarrow$	c

A	$\leftrightarrow$	C
a	$\leftrightarrow$	c

$\square$

Just like we defined $\mathbb{N}_{\geq m}$ which will be used to reason about infinite sets, we will define their relative sets which will be used to reason about finite sets.

Just like we defined $\mathbb{N}_{\geq m}$, we will define

$$\mathbb{N}_{<m} = \{n \in \mathbb{N} : n < m\} \text{ where } m \in \mathbb{N}.$$

For example, $\mathbb{N}_{<0} = \emptyset$ $\qquad \mathbb{N}_{<1} = \{0\},$ $\qquad \mathbb{N}_{<2} = \{0, 1\}$, and so on.

Note that for each $n \in \mathbb{N}$, the set $\mathbb{N}_{<n}$ has n elements and moreover, $\mathbb{N}_{<n} = \{0, 1, \cdots n-1\}$ can be put in one-to-one correspondence with $\{1, 2, \cdots, n\}$, for example via the correspondence:

$\mathbb{N}_{<n}$	$\leftrightarrow$	$\{1, 2, \cdots, n\}$
0	$\leftrightarrow$	1
1	$\leftrightarrow$	2
2	$\leftrightarrow$	3
$\vdots$		
$n-1$	$\leftrightarrow$	n

We have the following lemma which is shown using the induction axiom for $\mathbb{N}$. Its proof is left as an exercise (see Exercise 4.4).

Lemma 4.1.5. *For every* $m, n \in \mathbb{N}$, *the following hold:*

1. $\mathbb{N}_{<m}$ *is in one-to-one correspondence with* $\mathbb{N}_{<n}$ *iff* $m = n$.

2. *If we let every element of* $\mathbb{N}_{<m}$ *correspond to a unique element of* $\mathbb{N}_{<n}$ *such that no two different elements of* $\mathbb{N}_{<m}$ *correspond to the same element of* $\mathbb{N}_{<n}$ *then* $m \leq n$.

4.1.4   Countable and Uncountable sets

Cantor classified a set that can be put in one-to-one correspondence with the set of nonzero natural numbers to be *infinitely countable* and stated that all infinitely countable sets have the same size as $\mathbb{N}$. So he took Galileo's point here to mean that the infinite set of perfect squares is countable. Thus, a (proper) part of a set may have the same size as the set. We already saw above many proper subsets of $\mathbb{N}$ (e.g., E, O, $\mathbb{N}_{\geq m}$ for $m \in \mathbb{N}$, etc.) to be in one-to-one correspondence with $\mathbb{N}$ and hence according to Cantor, they would be of the same size as $\mathbb{N}$. In this section, we lay the foundations to this result of Cantor.

Definition 4.1.6 (Finite, Infinite, Size of Finite Sets, Countable, Uncountable).

> 1. *A set S is* finite *iff it can be put into one-to-one correspondence with $\mathbb{N}_{<n}$ for some $n \in \mathbb{N}$, in which case, we also say that the set S has n elements (or that the size of the set S is n) and write $|S| = n$. A set is* infinite *iff it is not finite.*
>
> 2. *A set is* countable *iff it is either finite or can be put into one-to-one correspondence with the nonzero natural numbers $\mathbb{N}^+$. In the latter case, the set is said to be* infinitely countable. *A set which is not countable is said to be* uncountable.

Note that $|\mathbb{N}_{<n}| = n$ for every $n \in \mathbb{N}$ and hence $|\emptyset| = 0$. Note also by Lemma 4.1.5.1, the size (i.e., the number of elements) of a finite set S is unique.

Since there is a one-to-one correspondence between $\mathbb{N}^+$ and $\mathbb{N}$ which takes any n in $\mathbb{N}^+$ to $n - 1$ in $\mathbb{N}$, then any one-to-one correspondence between $\mathbb{N}$ and a set S can be taken to a one-to-one correspondence between $\mathbb{N}^+$ and S.

$\mathbb{N}^+$	$\leftrightarrow$	$\mathbb{N}$	$\mathbb{N}$	$\leftrightarrow$	S		$\mathbb{N}^+$	$\leftrightarrow$	S
n	$\leftrightarrow$	$n-1$	$n-1$	$\leftrightarrow$	c		n	$\leftrightarrow$	c

Hence we can also give the definition of infinitely countable set S in terms of a one-to-one correspondence between $\mathbb{N}$ and S instead of a one-to-one correspondence between $\mathbb{N}^+$ and S. Similarly, on Page 91, we saw that

$$\mathbb{N}_{<n} \quad \leftrightarrow \quad \{1, 2, \cdots, n\}$$

and since a set S is finite iff it can be put into a one-to-one correspondence with $\mathbb{N}_{<n}$ for some n, then a set S is finite iff it can be put into a one-to-one

correspondence with $\{1, 2, \cdots, n\}$ for some n and in which case, we write $|S| = n$. We also have the following basic lemma about finite/infinite sets which states that there is no one-to-one correspondence between two finite sets of different sizes, or between a finite and an infinite set and that the set of natural numbers is infinite.

Lemma 4.1.7. *The following hold:*

1. *If $|S| = m$ and $|T| = n$ where $m, n \in \mathbb{N}$ and $n \neq m$ then there is no one-to-one correspondence between S and T.*

2. *If S is finite and T is infinite then there is no one-to-one correspondence between S and T.*

3. *$\mathbb{N}$ is infinite.*

4. *If A is infinitely countable, then A is infinite.*

Proof. 1. Since $|S| = m$ and $|T| = n$, then by Definition 4.1.6, S (resp. T) can be put into one-to-one correspondence with $\mathbb{N}_{<m}$ (resp. $\mathbb{N}_{<n}$). Now, if S can be put into one-to-one correspondence with T then by Lemma 4.1.4, $\mathbb{N}_{<m}$ can be put into one-to-one correspondence with $\mathbb{N}_{<n}$ and by Lemma 4.1.5, $m = n$. Contradiction.

2. Assume there is a one-to-one correspondence f between S and T. Since S is finite, then let $n \in \mathbb{N}$ be such that S can be put into one-to-one correspondence with $\mathbb{N}_{<n}$. Hence, by Lemma 4.1.4, T can be put into one-to-one correspondence with $\mathbb{N}_{<n}$. This means, T is finite, contradiction.

3. Assume $\mathbb{N}$ is finite and let $n \in \mathbb{N}$ such that $\mathbb{N}$ is in one-to-one correspondence with $\mathbb{N}_{<n}$. Let $m > n$. Since $\mathbb{N}_{<m} \subseteq \mathbb{N}$, we can use the one-to-one correspondence between $\mathbb{N}$ and $\mathbb{N}_{<n}$ to let every element in $\mathbb{N}_{<m}$ correspond to a unique element of $\mathbb{N}_{<n}$ such that no two different elements of $\mathbb{N}_{<m}$ correspond to the same element of $\mathbb{N}_{<n}$. Then, by Lemma 4.1.5 $m \leq n$. This contradicts $m > n$. Hence, $\mathbb{N}$ is infinite.

4. If A is infinitely countable, then A can be put in one-to-one correspondence with $\mathbb{N}$. If A is finite, then A can be put in one-to-one correspondence with $\mathbb{N}_{<n}$ for some $n \in \mathbb{N}$. Hence, $\mathbb{N}$ can be put in one-to-one correspondence with $\mathbb{N}_{<n}$ which is a contradiction by 2 above since $\mathbb{N}_{<n}$ is finite and $\mathbb{N}$ is infinite.

□

The proof of the following lemma is left as an exercise (see Exercise 4.5).

Lemma 4.1.8. *The following hold:*

1. *Let A be a finite non empty set and $x \in A$. Let B be the set that contains all and only the elements of A which differ from x[4]. Then B is finite and $|B| = |A| - 1$.*

2. *Let A be a finite set and $x \notin A$. Let B be the set that contains all and only the elements of A in addition to x[5]. Then B is finite and $|B| = |A| + 1$.*

3. *Let $n \in \mathbb{N}$. If we let every element of set A correspond to a unique element of $\mathbb{N}_{<n}$ such that no two different elements of A correspond to the same element of $\mathbb{N}_{<n}$ then A is finite and $|A| \leq n$.*

4. *If $A \subseteq B$ and B is finite then A is finite and $|A| \leq |B|$.*

5. *If $A \subseteq B$ and A is infinite then B is infinite.*

Up to here, we have seen in Definition 4.1.6 finite sets and infinite sets but we have only spoken about the size of a finite set. We still have not said anything about the size or number of elements of an infinite set. We did mention however that some infinite sets can be infinitely countable. That is, some infinite sets have the property that their elements can be associated on a one-to-one basis to the nonzero natural numbers. In other words, the elements of an infinitely countable set can be counted even though there are infinitely many of them. Now, what about sets that seem to be larger than the set of nonzero natural numbers? They certainly should be infinite, but are they all uncountable? The answer is No. Cantor proved that some sets that would seem to be larger than the set of nonzero natural numbers are also countable. In this section we show that the sets $\mathbb{Z}$, $\mathbb{Q}$ and the set of algebraic numbers are all infinitely countable and moreover that an infinite subset of a countable set is infinitely countable. But then, we close the section by showing that the set of all real numbers x in the interval $0 < x \leq 1$ is uncountable.

Theorem 4.1.9 ($\mathbb{Z}$ is infinitely countable). *The set of all integers, positive, negative, and zero, is infinitely countable.*

Proof. Write the integers in the following order:

$$0, 1, -1, 2, -2, \ldots$$

---

[4]Note here that $B = A \setminus \{x\}$, a set difference which we will define in the next section.
[5]Note here that $B = A \cup \{x\}$, a set union which we will define in the next section.

This implies the following one-to-one correspondence with the nonzero natural numbers:

$$
\begin{array}{ccc}
1 & \leftrightarrow & 0 \\
2 & \leftrightarrow & 1 \\
3 & \leftrightarrow & -1 \\
4 & \leftrightarrow & 2 \\
5 & \leftrightarrow & -2 \\
& \vdots &
\end{array}
$$

□

Theorem 4.1.10 ($\mathbb{Q}$ is infinitely countable). *The set of all rational numbers is infinitely countable.*

Proof. Let us begin by proving that the set $\mathbb{Q}^+$ of all *positive* rational numbers is infinitely countable. Consider the following array of numbers:

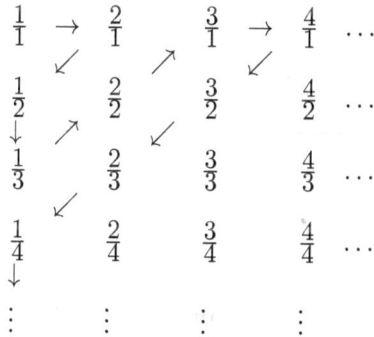

$$
\begin{array}{cccc}
\frac{1}{1} \rightarrow & \frac{2}{1} & \frac{3}{1} \rightarrow & \frac{4}{1} \cdots \\[4pt]
\frac{1}{2} & \frac{2}{2} & \frac{3}{2} & \frac{4}{2} \cdots \\[4pt]
\frac{1}{3} & \frac{2}{3} & \frac{3}{3} & \frac{4}{3} \cdots \\[4pt]
\frac{1}{4} & \frac{2}{4} & \frac{3}{4} & \frac{4}{4} \cdots \\[4pt]
\vdots & \vdots & \vdots & \vdots
\end{array}
$$

If we follow the arrows and delete any rational fractions equal to fractions that have appeared earlier, we get a *listing* $r_1, r_2, \ldots$ of positive rational numbers as follows:

$$
\begin{array}{cccccccccccccc}
r_1 & r_2 & r_3 & r_4 & & r_5 & r_6 & r_7 & r_8 & r_9 & r_{10} & & & & r_{11} & \cdots \\
\frac{1}{1} & \frac{2}{1} & \frac{1}{2} & \frac{1}{3} & \cancel{\tfrac{2}{2}} & \frac{3}{1} & \frac{4}{1} & \frac{3}{2} & \frac{2}{3} & \frac{1}{4} & \frac{1}{5} & \cancel{\tfrac{2}{4}} & \cancel{\tfrac{3}{3}} & \cancel{\tfrac{4}{2}} & \frac{5}{1} & \cdots
\end{array}
$$

This listing $r_1, r_2, \ldots$ includes all the positive rational numbers and is in one-to-one correspondence with the nonzero natural numbers $1, 2, 3, \cdots$ where $r_i \leftrightarrow i$ for each $i \in \mathbb{N}^+$. Thus, by Definition 4.1.6, the positive rational numbers are infinitely countable. If we then follow the procedure we saw above for the integers and write the rational numbers as

$$
0, r_1, -r_1, r_2, -r_2, \ldots,
$$

we get a one-to-one correspondence between all the rational numbers and the nonzero natural numbers. Hence by Definition 4.1.6, the rational numbers are infinitely countable.   □

And perhaps even more surprising, there is a larger set of numbers which is countable. This is the set of *algebraic* numbers.

Definition 4.1.11 (Polynomials, Algebraic Numbers). *A polynomial of degree $n \geq 0$ is a function of the form $f(x) = a_n x^n + a_{n-1} x^{n-1} + \cdots + a_2 x^2 + a_1 x + a_0$,[6] where the a_i's are real numbers which are called the coefficients of the polynomial.*

An algebraic number is a number (which may be real or complex) which is a root of a polynomial of the form

$$f(x) = a_n x^n + a_{n-1} x^{n-1} + \ldots a_1 x + a_0,$$

where $a_n \neq 0$ and each a_i is an integer (positive, negative, or 0).

The following lemma shows that all rational numbers (and much more) are algebraic.

Lemma 4.1.12. *Algebraic numbers include all rational numbers and all of their roots. I.e., for all $\frac{a_1}{a_2} \in \mathbb{Q}$ and for all $n \in \mathbb{N}^+$, $\sqrt[n]{\frac{a_1}{a_2}}$ is algebraic.*

The proof is left as an exercise (see Exercise 4.6).

Even though in the next theorem, we include the complex algebraic numbers, the same proof can be slightly adapted to apply for the real algebraic numbers (and hence excluding the complex ones).

Theorem 4.1.13 (Algebraic numbers are infinitely countable). *The set of all algebraic numbers is infinitely countable.*

Proof. Let a polynomial $f(x) = a_n x^n + a_{n-1} x^{n-1} + \ldots a_1 x + a_0$. Without loss of generality we may assume that $a_n > 0$. Define the "height" of the polynomial as the positive number

$$h = n + a_n + |a_{n-1}| + \ldots + |a_1| + |a_0|.$$

Clearly, h is a nonzero natural number. Furthermore, a given height is possessed by only a finite number of polynomials, because $n \leq h$ and for every n, $|a_n| \leq h$. It follows that there are only a finite number of algebraic numbers which are roots of polynomials of a given height. This makes it possible to write all the algebraic numbers as a sequence. First, we write all algebraic numbers associated with polynomials according to heights:

---

[6]If $f(x) = a_0$ where $a_0 \neq 0$, then the degree of the polynomial $f(x)$ is 0. If however $f(x) = 0$ then the degree of $f(x)$ is undefined.

- Height is 2: The only polynomials of height 2 are x and 2. Hence, the number 0 is the only algebraic number associated with polynomials of height 2.

- Height is 3: The polynomials of height 3 are $x^2, 2x, x+1, x-1$, and 3. These give the new roots -1 and $+1$.

- Height is 4: The new roots arising from polynomials of height 4 are, the reals in order of magnitude, $-2, -\frac{1}{2}, +\frac{1}{2}$, and $+2$ as well as the remaining complex numbers $-i$ and i.

- Height is 5: The new numbers we get from polynomials of height 5 are the reals $-3, -\frac{1}{2} - \frac{1}{2}\sqrt{5}, -\sqrt{2}, -\frac{1}{2}\sqrt{2}, \frac{1}{2} - \frac{1}{2}\sqrt{5}, -\frac{1}{3}, \frac{1}{3}, -\frac{1}{2} + \frac{1}{2}\sqrt{5}, \frac{1}{2}\sqrt{2}, \sqrt{2}, \frac{1}{2} + \frac{1}{2}\sqrt{5}$, and 3, as well as the complex numbers $-\frac{1}{2} - \frac{i}{2}\sqrt{3}, -\frac{1}{2} + \frac{i}{2}\sqrt{3}, \frac{1}{2} - \frac{i}{2}\sqrt{3}, \frac{1}{2} + \frac{i}{2}\sqrt{3}$.

- etc...

By allowing the height to run through all the natural numbers and writing down for each such height the finitely many newly arising algebraic numbers corresponding to each value of the height, we obtain a sequence of distinct algebraic numbers. Since every polynomial has a height, all algebraic numbers appear in this sequence. □

Theorem 4.1.14. *Every infinite subset of a countable set is infinitely countable.*

Proof. Let $S \subseteq T$ where S is infinite and T is countable. By Lemma 4.1.8, T is also infinite. Let $a_1, a_2, \ldots, a_n, \ldots$ give the correspondence between the elements of T and the nonzero natural numbers (which exists because T is countable and not finite). Start with a_1 and go through this sequence finding the elements of S. Since $S \neq \emptyset$, there will be a first such element of this sequence; call that b_1. Continue this process, where, in each case, b_n be the first a_i after b_{n-1} which is an element of S. This way, we find a one-to-one correspondence between S and $\mathbb{N}^+$ and hence, S is countable. □

Corollary 4.1.15. *Every subset of a countable set is countable.*

Proof. Let $S \subseteq T$ and T countable. If S is finite then S is countable. If S is infinite then by Theorem 4.1.14, S is infinitely countable and hence countable. □

Theorem 4.1.14 strongly suggests that countable sets are the "smallest" infinite sets.

It may appear from these results that all infinite sets are countable, but this is not the case. First, we give some definitions of intervals:

Definition 4.1.16. *An* open interval *is a set of values x of the form $a < x < b$, where $a < b$, which does not include the endpoints. Such an open interval is denoted (a, b). A* closed interval *is a set of values x of the form $a \leq x \leq b$, which does include the endpoints. Such a closed interval is denoted $[a, b]$. A* half-open interval *is an interval which includes one endpoint but not the other, so $(a, b]$ is the collection of values x such that $a < x \leq b$, which includes b but not a, and $[a, b)$ is the collection of values x such that $a \leq x < b$, which includes a but not b.*

Theorem 4.1.17 ($(0, 1]$ **is uncountable**)**.** *The set of all real numbers x in the interval $0 < x \leq 1$ is uncountable.*

Proof. First note that every real number x in the interval $0 < x \leq 1$ can be written as an infinite decimal of the form $0.a_1 a_2 a_3 \ldots$. We will agree that any decimal that ends in a sequence of all 0s, such as $0.500\ldots$ will be written as the equivalent decimal ending in all 9s, in this case $0.4999\ldots$, then no two of these infinite decimals will represent the same real number. (Note that we are writing 1 as $0.999\ldots$.)

Now suppose that the set of real numbers x in the interval $0 < x \leq 1$ were countable, we would have a correspondence of the form

$$1 \leftrightarrow 0.a_{11} a_{12} a_{13} \ldots$$
$$2 \leftrightarrow 0.a_{21} a_{22} a_{23} \ldots$$
$$3 \leftrightarrow 0.a_{31} a_{32} a_{33} \ldots$$
$$\vdots \quad \vdots \quad \vdots$$

To complete the proof we need only define an infinite decimal d in the interval $0 < x < 1$ which does not correspond to a nonzero natural number under this correspondence. So we define $d = 0.b_1 b_2 b_3$ as follows:

$$b_n = \begin{cases} 3, & \text{if } a_{nn} = 2; \\ 2, & \text{otherwise.} \end{cases}$$

Then d cannot be in the indicated correspondence because its nth digit b_n differs from a_{nn}, the nth digit of the nth decimal in the correspondence. Hence, d does not correspond to a nonzero natural numbers under this correspondence. So, our assumption that the set of real numbers x in the interval $0 < x \leq 1$ is countable, is incorrect. It must be uncountable. $\square$

It may appear with this proof that something is missing, since this proof only gives one real number in the interval that does not correspond to a nonzero natural numbers. But note that we have infinitely many ways of finding this real number in the interval: we constructed d by using the main diagonal going down to the right from a_{11}. But we could have used any path through the array to construct a different d, and so there are infinitely many real numbers that to not correspond to nonzero natural numbers.

Theorem 4.1.18 ($\mathbb{R}$ is uncountable). *The set of the real numbers $\mathbb{R}$ is uncountable.*

Proof. If $\mathbb{R}$ were countable, then since the set of all real numbers x in the interval $0 < x \leq 1$ is a subset of $\mathbb{R}$, then by Theorem 4.1.14, that set would also be countable. But, by Theorem 4.1.17, that set is uncountable. Hence, our assumption that $\mathbb{R}$ is countable cannot hold and $\mathbb{R}$ is uncountable. $\square$

4.1.5 Exercises

Exercise 4.1. Prove Lemma 4.1.2.

Exercise 4.2. Show that $\emptyset \in \{\emptyset\}$ and $\emptyset \subset \{\emptyset\}$.
Similarly, show that $\{\emptyset\} \in \{\{\emptyset\}\}$ and $\{\emptyset\} \subset \{\{\emptyset\}\}$.

Exercise 4.3. Assume that $0 = \emptyset$, $1 = \{\emptyset\}$, $2 = \{\{\emptyset\}\}$, $\cdots$. Let $S = \{\emptyset, \{\emptyset\}, \{\{\emptyset\}\}, \cdots\}$. Show that $S = \mathbb{N}$.

Exercise 4.4. Prove Lemma 4.1.5.

Exercise 4.5. Prove Lemma 4.1.8.

Exercise 4.6. Prove Lemma 4.1.12.

Exercise 4.7. 1. Give all the polynomials for height 4 as in the proof of Theorem 4.1.13.

2. Give the polynomials for height 5 which give all the new numbers listed under height 5 in the proof of Theorem 4.1.13.

Exercise 4.8. For each of the following sets, show that it is countable by giving the relevant one-to-one correspondences with the nonzero natural numbers.

1. $E = \{x \in \mathbb{Z} : x = 2y \text{ for some } y \in \mathbb{Z}\}$.

2. $O = \{x \in \mathbb{Z} : x = 2y + 1 \text{ for some } y \in \mathbb{Z}\}$.

4.2   Set Operations and Functions

We are used to operating on numbers. We can also operate on sets. This section introduces the basic operations on sets and also define functions which will be crucial for establishing properties on sets and their sizes. In fact, if you think about it, we have already used the function notion in this chapter: the one-to-one correspondence that was crucial to the size of a set is nothing else than the notion of a bijective function that we will define below. So far, we have been informal in our description of one-to-one correspondences and in our proofs that something is indeed a one-to-one correspondence. There comes a stage where we need to have a more precise definition in order to either construct or prove that something is a one-to-one correspondence. That is why we need the details on functions we introduce here.

Although we introduce formal logic later on in this book (see Chapter 6), we use basic logic to carry out some proofs in this chapter. We will be as clear as possible, and the student can refer to Chapter 6.

4.2.1   Basic Set Operations

Definition 4.2.1. *If S and T are sets, then the* union *of the sets S and T, in symbols $S \cup T$, is the set which consists of all the elements of S together with all the elements of T. We can also write this as*

$$S \cup T = \{x \in U : x \in S \text{ or } x \in T\}.$$

Example 4.2.2. *If $S = \{1, 2, 3\}$ and $T = \{2, 3, 4, 5\}$, then $S \cup T = \{1, 2, 3, 4, 5\}$.*

If $S = \{x : x \text{ is a rational number}\}$ and $T = \{x : x \text{ is an irrational number}\}$, then $S \cup T = \{x : x \text{ is a real number}\}$. Here, the universe U is not written (see Page 87), but it could be the set of real numbers $\mathbb{R}$ or the set of complex numbers $\mathbb{C}$, etc.

As we see, the union of sets amounts to putting them together.

Theorem 4.2.3. *The operation of union on sets has the following properties:*

1. *(Idempotence) For any set S, $S \cup S = S$.*

2. *(Identity) $\emptyset$ is a identity for sets under $\cup$; i.e., $\emptyset \cup S = S = S \cup \emptyset$.*

3. *(Commutativity) For any sets S and T, $S \cup T = T \cup S$.*

4. *(Associativity) For any sets S, T, and R, $S \cup (T \cup R) = (S \cup T) \cup R$.*

Proof. We only prove the last item and leave the first 3 as exercises (see exercise 4.9).

We need to prove that $x \in S \cup (T \cup R)$ if and only if $x \in (S \cup T) \cup R$. We have

$$
\begin{aligned}
x \in S \cup (T \cup R) \quad &\Leftrightarrow \quad x \in S \text{ or } x \in T \cup R \\
&\Leftrightarrow \quad x \in S \text{ or } (x \in T \text{ or } x \in R) \\
&\Leftrightarrow \quad (x \in S \text{ or } x \in T) \text{ or } (x \in R) \\
&\Leftrightarrow \quad x \in S \cup T \text{ or } x \in R \\
&\Leftrightarrow \quad x \in (S \cup T) \cup R.
\end{aligned}
$$

$\square$

Part 4 of this theorem suggests that we can take the union of more than two sets.

Definition 4.2.4. *The union of many sets consists of the set of all elements that are in at least one of the sets. This is denoted as follows:*

- *If there are finitely many sets $S_1, S_2, \ldots, S_n$:*

$$
\bigcup_{i=1}^{n} S_i
$$

Hence, $\bigcup_{i=1}^{n} S_i = \{x \in U : x \in S_i \text{ for some } 1 \le i \le n\}$.

- *If there are countably infinite many sets, $S_1, S_2, \ldots$:*

$$
\bigcup_{i=1}^{\infty} S_i.
$$

Hence, $\bigcup_{i=1}^{\infty} S_i = \{x \in U : x \in S_i \text{ for some } i \in \mathbb{N}^+\}$.

- *If there are uncountably many sets S where all the sets S are elements of a set T:*

$$
\bigcup_{S \in T} S.
$$

Hence, $\bigcup_{S \in T} S = \{x \in U : x \in S \text{ for some } S \in T\}$.

The operation of union of sets tends to make the sets bigger. There is another operation, intersection which makes them smaller.

Definition 4.2.5. *The* intersection *of the sets S and T, in symbols $S \cap T$, is the set of all elements that are in both S and T at the same time. We can write this as*

$$S \cap T = \{x \in U : x \in S \text{ and } x \in T\}.$$

Example 4.2.6. *If $S = \{1, 2, 3\}$ and $T = \{2, 3, 4, 5\}$, then $S \cap T = \{2, 3\}$.*

If $S = \{x : x \text{ is a rational number}\}$ and $T = \{x : x \text{ is an irrational number}\}$, then $S \cap T = \emptyset$.

Theorem 4.2.7. *The intersection of sets has the following properties:*

1. *(Idempotence) For any set S, $S \cap S = S$.*

2. *(Zero) The set $\emptyset$ acts like 0: for any set S, $\emptyset \cap S = \emptyset = S \cap \emptyset$.*

3. *(Commutativity) For any two set S and T, $S \cap T = T \cap S$.*

4. *(Associativity) For any three sets S, T, and R, $S \cap (T \cap R) = (S \cap T) \cap R$.*

Proof. We only prove the last item and leave the first 3 as exercises (see exercise 4.10).
We need to prove that $x \in S \cap (T \cap R)$ if and only if $x \in (S \cap T) \cap R$. We have

$$
\begin{aligned}
x \in S \cap (T \cap R) \quad &\Leftrightarrow \quad x \in S \text{ and } x \in T \cap R \\
&\Leftrightarrow \quad x \in S \text{ and } (x \in T \text{ and } x \in R) \\
&\Leftrightarrow \quad (x \in S \text{ and } x \in T) \text{ and } x \in R \\
&\Leftrightarrow \quad x \in S \cap T \text{ and } x \in R \\
&\Leftrightarrow \quad x \in (S \cap T) \cap R.
\end{aligned}
$$

$\square$

Definition 4.2.8. *The intersection of more than two sets consists of those elements in all of the sets involved. This is denoted as follows:*

- *If there are finitely many sets $S_1, S_2, \ldots, S_n$:*

$$\bigcap_{i=1}^{n} S_i$$

Hence, $\bigcap_{i=1}^{n} S_i = \{x \in U : x \in S_i \text{ for every } 1 \leq i \leq n\}$.

- *If there are countably infinite many sets, $S_1, S_2, \ldots$:*

$$\bigcap_{i=1}^{\infty} S_i$$

Hence, $\bigcap_{i=1}^{\infty} S_i = \{x \in U : x \in S_i \text{ for every } i \in \mathbb{N}^+\}$.

- *If there are uncountably many sets S where all the sets S are elements of a set T:*

$$\bigcap_{S \in T} S.$$

Hence, $\bigcap_{S \in T} S = \{x \in U : x \in S \text{ for every } S \in T\}$.

Definition 4.2.9. *If the intersection of two or more sets is the empty set, then the sets are said to be* disjoint.

Example 4.2.10. *Assume for each $n \in \mathbb{N}^+$, $S_n = \{n\}$. Then, $\bigcup_{i=1}^{\infty} S_i = \mathbb{N}^+$ and $\bigcap_{i=1}^{\infty} S_i = \emptyset$. Additionally, $\bigcup_{i=1}^{n} S_i = \{1, 2, \cdots, n\}$ and $\bigcap_{i=1}^{n} S_i = \emptyset$.*

As shown by part 2 of Theorem 4.2.7, the intersection of sets one of which is empty is always empty.

In ordinary algebra, we are used to the fact that multiplication distributes over addition:

$$a(b + c) = ab + ac.$$

But addition does not distribute over multiplication; we do not have

$$a + bc = (a + b)(a + c).$$

In set theory we have that each of $\cup$ and $\cap$ distributes over the other:

Theorem 4.2.11. *Each of the operations $\cup$ and $\cap$ distributes over the other. That is, we have both*

$$S \cap (T \cup R) = (S \cap T) \cup (S \cap R)$$

and

$$S \cup (T \cap R) = (S \cup T) \cap (S \cup R).$$

Proof. For the first part, we have

$$
\begin{aligned}
x \in S \cap (T \cup R) \quad &\Leftrightarrow \quad x \in S \text{ and } (x \in T \text{ or } x \in R) \\
&\Leftrightarrow \quad (x \in S \text{ and } x \in T) \text{ or } (x \in S \text{ and } x \in R) \\
&\Leftrightarrow \quad (x \in S \cap T) \text{ or } (x \in S \cap R) \\
&\Leftrightarrow \quad x \in (S \cap T) \cup (S \cap R)
\end{aligned}
$$

The second part is similar. We leave it as an exercise (see Exercise 4.11).

□

Definition 4.2.12. *The* difference *between sets S and T, in symbols $S \setminus T$, consists of those elements of S which are* not *in T. This can be written as*

$$S \setminus T = \{x \in U : x \in S \text{ and } x \notin T\}.$$

The assumption here is that $T \subseteq S$. If S is the universe of discourse U, then $S \setminus T$ is $U \setminus T$, and is called the complement *of T, or T^c.*

Example 4.2.13.   • *If $S = \mathbb{Z}$ and $T = \mathbb{N}$ then $S \setminus T = \{-n : n \in \mathbb{N}^+\}$.*

   • *If $S = \mathbb{R}$ and $T = \mathbb{Q}$ then $S \setminus T = \{x : x \text{ is an irrational number}\}$.*

   • *If $S = \mathbb{N}$ and $T = \{x \in \mathbb{N} : x = 2y \text{ for some } y \in \mathbb{N}\}$, then $S \setminus T = \{x \in \mathbb{N} : x = 2y + 1 \text{ for some } y \in \mathbb{N}\}$.*

Theorem 4.2.14. *For sets S, T, and R, we have that*

$$S \setminus (T \cap R) = (S \setminus T) \cup (S \setminus R)$$

and

$$S \setminus (T \cup R) = (S \setminus T) \cap (S \setminus R).$$

Proof. For the first result, we have

$$
\begin{aligned}
x \in S \setminus (T \cap R) \quad &\Leftrightarrow \quad x \in S \text{ and } x \notin (T \cap R) \\
&\Leftrightarrow \quad x \in S \text{ and not } x \in (T \cap R) \\
&\Leftrightarrow \quad x \in S \text{ and not both } (x \in T) \text{ and } (x \in R) \\
&\Leftrightarrow \quad (x \in S \text{ and } x \notin T) \text{ or } (x \in S \text{ and } x \notin R) \\
&\Leftrightarrow \quad (x \in S \setminus T) \text{ or } x \in S \setminus R) \\
&\Leftrightarrow \quad x \in (S \setminus T) \cup (S \setminus R).
\end{aligned}
$$

For the second result,

$$
\begin{aligned}
x \in S \setminus (T \cup R) \quad &\Leftrightarrow \quad x \in S \text{ and } x \notin (T \cup R) \\
&\Leftrightarrow \quad x \in S \text{ and not } x \in (T \cup R) \\
&\Leftrightarrow \quad x \in S \text{ and not } ((x \in T) \text{ or } (x \in R)) \\
&\Leftrightarrow \quad x \in S \text{ and not } (x \in T) \text{ and not } (x \in R) \\
&\Leftrightarrow \quad x \in S \text{ and } (x \notin T) \text{ and } (x \notin R) \\
&\Leftrightarrow \quad x \in S \text{ and } x \in S \text{ and } (x \notin T) \text{ and } (x \notin R) \\
&\Leftrightarrow \quad (x \in S \text{ and } x \notin T) \text{ and } (x \in S \text{ and } x \notin R) \\
&\Leftrightarrow \quad (x \in S \setminus T) \text{ and } (x \in S \setminus R) \\
&\Leftrightarrow \quad x \in (S \setminus T) \cap (S \setminus R).
\end{aligned}
$$

□

Note in the above proof that we used a famous logic rule known as De Morgan's law, which can be written as follows for A, B, C propositions, $\vee$ standing for disjunction "or", $\wedge$ standing for conjunction "and", and $\neg$ standing for negation "not":

$$\neg(A \wedge B) = \neg A \vee \neg B \qquad \neg(A \vee B) = \neg A \wedge \neg B$$

We will see De Morgan's law in Chapter 6.

Corollary 4.2.15. *For sets S and T,*

$$(S \cap T)^c = S^c \cup T^c$$

and

$$(S \cup T)^c = S^c \cap T^c.$$

There is another operation that is important for us: the *Cartesian product*.

Definition 4.2.16 (Cartesian product). *The* Cartesian product *of sets S and T is defined to be*

$$S \times T = \{(x, y) : x \in S \text{ and } y \in T\}.$$

If there are finitely many sets $S_1, S_2, \ldots, S_n$, then the Cartesian product *of sets $S_1, S_2, \ldots, S_n$, is defined to be*

$$S_1 \times S_2 \cdots \times S_n = \{(x_1, x_2, \cdots, x_n) : x_i \in S_i \text{ for all } 1 \leq i \leq n\}.$$

Example 4.2.17. *If $S = \{1, 2, 3\}$ and $T = \{2, 3, 4, 5\}$, then $S \times T =$*

$$\{(1,2), (1,3), (1,4), (1,5), (2,2), (2,3), (2,4), (2,5), (3,2), (3,3), (3,4), (3,5)\}.$$

The set of points of the Euclidean plane as used in analytic geometry is the set $\mathbb{R} \times \mathbb{R}$. This is why this product is called the "Cartesian" product.[7]

Another definition that is important to us is the notion of *power set*.

Definition 4.2.18 (Powerset). *If S is a set, the* power set of S, *or $\mathcal{P}S$, is the set of all subsets of S.*

Example 4.2.19. *If $S = \{1, 2, 3\}$ then*

$$\mathcal{P}S = \{\emptyset, \{1\}, \{2\}, \{3\}, \{1,2\}, \{1,3\}, \{2,3\}, \{1,2,3\}\}.$$

---

[7]The idea is due to Rene Descartes (1596–1650), a French philosopher and mathematician. He invented what we now call *analytic geometry*.

4.2.2   Basic Properties of Functions

In what follows, we give the main definitions needed for functions and their properties. Note that the definition of a bijective function captures exactly the definition of a one-to-one correspondence we gave in Section 4.1.3 and hence everything we said about one-to-one correspondences there holds here. Now of course, if you like, you can carry out many of the proofs we did in Section 4.1.3 more precisely using the more explicit definition of function and its properties.

Definition 4.2.20 (Functions, Injection, Surjection, Bijection, $=$).
If S and T are sets we write $f : S \mapsto T$ for the function that maps each element $a \in S$ to a unique element $f(a) \in T$.

Let $f : S \mapsto T$. We define the following:

- *f is injective (or one-to-one) iff (for every $a, b \in S$, if $f(a) = f(b)$ then $a = b$).*

- *f is surjective (or onto) iff (for every $b \in T$, there is an $a \in S$ such that $f(a) = b$).*

- *f is bijective (or one-to-one correspondence) iff (for every $b \in T$, there is a unique $a \in S$ such that $f(a) = b$).*

- *Function Extensionality: If also $g : S \mapsto T$, we define $f = g$ iff for all $x \in S$, $f(x) = g(x)$.*

Note here that if $f : S \mapsto T$, then we are assuming that every element a of S is mapped to an element $f(a)$ of T so no element of S is left without being mapped into an element in T. Furthermore, we are requiring that each element a of S is mapped to a unique element $f(a)$ of T. So, it is impossible that a gets two different mappings in T. So in the lefthand side figure below, we do not have a function because two arrows are coming out of 1. In the righthand side figure below, again we do not have a function because no arrow is coming out of 2.

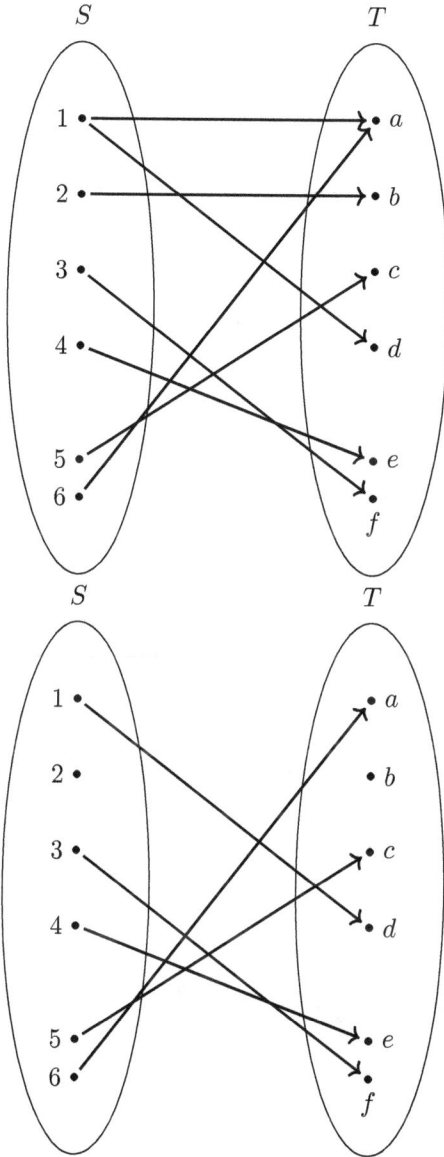

Now, both figures on the left and right below do represent functions. However, for injectivity, the figure on the left below does not represent an injective function since two arrows are coming into a, but the figure on the right below does indeed represent an injective function.

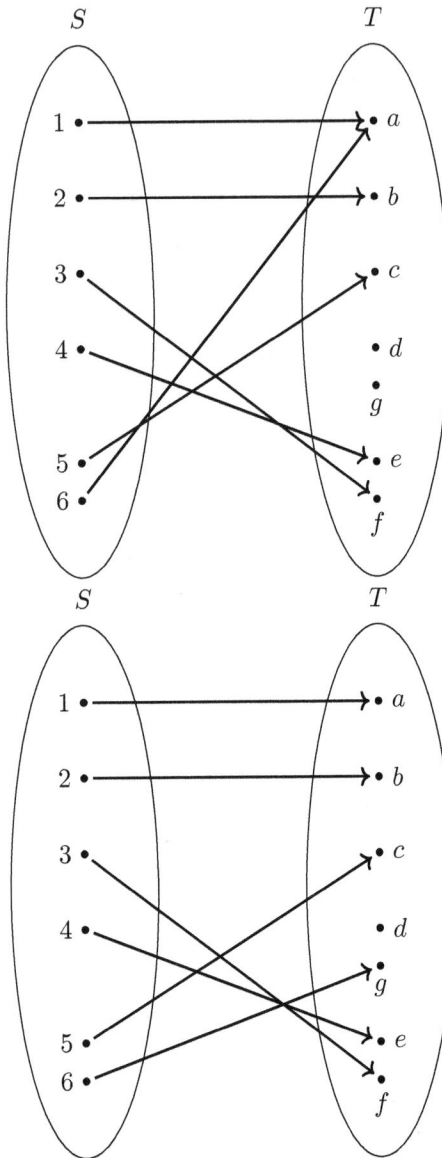

Now, these functions represented just here above on the left and right are not surjective since there are elements in T to which no arrow is coming. However, the functions below on both left and right, are surjective. Of the 2 surjective functions below, only the one on the right is also injective, and only the one on the right is bijective.

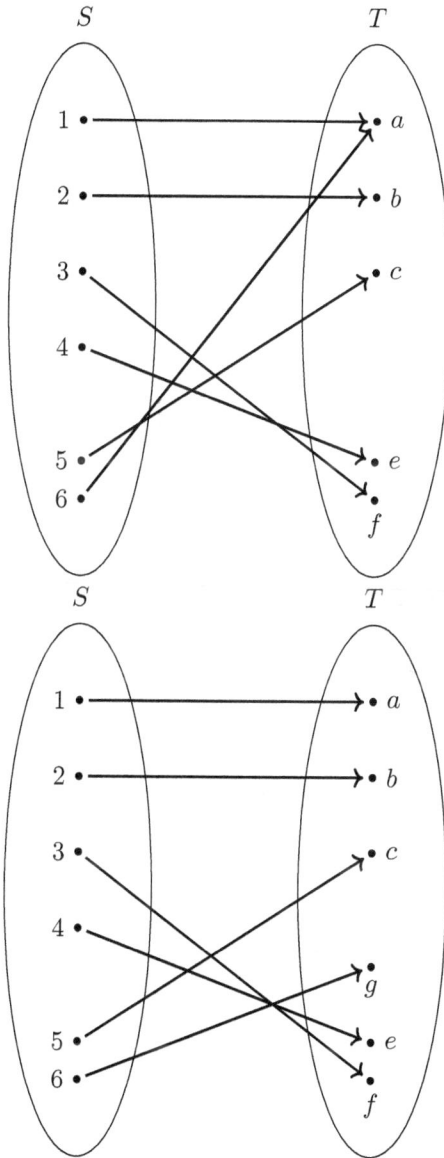

Definition 4.2.21 (Image, pre-image, composition, inverse, restriction). *Let f be a function such that f : S ↦ T.*

- If $A \subseteq S$, we define the image of A by f, $f[A] = \{f(a) : a \in A\}$.

- If $B \subseteq T$, we define pre-image of B by f, $f^{-1}[B] = \{a \in S : f(a) \in B\}$.

- If also $g : T \mapsto V$ is a function then $g \circ f : S \mapsto V$ such that $g \circ f(a) = g(f(a))$ is a function from S to V.

- If f is bijective, we define $f^{-1} : T \mapsto S$ such that $(f^{-1}(y) = x$ iff $f(x) = y)$.

- If $S' \subseteq S$, we define $f_{|S'}$, the restriction of f to S' for the function $f_{|S'} : S' \mapsto T$. So, for all $x \in S'$, $f_{|S'}(x) = f(x)$.

Example 4.2.22. Let $f : \mathbb{N} \mapsto \mathbb{N}$ be such that $f(x) = x^2$. Then,

- If $A = \{1, 2, 3, 4\}$ then $f[A] = \{1, 4, 9, 16\}$.

- $f^{-1}[\mathbb{N}_{<26}] = \{1, 2, 3, 4, 5\}$. Note that there are elements y in $\mathbb{N}_{<26}$ such that for every $x \in \mathbb{N}$, $f(x) \neq y$.

- If $g : \mathbb{N} \mapsto \mathbb{N}$ such that $g(x) = x + 1$. Then, $g \circ f :: \mathbb{N} \mapsto \mathbb{N}$ such that $g \circ f(x) = x^2 + 1$.

- $f_{|\mathbb{N}_{<26}} : \mathbb{N}_{<26} \mapsto \mathbb{N}$ such that for all $x \in \mathbb{N}_{<26}$, $f_{|\mathbb{N}_{<26}}(x) = f(x) = x^2$.

The next lemma whose proof is left as an exercise (see Exercise 4.15), sets the properties we need about functions.

Lemma 4.2.23. Let $f : S \mapsto T$. The following hold:

1. f is bijective iff (f is both injective and surjective).

2. If f is bijective then $f^{-1} : T \mapsto S$ is a function which is also bijective and $f \circ f^{-1} = 1_T : T \mapsto T$ and $f^{-1} \circ f = 1_S : S \mapsto S$ where $1_S(x) = x$ for any $x \in S$ and $1_T(y) = y$ for any $y \in T$.

3. Let $g : T \mapsto V$. The following hold:

    (a) $g \circ f : S \mapsto V$ is also a function.

    (b) If f and g are injective, then $g \circ f$ is also injective.

    (c) If f and g are surjective, then $g \circ f$ is also surjective.

    (d) If f and g are bijective, then $g \circ f$ is also bijective.

4. f is injective iff f is one-to-one correspondence between f and $f[S]$.

5. If f is bijective then there is also a bijection between $\mathcal{P}S$ and $\mathcal{P}T$.

6. If f is injective and $S' \subseteq S$ then $f_{|S'} : S' \mapsto T$ is also injective.

7. If $S' \subseteq S$ then there is an injection from S' to S.

Sometimes, it is hard to find a bijection between two sets S and T. An easier step is to find an injection from S to T and another injection from T to S. This is guaranteed by the following theorem which was conjectured by Cantor and independently proved by Schröder and Bernstein in the 1890s:

Theorem 4.2.24 (Cantor-Schröder-Bernstein). *Let S and T be sets. If there are two injections $f : S \mapsto T$ and $g : T \mapsto S$, then there is a bijection between S and T.*

Proof. For $n \geq 1$, define S_n and T_n as follows:

$$S_1 = S \qquad\qquad T_1 = T$$
$$S_{n+1} = g[f[S_n]] \qquad\qquad T_{n+1} = f[g[T_n]]$$
$$S^* = \bigcap_{n=1}^{\infty} S_n \qquad\qquad T^* = \bigcap_{n=1}^{\infty} T_n$$

By the easy Exercise 4.20, we have for all $n \geq 1$,

- $\cdots \subseteq S_{n+2} \subseteq g[T_{n+1}] \subseteq S_{n+1} \subseteq g[T_n] \subseteq S_n \cdots \subseteq g[T_3] \subseteq S_3 \subseteq g[T_2] \subseteq S_2 \subseteq g[T_1] \subseteq S_1$ and

- $\cdots \subseteq T_{n+2} \subseteq f[S_{n+1}] \subseteq T_{n+1} \subseteq f[S_n] \subseteq T_n \cdots \subseteq f[S_3] \subseteq T_3 \subseteq f[S_2] \subseteq T_2 \subseteq f[S_1] \subseteq T_1$

- $T^* = f[S^*]$ and $f_{|S*} : S^* \mapsto T^*$ is a bijection.

- $U'_n = f[S_n] \setminus T_{n+1} = f[S_n \setminus g[T_n]]$ and $V'_n = g[T_n] \setminus S_{n+1} = g[T_n \setminus f[S_n]]$. Let $U_n = S_n \setminus g[T_n]$ and $V_n = T_n \setminus f[S_n]$.

By the above items, you can establish easily that:

- For all $n \geq 1$, $U_n \subseteq S$, $V_n \subseteq T$, $f_{|U_n}[U_n] = f[U_n] = U'_n$ and $g_{|V_n}[V_n] = g[V_n] = V'_n$. Since f and g are injections, then by Lemma 4.2.23, $f_{|U_n} : U_n \mapsto U'_n$ and $g_{|V_n} : V_n \mapsto V'_n$ are bijections.

- For all $n, m \geq 1$, S^*, U_n, V'_n, U_m, V'_m are all mutually exclusive (they have no elements in common) and $S = S^* \cup U_1 \cup V'_1 \cup U_2 \cup V'_2 \cup \cdots$.

- For all $n, m \geq 1$, T^*, V_n, U'_n, V_m, U'_m are all mutually exclusive and $T = T^* \cup U'_1 \cup V_1 \cup U'_2 \cup V_2 \cup \cdots$.

Now look at the following one-to-one correspondences we have established so far:

$$
\begin{array}{ccc}
S^* & \overset{f_{|S^*}}{\leftrightarrow} & T^* \\[4pt]
U_1 & \overset{f_{|U_1}}{\leftrightarrow} & U_1' \\[4pt]
V_1' & \overset{(g_{|V_1})^{-1}}{\leftrightarrow} & V_1 \\[4pt]
U_2 & \overset{f_{|U_2}}{\leftrightarrow} & U_2' \\[4pt]
V_2' & \overset{(g_{|V_2})^{-1}}{\leftrightarrow} & V_2 \\[4pt]
\vdots & \vdots & \vdots
\end{array}
$$

Since S is the union of the mutually exclusive sets S^*, U_1, V_1', U_2, V_2', $\cdots$, we can build a function $h : S \mapsto T$ as follows:

$$
h(x) = \begin{cases}
f_{|S^*}(x) & \text{if } x \in S^* \\
f_{|U_n}(x) & \text{if } x \in U_n \text{ for some } n \geq 1 \\
(g_{|V_n})^{-1}(x) & \text{if } x \in V_n' \text{ for some } n \geq 1
\end{cases}
$$

Since also T is the union of the mutually exclusive sets T^*, U_1', V_1, U_2', V_2, $\cdots$, and all the functions $f_{|S^*}$, $f_{|U_n}$ and $(g_{|V_n})^{-1}$, for $n \geq 1$ are bijections, we can show that h is a bijection (see Exercises 4.17 and 4.18). □

Corollary 4.2.25. *Let S and T be sets. There is a bijection between S and T iff there is an injection from S to T and there is an injection from T to S.*

Proof. If $f : S \mapsto T$ is bijective then $f : S \mapsto T$ and $f^{-1} : T \mapsto S$ are injective (see Lemma 4.2.23).
If there are injections $f : S \mapsto T$ and $g : T \mapsto S$ then by Cantor-Schröder-Bernstein theorem there is a bijections between S and T. □

4.2.3   Exercises

Exercise 4.9. Prove the first three items of Theorem 4.2.3.

Exercise 4.10. Prove the first three items of Theorem 4.2.7.

Exercise 4.11. Prove the second part of Theorem 4.2.11.

Exercise 4.12.   1. Let $A \subseteq B$. Show that $A \cap B = A$ and $A \cup B = B$.

2. Let $A \subseteq B$ and $C \subseteq D$ Show that $A \cap C \subseteq B \cap D$ and $A \cup C \subseteq B \cup D$.

3. Let a set T and for every $n \geq 1$, a set T_n such that $T_n \subseteq T$. Show that for all $i \geq 1$, $\bigcap_{n=1}^{\infty} T_n \subseteq T_i \subseteq \bigcup_{n=1}^{\infty} T_n \subseteq T$.

Exercise 4.13. Let $n \in \mathbb{N}^+$.

1. Give 2 different elements of $\mathcal{P}\mathbb{N}_{<n}$ which are finite sets and give their sizes.

2. Give 2 different elements of $\mathcal{P}\mathbb{N}_{\geq n}$ which are finite sets and give their sizes.

3. Give 2 different elements of $\mathcal{P}\mathbb{N}_{\geq n}$ which are infinite sets.

4. Let $S \in \mathcal{P}\mathbb{N}_{\geq n}$. Is S countable or uncountable? Justify your answer.

Exercise 4.14. Is $\mathbb{R} \setminus \mathbb{N}_{n \geq 2}$ countable or uncountable? Justify your answer.

Exercise 4.15. Prove Lemma 4.2.23.

Exercise 4.16. Let S and T be two sets let $f : S \mapsto T$ be a function. Assume $A \subseteq B \subseteq S$ and $C \subseteq D \subseteq T$. Show that $f[A] \subseteq f[B]$ and $f^{-1}[C] \subseteq f^{-1}[D]$.

Exercise 4.17. Let $f : A_1 \mapsto B_1$ and $g : A_2 \mapsto B_2$ be two bijections such that $A_1 \cap A_2 = \emptyset$ and $B_1 \cap B_2 = \emptyset$. Define $h : A_1 \cup A_2 \mapsto B_1 \cup B_2$ such that

$$h(x) = \begin{cases} f(x) & \text{if } x \in A_1 \\ g(x) & \text{if } x \in A_2. \end{cases}$$

Show that h is a bijection.

Exercise 4.18. For each $n \geq 1$, let A_n and B_n be sets such for all $n \neq m$, $A_n \cap A_m = \emptyset$ and $B_n \cap B_m = \emptyset$. For each $n \geq 1$, let $f_n : A_n \mapsto B_n$ be a bijection and define $f : \bigcup_{n \geq 1} A_n \mapsto \bigcup_{n \geq 1} B_n$ such that $f(x) = f_n(x)$ for $x \in A_n$. Show that h is a bijection.

Exercise 4.19. Let $f : S \mapsto T$ and $S_i \subseteq S$ and $T_i \subseteq T$ for $i \in \mathbb{N}^+$. Prove the following:

1. $f[S_1 \cup S_2] = f[S_1] \cup f[S_2]$ and $f[\bigcup_{i=1}^{\infty} S_i] = \bigcup_{i=1}^{\infty} f[S_i]$.

2. If f is injective then $f[S_1 \cap S_2] = f[S_1] \cap f[S_2]$, $f[S_1 \setminus S_2] = f[S_1] \setminus f[S_2]$ and $f[\bigcap_{i=1}^{\infty} S_i] = \bigcap_{i=1}^{\infty} f[S_i]$. Give examples where these equalities fail when f is not injective.

3. $f^{-1}[T_1 \cup T_2] = f^{-1}[T_1] \cup f^{-1}[T_2]$ and $f^{-1}[\bigcup_{i=1}^{\infty} T_i] = \bigcup_{i=1}^{\infty} f^{-1}[T_i]$.

4. $f^{-1}[T_1 \cap T_2] = f^{-1}[T_1] \cap f^{-1}[T_2]$ and $f^{-1}[\bigcap_{i=1}^{\infty} T_i] = \bigcap_{i=1}^{\infty} f^{-1}[T_i]$.

Exercise 4.20. Let S and T be two sets and let f and g be two functions such that $f : S \mapsto T$ and $g : T \mapsto S$. For $n \geq 1$, define S_n and T_n as follows:

$$S_1 = S \qquad\qquad S_{n+1} = g[f[S_n]]$$
$$T_1 = T \qquad\qquad T_{n+1} = f[g[T_n]]$$

Show that for every $n \geq 1$, we have $S_{n+1} \subseteq g[T_n] \subseteq S_n$ and $T_{n+1} \subseteq f[S_n] \subseteq T_n$.

Conclude that

1. $\cdots \subseteq S_{n+2} \subseteq g[T_{n+1}] \subseteq S_{n+1} \subseteq g[T_n] \subseteq S_n \cdots \subseteq g[T_3] \subseteq S_3 \subseteq g[T_2] \subseteq S_2 \subseteq g[T_1] \subseteq S_1$ and

2. $\cdots \subseteq T_{n+2} \subseteq f[S_{n+1}] \subseteq T_{n+1} \subseteq f[S_n] \subseteq T_n \cdots \subseteq f[S_3] \subseteq T_3 \subseteq f[S_2] \subseteq T_2 \subseteq f[S_1] \subseteq T_1$

3. If f is injective then for $S^* = \bigcap_{n=1}^{\infty} S_n$ and $T^* = \bigcap_{n=1}^{\infty} T_n$, we have $T^* = f[S^*]$ and $f_{|S*} : S^* \mapsto T^*$ is a bijection.

4. If f and g are injective, then for all $n \geq 1$, $f[S_n \setminus g[T_n]] = f[S_n] \setminus T_{n+1}$ and $g[T_n \setminus f[S_n]] = g[T_n] \setminus S_{n+1}$.

Exercise 4.21. Give a compact definition of the one-to-one correspondence given in the proof of Theorem 4.1.9

Exercise 4.22. Show that the function $f : (-1, 1) \mapsto \mathbb{R}$ defined by $f(x) =$
$$\begin{cases} 0 & \text{if } x = 0 \\ \frac{1}{x} - 1 & \text{if } x > 0 \\ \frac{1}{x} + 1 & \text{if } x < 0 \end{cases}$$
is a bijection.

Exercise 4.23. Let S be a set and $f : \mathcal{P}S \mapsto \mathcal{P}S$ such that for any $A, B \in \mathcal{P}S$, if $A \subseteq B$ then $f(A) \subseteq f(B)$. Show that there is a set T such that $f(T) = T$.

4.3　Further Results on Countable sets

In section 4.2, we defined a number of set operations including finite and infinite union or intersection, set difference, finite cartesian products and powersets. In this section we will discuss how sets formed using these set operations preserve countability. We will also discuss how injections/surjections/bijections between sets affect countability.

First, we revisit the definition of finite and infinite sets and use functions to establish further properties of these sets.

4.3.1 Properties of Functions and Sizes of Sets

Lemma 4.3.1. *Let sets S and T. The following hold:*

1. *Assume there is a one-to-one correspondence between S and T. Then either S and T are both finite and have the same size or they are both infinite.*

2. *Let $n \in \mathbb{N}$. (S is finite and $|S| \leq n$) iff there is an injection from S to $\mathbb{N}_{<n}$.*

3. *S is countable iff there is an injection from S to $\mathbb{N}$.*

4. *Let $S \neq \emptyset$. Then, S is countable iff there is a surjection from $\mathbb{N}$ to S.*

5. *S is infinite iff there is $S' \subseteq S$ such that S' and $\mathbb{N}$ are in one-to-one correspondence.*

6. *Let $f : S \mapsto T$. The following hold:*

    (a) *If f is injective and S is uncountable then T is uncountable.*

    (b) *If f is surjective and T is uncountable then S is uncountable.*

    (c) *If f is bijective then either both S and T are countable or they are both uncountable.*

Proof. 1. Assume S is finite, then there is a one-to-one correspondence (i.e., a bijection) between $\mathbb{N}_{<n}$ and S for some $n \in \mathbb{N}$ and $|S| = n$. Since there is also a one-to-one correspondence between S and a set T, then by Lemma 4.1.4, there is a one-to-one correspondence between $\mathbb{N}_{<n}$ and T and $|T| = n$. The same proof holds if we started with T being finite. Hence either both S and T are both finite and have the same size or they are both infinite.

2. Assume S is finite and $|S| \leq n$, then there is a one-to-one correspondence f (i.e., a bijection) between S and $\mathbb{N}_{<m}$ for some $m \in \mathbb{N}$ and $|S| = m \leq n$. Since $\mathbb{N}_{<m} \subseteq \mathbb{N}_{<n}$ then by Lemma 4.2.23.7, there is an injection $g : \mathbb{N}_{<m} \mapsto \mathbb{N}_{<n}$ and by Lemma 4.2.23.3, there is an injection $g \circ f : S \mapsto \mathbb{N}_{<n}$.

On the other hand, if there is an injection f from S to $\mathbb{N}_{<n}$ for some $n \in \mathbb{N}$, then by Lemma 4.2.23.4, f is a one-to-one correspondence between S and $f[S]$. Since $f[S] \subseteq \mathbb{N}_{<n}$ and $\mathbb{N}_{<n}$ is finite, by Lemma 4.1.8.4, $f[S]$ is finite and $|f[S]| \leq |\mathbb{N}_{<n}| = n$. Since $f[S]$ is finite then there is a one-to-one correspondence between $f[S]$ and $\mathbb{N}_{<m}$ for some $m \in \mathbb{N}$ and $|f[S]| = m$. By 1, S is finite and $|S| = |f[S]| = m \leq n$.

3. Assume S is countable. Then either there is a one-to-one correspondence $f : S \mapsto \mathbb{N}_{<n} \subseteq \mathbb{N}$ for some $n \in \mathbb{N}$ or a one-to-one correspondence $f : S \mapsto \mathbb{N}$. In either case, this f is an injection from S to $\mathbb{N}$.

   On the other hand, if S is finite then S is countable, else, if S is infinite and there is an injection f from S to $\mathbb{N}$ then f is a bijection from S to $f[S]$ and hence by 1 above, $f(S) \subseteq \mathbb{N}$ is infinite. By Theorem 4.1.14, $f[S]$ is infinitely countable. Since there is a one-to-one correspondence from S to $f[S]$ and a one-to-one correspondence from $f[S]$ to $\mathbb{N}$ then by Lemma 4.1.4 there is a one-to-one correspondence from S to $\mathbb{N}$ and S is infinitely countable.

4. Since $S \neq \emptyset$, let $a \in S$. Assume S is countable. Then by Lemma 4.1.4, either there is a one-to-one correspondence $f : \mathbb{N}_{<n} \mapsto S$ for some $n \in \mathbb{N}$ or a one-to-one correspondence $f : \mathbb{N} \mapsto S$. In the latter case, by Lemma 4.2.23.1, there is a surjection $f : \mathbb{N} \mapsto S$. In the former case, let $g : \mathbb{N} \mapsto S$ such that $g(m) = \begin{cases} f(m) & \text{if } m < n \\ a & \text{if } m \geq n. \end{cases}$

   It is easy to show that g is a surjection.

   On the other hand, if there is a surjection from $f : \mathbb{N} \mapsto S$ then let $g : S \mapsto \mathbb{N}$ such that for every $b \in S$, $g(b) = minimum\{n : f(n) = b\}$. This minimum exists by the least number principle given on page 89. It is easy to show that g is an injective function. Hence, by 3 above, S is countable.

   Therefore, if $S \neq \emptyset$ then, S is countable iff there is a surjection from $\mathbb{N}$ to S.

5. Assume there is $S' \subseteq S$ such that S' and $\mathbb{N}$ are in one-to-one correspondence. By Lemma 4.1.7.3, $\mathbb{N}$ is infinite and by Lemma 4.1.7.2, S' is infinite. Hence, by Lemma 4.1.8.5, S is infinite.

   If on the other hand S is infinite then let the function $f : \mathbb{N} \mapsto S$ be defined as follows: $f(0) = a_0 \in S$, $f(1) = a_1 \in S \setminus \{a_0\}$, $f(2) = a_2 \in S \setminus \{a_0, a_1\}$, $\cdots$, $f(n) = a_n \in S \setminus \{a_0, a_1, \cdots, a_{n-1}\}$, $\cdots$.

   Since S is infinite we can find these $a_n \in S$ for any $n \in \mathbb{N}$. Now, it is easy to show that f is an injection. Let $S' = f(\mathbb{N})$. Obviously $S' \subseteq S$. By Lemma 4.2.23.4, $f : \mathbb{N} \mapsto S'$ is a bijection. Hence, there is $S' \subseteq S$ such that S' and $\mathbb{N}$ are in one-to-one correspondence.

6. (a) If T was countable then by 3 above, there is an injection $g : T \mapsto \mathbb{N}$. By Lemma 4.2.23.3, $g \circ f : S \mapsto \mathbb{N}$ is an injection which by 3 above contradicts that S is uncountable.

(b) Note that $S \neq \emptyset$ since T is uncountable and hence infinite and so non empty and each element of T must be the image of an element of S by the surjection f. If S was countable then by 4 above, there is a surjection $g : \mathbb{N} \mapsto S$. By Lemma 4.2.23.3, $f \circ g : \mathbb{N} \mapsto T$ is a surjection which by 4 above contradicts that T is uncountable.

(c) This is a corollary of the above 2 items.

$\square$

Now we move to establishing how set building operators preserve countability.

4.3.2 Set Building Operations and Countability

For set difference, we refer the reader to Exercise 4.28. For the intersection of sets to be countable, it suffices that one of the sets be countable.

Theorem 4.3.2. *The following hold:*

1. *If for some $1 \leq j \leq n$, S_j is countable then $\bigcap_{i=1}^{n} S_i$ is also countable.*

2. *If for some $j \in \mathbb{N}^+$, S_j is countable then $\bigcap_{i=1}^{\infty} S_i$ is also countable.*

Proof. 1. By Corollary 4.1.15, every subset of a countable set is countable. Since $\bigcap_{i=1}^{n} S_i \subseteq S_j$, then $\bigcap_{i=1}^{n} S_i$ is also countable. The proof of 2., is similar. $\square$

The following theorem shows that the union of two countable sets is also a countable set.

Theorem 4.3.3. *If S and T are two countable sets then $S \cup T$ is also countable.*

Proof. Assume S and T are countable sets. If one of S or T is empty, then this is trivial. We assume that both S and T are non empty. By Definition 4.1.6, each of S and T is either finite or in one-to-one correspondence with $\mathbb{N}$.

- If S and T are both finite, then for some n, m, there are one-to-one correspondences $f : \{1, \cdots n\} \mapsto S$ and $g : \{1, \cdots m\} \mapsto T$. Without loss of generality we can assume $n \geq m$. Let p be the number of elements of $T \setminus S$. Since g is bijective then g^{-1} is also bijective and $g^{-1}(T \setminus S) \subseteq \{1, \cdots m\}$ and hence by the least number principle, we can order $g^{-1}(T \setminus S)$ as follows: $k_1 < k_2 < \cdots < k_p$ such that $\forall 1 \leq i \leq p$,

we have $1 \leq k_i \leq m$ and $g(k_i) \in T \setminus S$. We define the function
$h : \{1, \cdots, n+p\} \mapsto S \cup T$ by: $h(l) = \begin{cases} f(l) & \text{if } 1 \leq l \leq n \\ g(k_{l-n}) & \text{if } n+1 \leq l \leq n+p. \end{cases}$

We can easily show that h is a one-to-one correspondence and hence $S \cup T$ is countable.

- If S is finite and T is countably infinite, then there are one-to-one correspondences $f : \{1, \cdots n\} \mapsto S$ and $g : \mathbb{N}^+ \mapsto T$. Let $h : \mathbb{N}^+ \mapsto S \cup T$ be defined by: $h(l) = \begin{cases} f(l) & \text{if } 1 \leq l \leq n \\ g(l-n) & \text{if } l > n. \end{cases}$

  It is easy to show that h is surjective and hence by Lemma 4.3.1.4, $S \cup T$ is countable and hence it is $S \cup T$ is infinitely countable.

- If S and T are countably infinite, then there are one-to-one correspondences $f : \mathbb{N}^+ \mapsto S$ and $g : \mathbb{N}^+ \mapsto T$. Let $h : \mathbb{N}^+ \mapsto S \cup T$ be defined by:
  $$h(l) = \begin{cases} f(\frac{l}{2}) & \text{if } l \text{ is even} \\ g(\frac{l+1}{2}) & \text{if } l \text{ is odd}. \end{cases}$$

  It is easy to show that h is surjective and hence by Lemma 4.3.1.4, $S \cup T$ is countable and hence $S \cup T$ is infinitely countable.

  $\square$

A moment reflection on the functions h defined in the proof of the above theorem and especially on the function h defined when S and T are both infinitely countable gives the idea to *zip* functions. That is:

$1 \overset{g}{\mapsto} t_1$		$1 \overset{h}{\mapsto} t_1 = g(1)$
	$1 \overset{f}{\mapsto} s_1$	$2 \overset{h}{\mapsto} s_1 = f(1)$
$2 \overset{g}{\mapsto} t_2$		$3 \overset{h}{\mapsto} t_2 = g(2)$
	$2 \overset{f}{\mapsto} s_2$	$4 \overset{h}{\mapsto} s_2 = f(2)$
$3 \overset{g}{\mapsto} t_3$		$5 \overset{h}{\mapsto} t_3 = g(3)$
	$3 \overset{f}{\mapsto} s_3$	$6 \overset{h}{\mapsto} s_3 = f(3)$
$\vdots$	$\vdots$	$\vdots$

We have used this zip idea already in Theorem 4.1.9 where we zipped $\mathbb{N}^+$ together with the set $-\mathbb{N}$ to show that $\mathbb{Z}$ is countable.

This is easy to do when the two sets S and T are infinite and so both f and g are functions from the entire $\mathbb{N}^+$. But, we can also do this even

if the sets are finite. In fact, a surjective function $f : \{1, 2, \cdots, n\} \mapsto S$ can also be made into a surjective function $g : \mathbb{N}^+ \mapsto S$ as follows: $g(m) =$
$$\begin{cases} f(m) & \text{if } m \leq n \\ f(n) & \text{if } m > n. \end{cases}$$
Here, an element $f(n)$ in S can be the image of numerous elements in $\mathbb{N}^+$. Why not, all we are interested in is surjectivity.

So, we could start from a *listing* $s_1, s_2, \cdots s_n$ of the elements of S where all the s_i's are different and build a listing $s_1, s_2, \cdots, s_n, s_n, s_n \cdots$ where now we can see that some elements of S are repeated in the listing. This is acceptable. All that matters is that *all the elements* of S must appear in the listing. No element must be missing. So, we tolerate repetitions but we cannot tolerate missing elements. Hence we give the following definition:

Definition 4.3.4. *We call $s_1, s_2, s_3, \cdots$ a listing of a set S iff all the elements of S appear in the list $s_1, s_2, \cdots$ and for each $i \in \mathbb{N}^+$, it holds that $s_i \in S$.*

Lemma 4.3.5. *Let S be a non empty set. Then, S is countable iff S has a listing.*

Proof. Assume S is countable. By Lemma 4.3.1.4, since S is countable, there is a surjection from $\mathbb{N}$ to S and hence there is a surjection f from $\mathbb{N}^+$ to S. Therefore, $f[\mathbb{N}^+] = S$. For each $i \in \mathbb{N}^+$, we write $f(i)$ as s_i. Now, $s_1, s_2, \cdots$ is a listing of S because all the elements of S appear in the list $s_1, s_2, \cdots$ and for each $i \in \mathbb{N}^+$, it holds that $s_i \in S$.

$$\begin{array}{ccl} \mathbb{N}^+ & \mapsto & S \\ \hline 1 & \rightarrow & s_1 = f(1) \\ 2 & \rightarrow & s_2 = f(2) \\ 3 & \rightarrow & s_3 = f(3) \\ & \vdots & \\ n & \rightarrow & s_n = f(n) \\ & \vdots & \end{array}$$

On the other hand, assume S has a listing $s_1, s_2, \cdots$. Let $f : \mathbb{N}^+ \mapsto S$ be defined such that $f(n) = s_n$ for each $n \in \mathbb{N}$. We can easily prove that f is a surjection and hence by Lemma 4.3.1.4, S is countable. $\square$

From now on, whenever we have a countable set S, we can list its elements as $s_1, s_2, \cdots$. Now, we can give a more compact proof of the countability of the union of two countable sets:

Proof. Let S and T be countable and non empty. By Lemma 4.3.5, let $s_1, s_2, \cdots$ and $t_1, t_2, \cdots$ be listings of S respectively T.

$$
\begin{array}{llllll}
S: & s_1, & s_2, & s_3, & \cdots, \\
T: & & t_1, & t_2, & t_3, & \cdots, \\
S \cup T: & s_1, & t_1, & s_2, & t_2, & s_3, & t_3, & \cdots,
\end{array}
$$

We can easily prove that $s_1, t_1, s_2, t_2, \cdots$ is a listing of $S \cup T$ and hence by Lemma 4.3.5, $S \cup T$ is countable. $\qquad\square$

We can also apply a similar method to show that the product of two countable sets is also countable.

Theorem 4.3.6. *If S and T are countable then $S \times T$ is countable.*

Proof. Let S and T be countable. If either S or T is empty, then this is trivial. We assume that neither of S or T is empty. By Lemma 4.3.5, let $s_1, s_2, \cdots$ and $t_1, t_2, \cdots$ be listings of S respectively T. Let us now consider for each $n \in \mathbb{N}^+$, $\text{block}_n = \{(s_i, t_j) : i + j - 1 = n\}$ and let us write the elements of block_1, block_2, etc., as follows one after another:

$$\underbrace{(s_1, t_1),}_{1} \underbrace{(s_1, t_2), (s_2, t_1),}_{2} \underbrace{(s_1, t_3), (s_2, t_2), (s_3, t_1),}_{3}$$
$$\underbrace{(s_1, t_4), (s_2, t_3), (s_3, t_2), (s_4, t_1),}_{4} \cdots$$

Here, we have grouped all pairs (s_i, s_j) and $(s_{i'}, s_{j'})$ such that $i+j = i'+j'$ under the same block labeled $i+j-1$. So for example (s_1, t_1) falls into block_1 and all of $(s_1, t_3), (s_2, t_2), (s_3, t_1)$ fall into block_3. Moreover, within the same block, (s_i, s_j) comes before $(s_{i'}, s_{j'})$ if $i < i'$.

So to summarise: for all pairs (s_i, s_j) and $(s_{i'}, s_{j'})$ in the listing, if (s_i, s_j) appears before $(s_{i'}, s_{j'})$ then either $i + j < i' + j'$ or ($i + j = i' + j'$ and $i < j$).

Now, we can easily prove that this is indeed a listing of $S \times T$. In fact, if (s_i, t_j) is in one of the blocks then it is in $S \times T$. Moreover, if (s_i, t_j) is in $S \times T$, then it is in block_{i+j-1}.

Hence by Lemma 4.3.5, $S \times T$ is countable. $\qquad\square$

Remark 4.3.7. *We could mark the listing in the above proof of Theorem 4.3.6 more visually by the following diagram where we can see the zipping more easily:*

$(s_1, t_1) \longrightarrow (s_1, t_2) \qquad (s_1, t_3) \qquad (s_1, t_4) \qquad \cdots$

$(s_2, t_1) \qquad (s_2, t_2) \qquad (s_2, t_3) \qquad (s_2, t_4) \qquad \cdots$

$(s_3, t_1) \qquad (s_3, t_2) \qquad (s_3, t_3) \qquad (s_3, t_4) \qquad \cdots$

$(s_4, t_1) \qquad (s_4, t_2) \qquad (s_4, t_3) \qquad (s_4, t_4) \qquad \cdots$

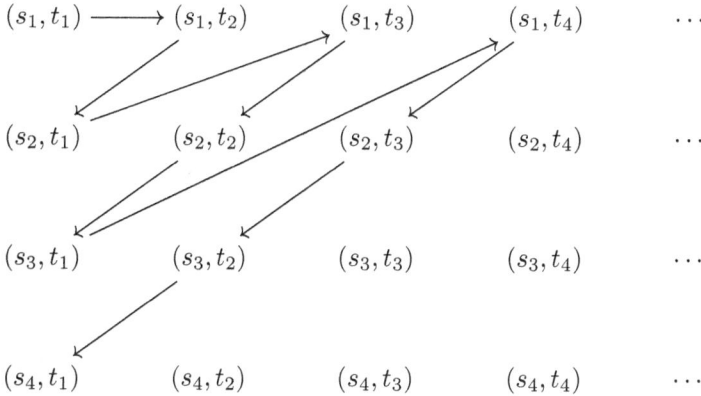

Of course, the listing given above is not unique. We could think of other ways to list $S \times T$. For example, the following is another possible listing:

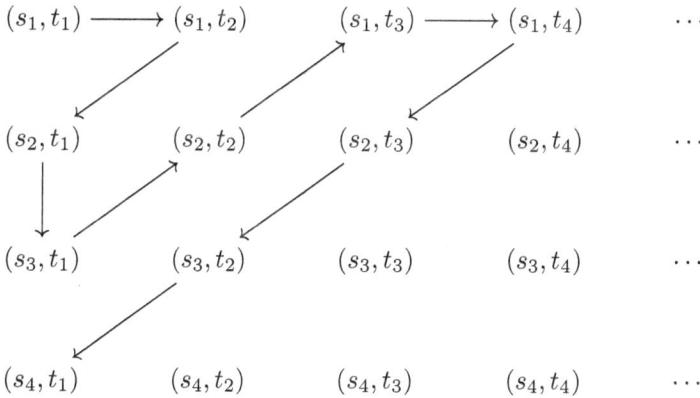

$(s_1, t_1) \longrightarrow (s_1, t_2) \qquad (s_1, t_3) \longrightarrow (s_1, t_4) \qquad \cdots$

$(s_2, t_1) \qquad (s_2, t_2) \qquad (s_2, t_3) \qquad (s_2, t_4) \qquad \cdots$

$(s_3, t_1) \qquad (s_3, t_2) \qquad (s_3, t_3) \qquad (s_3, t_4) \qquad \cdots$

$(s_4, t_1) \qquad (s_4, t_2) \qquad (s_4, t_3) \qquad (s_4, t_4) \qquad \cdots$

In all these listings, we are avoiding going through an entire row or an entire column (because these are infinite) and instead, we are traversing one diagonal at a time. First we traverse the diagonal which contains one element only (s_1, t_1), then the diagonal which contains (s_1, t_2) and (s_2, t_1). Then, the diagonal which contains (s_1, t_3), (s_2, t_2) and (s_3, t_1), etc. Each such diagonal is finite and we can pass to the following diagonal.

We will now show that both the union and the product of a finite number of countable sets is also countable.

Theorem 4.3.8. Let $n \in \mathbb{N}$. If $S_1, S_2, \cdots, S_n$ are countable sets then $\bigcup_{i=1}^{n} S_i$ and $S_1 \times S_2 \times \cdots \times S_n$ are countable.

Proof. Let $I = \{n \in \mathbb{N} : \text{for all countable } S_1, S_2, \cdots S_n,$

$\bigcup_{i=1}^{n} S_i$ and $S_1 \times S_2 \times \cdots \times S_n$ are countable}.
We will show by the induction axiom that $I = \mathbb{N}$. Obviously $0 \in I$. Assume $n \in I$. We will show that $n + 1 \in I$. Let $S_1, S_2, \cdots S_n, S_{n+1}$ be countable sets. By hypothesis, since $n \in I$, $\bigcup_{i=1}^{n} S_i$ and $S_1 \times S_2 \times \cdots \times S_n$ are countable.

- By Theorem 4.3.3, $\bigcup_{i=1}^{n+1} S_i = (\bigcup_{i=1}^{n} S_i) \cup S_{n+1}$ is also countable.

- By Theorem 4.3.6, $(S_1 \times S_2 \times \cdots \times S_n) \times S_{n+1}$ is also countable. We can assume that all the S_i's are non empty since if one is empty, the entire product is empty and is countable. Let us define the function $f : (S_1 \times S_2 \times \cdots \times S_n) \times S_{n+1} \mapsto S_1 \times S_2 \times \cdots \times S_n \times S_{n+1}$ such that $f((x_1, x_2, \cdots, x_n), x_{n+1}) = (x_1, x_2, \cdots, x_n, x_{n+1})$. It is easy to show that f is a bijection and hence by Lemma 4.3.1.6, $S_1 \times S_2 \times \cdots \times S_n \times S_{n+1}$ is countable.

Hence, $n + 1 \in I$. By the induction axiom for $\mathbb{N}$, $I = \mathbb{N}$. □

We can also show that the union of a countably infinite number of countable sets is countable.

Theorem 4.3.9. *Let $S_1, S_2, S_3, \ldots$ be a countably infinite number of countable sets. Then $\bigcup_{i=1}^{\infty} S_i$ is countable.*

Proof. We can assume that all the S_i's are non empty since the empty ones can be discarded without impacting the union. Now, by Lemma 4.3.5, each S_i can be listed as follows:

$S_1 : s_{11}, s_{12}, \cdots s_{1n}, \cdots$
$S_2 : s_{21}, s_{22}, \cdots s_{2n}, \cdots$
$S_3 : s_{31}, s_{32}, \cdots s_{3n}, \cdots$
$\vdots$
$S_m : s_{m1}, s_{m2}, \cdots s_{mn}, \cdots$
$\vdots$

We can then form the listing of $\bigcup_{i=1}^{\infty} S_i$ just like we developed the listing of $S_1 \times S_2$ in Theorem 4.3.6 as follows:

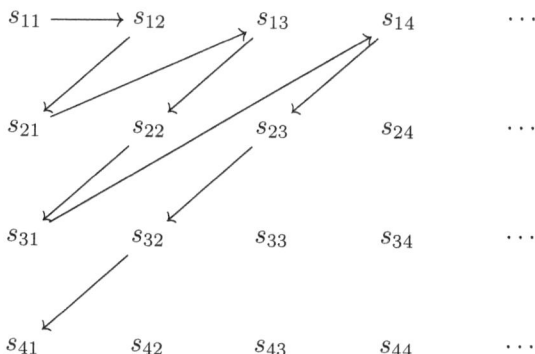

$$s_{11} \longrightarrow s_{12} \qquad s_{13} \qquad s_{14} \qquad \cdots$$

$$s_{21} \qquad s_{22} \qquad s_{23} \qquad s_{24} \qquad \cdots$$

$$s_{31} \qquad s_{32} \qquad s_{33} \qquad s_{34} \qquad \cdots$$

$$s_{41} \qquad s_{42} \qquad s_{43} \qquad s_{44} \qquad \cdots$$

$\square$

Now, we have established how countability is preserved for all the sets formed using the set forming operations of Section 4.2 except for the power sets of sets. If S is countable, would $\mathcal{P}(S)$ also be countable? Before answering this question, we may think that actually so far, most of the infinite sets we worked with so far have been countable and so perhaps if S is countable, then $\mathcal{P}(S)$ would also be countable. For, we have seen that $\mathbb{N}$, $\mathbb{Z}$, $\mathbb{Q}$, and the algebraic numbers are all countable and also that the product of countable sets is countable and a countably infinite union of countable sets is also countable. The only uncountable sets we have seen so far are $(0, 1]$ and $\mathbb{R}$ (see Theorems 4.1.17 and 4.1.18). Cantor himself originally believed that $\mathbb{R}$ would be infinitely countable but then in the end, he developed the *diagonalisation argument* that we used to show that $(0, 1]$ is uncountable in Theorem 4.1.17.

So, we have finite sets (which are countable and we have even defined their size/number of elements). We have infinite sets which can be countable like $\mathbb{N}$, $\mathbb{Z}$, $\mathbb{Q}$, or uncountable like $(0, 1]$ and $\mathbb{R}$. We have not yet defined the size of infinite sets. It is certain that since $\mathbb{N}$ is countable and $\mathbb{R}$ is uncountable, they cannot have the same size. So, we already see that there are at least two infinite set sizes. In fact, there are infinitely many different infinite sizes. All the sets below have different sizes:

$$\mathbb{N}, \mathcal{P}(\mathbb{N}), \mathcal{P}(\mathcal{P}(\mathbb{N})), \mathcal{P}(\mathcal{P}(\mathcal{P}(\mathbb{N}))), \ldots$$

We will see this in the next chapter where we will also see where to place the set $\mathbb{R}$ in terms of its size with respect to the above increasing size hierarchy.

4.3.3 Exercises

Exercise 4.24. Let $f : S \mapsto T$. Show the following:

1. If f is injective and S is infinite, then T is infinite.

2. If f is injective and T is finite, then S is finite.

3. If f is surjective and T is infinite, then S is infinite.

4. If f is surjective and S is finite, then T is finite.

Exercise 4.25. Let S and T be infinite sets such that S is countable and $f : S \mapsto T$ is a surjection. Show that T is countable.

Exercise 4.26. Explicitly construct a bijection between $\mathbb{N}$ and S to show that if S is an infinite set and $f : \mathbb{N} \mapsto S$ is surjective then S is infinitely countable.

Exercise 4.27. Prove that the first function h defined in the proof of Theorem 4.3.3 is bijective and that each of the following two functions h in that proof are surjective.

Exercise 4.28. Show that if S is countable then $S \setminus T$ is also countable. Give an example of two sets S and T such that S and T are uncountable but $S \setminus T$ is countable. Also, give an example of two sets S and T such that S and T are uncountable and also $S \setminus T$ is uncountable.

Exercise 4.29. Recall that a polynomial of the form $a_n x^n + a_{n-1} x^{n-1} + \ldots a_1 x + a_0$ where $a_n \neq 0$ and each a_i is an integer, is a polynomial of degree n. It is known that a polynomial of degree n has at most n roots. Using the results of this section, can you give a more compact proof of the countability of the Algebraic numbers than that given in the proof of Theorem 4.1.13.

Exercise 4.30. In Remark 4.3.7, we saw two different listings of the cartesian product of two countable sets. Apply these two listings to give two different listings of the cartesian product $\mathbb{N} \times \mathbb{Q}$.

Exercise 4.31. Let S and T be sets. Show that (S and T are finite) iff $S \cup T$ is finite. Moreover, show that if S and T are finite then $|S \cup T| \leq |S| + |T|$.

Exercise 4.32. Let $f : \mathbb{N} \mapsto \mathbb{N} \times \mathbb{N}$ such that

$$
\begin{aligned}
f(0) &= (0,0) \\
f(n) &= \begin{cases} (0, k+1) & \text{if } n \neq 0 \text{ and } f(n-1) = (k,0) \\ (k+1, l-1) & \text{if } n \neq 0,\, l \neq 0 \text{ and } f(n-1) = (k,l) \end{cases}
\end{aligned}
$$

Show that f is a bijection and then write the elements $f(0), f(1), f(2), \cdots$ in this order and state where in the chapter you have seen this listing.

Exercise 4.33. Let $n, m \in \mathbb{N}^+$. Show that $\mathbb{N}_{<n} \times \mathbb{N}_{<m}$ is finite and give its size.

Exercise 4.34. Let S and T be non empty sets. Show that (S and T are finite) iff $S \times T$ is finite. Moreover, show that if S and T are finite then $|S \times T| = |S| \times |T|$.

Exercise 4.35. Let S be a set. Show that if S is finite then $\mathcal{P}S$ is also finite.

Exercise 4.36. Let S be a finite set whose elements are also finite sets. Show that $\bigcup S = \{x : x \in T \text{ for some } T \in S\}$ is finite.

Exercise 4.37. Let $m \in \mathbb{N}^+$ and $A = \{x \in \mathbb{R} : x = m^n \text{ for some } n \in \mathbb{N}\}$. Is A finite or infinitely countable or uncountable? Prove your answer.

Exercise 4.38. Let $f : \mathbb{N} \mapsto \mathbb{Q}$ be a bijection and denote $f(n)$ by q_n for any $n \in \mathbb{N}$. Use this bijection to give a bijection $g : \mathbb{N} \mapsto \mathbb{Q} \times \mathbb{Q}$.

Exercise 4.39. Let S be an infinitely countable set and let T be a set. In each of the following cases below, give a bijection from $S \cup T$ to $\mathbb{N}$.

1. T is finite and $S \cap T = \emptyset$.

2. T is finite and $S \cap T \neq \emptyset$.

3. T is infinitely countable and $S \cap T = \emptyset$.

4. T is infinitely countable and $S \cap T \neq \emptyset$.

Chapter 5

Cardinal Numbers

In the previous chapter we spoke about finite and infinite sets and about their sizes. We learned how to calculate the size of a finite set. However, for infinite sets, we mostly measured their size in terms of the size of the natural numbers. We saw many sets (like $\mathbb{Z}$, $\mathbb{Q}$, E, O, etc.) that appeared larger or smaller than $\mathbb{N}$ but we still treated them as having the same size as $\mathbb{N}$. We also saw some sets (like $\mathbb{R}$) that were truly larger than $\mathbb{N}$. We classified $\mathbb{N}$ as infinitely countable and $\mathbb{R}$ as uncountable.

By this stage you are familiar with calculating the sizes of finite sets, but there is still a lot to learn about calculating the sizes of infinite sets. For finite sets, calculating with set sizes is just like calculating with the natural numbers. The empty set has size 0, the size of the union of two finite *disjoint* sets is the sum of the sizes of the two sets, the size of the Cartesian product of two finite sets is the product of the sizes of the sets, etc. But, what exactly is the size of an infinite set and how do we calculate the size of the union or product of infinite sets? How do we add or multiply infinites?

In this chapter, we study the cardinal numbers which are sizes of sets (finite and infinite) and explain how to measure sizes of infinite sets. We show that all infinitely countable sets have the same size and that there are infinitely many infinite cardinal numbers the smallest of which is the size $\mathfrak{a}$ of $\mathbb{N}$ and of all the infinitely countable sets. The cardinal number $\mathfrak{a}$ is different from the cardinal $\mathfrak{c}$ which is the size of the real numbers. In particular we show that $\#\mathbb{N} = \#\mathbb{Z} = \#\mathbb{Q} = \mathfrak{a}$ and $\#\mathbb{R} = \#\mathcal{P}\mathbb{N} = \#\mathcal{P}\mathbb{Z} = \#\mathcal{P}\mathbb{Q} = \mathfrak{c} < \#\mathcal{P}\mathbb{R} \cdots$ where $\#S$ stands for the size of S (also called the cardinal number of S). We then move in the second section to give the arithmetic of cardinal numbers. You are already familiar with the arithmetic of the finite cardinal numbers which you know from the natural numbers since finite cardinals are just the natural numbers. But, when it comes to the arithmetic of cardinal numbers that involve infinite cardinals, there will be surprises. Finally, in the third section we reflect on the use of the universal set and on the paradoxes and then reflect further on the so-called *Continuum Hypothesis* and discuss why we chose to use $\mathfrak{a}$ for the smallest infinite cardinal number and $\mathfrak{c}$ for the size of the real numbers instead of $\aleph_0$ and $\aleph_1$ that are used in some sources.

5.1   More on Sizes of Sets

With finite sets, we are used to talking about the number of elements of a set and we have indeed introduced in Definition 4.1.6 of Chapter 4 the sizes of finite sets. For example, we showed in Chapter 4 that $|\{x,y\}| = 2$ and $|\{x,y,z\}| = 3$ and that $|\{x,y\}| < |\{x,y,z\}|$. But we have still not defined the size of an infinite set. The reason for this is that sizes of infinite sets were not really defined nor properly compared before the late 1800's when Cantor developed a system for comparing the sizes of infinite sets. Cantor gave every set S a cardinality $\#S$ representing its size, which is an ordinary nonnegative integer if S is finite, but is a new kind of number if S is infinite. In this chapter, we will study the sizes of sets in general (finite and infinite) and discuss the cardinal number of a set which represents the size of any set (finite and infinite). The cardinal number of a finite set will coincide with the size of the set as we defined in Definition 4.1.6. First, let us note the following lemma about the sizes of finite sets which will be ported to any set, finite or infinite:

Lemma 5.1.1. *Let S and T be finite sets. The following hold:*

1. *$|S| = |T|$ iff there is a bijection $f : S \mapsto T$.*

2. *$|S| \leq |T|$ iff there is an injection $f : S \mapsto T$.*

Proof. Since S and T are finite sets, then by Definition 4.1.6, let $n, m \in \mathbb{N}$ where S (resp. T) can be put into a one-to-one correspondence with $\mathbb{N}_{<n}$ (resp. $\mathbb{N}_{<m}$) and $|S| = n$ (resp. $|T| = m$).

1. We already know that we have a bijection from S to $\mathbb{N}_{<n}$ and another bijection from T to $\mathbb{N}_{<m}$ and that $|S| = n$ and $|T| = m$.
   Now, by the results we learned about bijections from Chapter 4, there is a bijection $f : S \mapsto T$ iff there is a bijection from $\mathbb{N}_{<n}$ to $\mathbb{N}_{<m}$ iff $|S| = n = |\mathbb{N}_{<n}| = |\mathbb{N}_{<m}| = m = |T|$.

2. By Lemma 4.3.1.2, $|S| \leq |T| = m$ iff there is an injection from S to $\mathbb{N}_{<m}$.

Since there is a bijection from T to $\mathbb{N}_{<m}$, then by Lemma 4.2.23.(3 and 2), there is an injection from S to $\mathbb{N}_{<m}$ iff there is an injection from S to T.

Hence, $|S| \leq |T| = m$ iff there is an injection from S to T.

$\square$

This lemma states that for finite sets, all bijective sets have the same size and that an order on set sizes could be made through injections. We will follow this idea for the size of all sets (finite or infinite). Before doing so, we will introduce the following notation to describe the existence of a bijection or an injection between sets:

Definition 5.1.2 (Set equivalence and embedding: $\sim$ and $\preceq$).

1. *Two sets S and T are said to be* equivalent *iff there is a one-to-one correspondence (i.e., bijection) between them. If S and T are equivalent, we write $S \sim T$.*

2. *A set S is said to have an* embedding *in a set T iff there is an injection from S to T. If S has an embedding in T, we write $S \preceq T$.*

By Definition 3.1.1 and Theorem 4.2.24 we have the following theorem whose proof is left as an exercise (see Exercise 5.1).

Theorem 5.1.3. *The following hold:*

1. *$\sim$ is an equivalence relation.*

2. *$\preceq$ is reflexive and transitive and moreover, for any sets S and T:*

$$(S \preceq T \text{ and } T \preceq S) \text{ iff } S \sim T.$$

$$S \sim \emptyset \text{ iff } S = \emptyset.$$

Theorem 5.1.3 means that, given a suitable universe of sets U, we have *equivalence classes* under $\sim$. The equivalence class of a set S under $\sim$ is the set of all sets equivalent to S; i.e., it is $\{T \in U : T \sim S\}$.[1]

---

[1] In fact, this equivalence class of a set S could be used as the definition of $\#S$ of the cardinal number of the set S, however, readers familiar with ZF set theory and its version ZFC with the axiom of Choice know that defining cardinal numbers as equivalence classes of the relation of similarity between sets will not work in ZFC because it cannot be proved there that all of the equivalence classes exist. For more details on this, see Suppes [39] where an axiom is added to take care of this. Another definition of a cardinal number could be assumed in terms of for example, the so-called initial ordinal, etc., but since we have not defined the full machinery to give an exact definition of the cardinal number, we will be content with comparing cardinal numbers which is all we need to develop the arithmetic of the cardinals. We already have defined all the needed machinery to compare cardinal numbers and hence, we will concentrate on the comparison of cardinal numbers and their arithmetic without defining exactly what a cardinal number is.

Even though we have still not defined the size of an infinite set like we defined the size of a finite set in Definition 4.1.6, we will introduce the notation $\#S$ to stand for the size of any set S (also called the cardinal number of S) whether the set is finite is infinite, and we will let $\#S$ coincide with $|S|$ for finite sets. We will define an equal relation $=$ (determined by $\sim$) and a comparison relation $\leq$ (determined by $\preceq$) on the cardinal numbers of sets (i.e., the sizes of sets) and we will give the arithmetic of the cardinal numbers which has similarities and also differences with the arithmetic of the natural numbers. This way, this chapter will concentrate on *how* cardinal numbers work and not on *what* cardinal numbers are.

> We always want any two equivalent sets to have the same size (i.e., cardinal number) and for any two sets S and T, if there is an injection from S to T then the cardinal number of S must be smaller or equal to the cardinal number of T.

In the case of finite sets, this was clearly established in Lemma 5.1.1.

The problem with the case of infinite sets is that even though one set might appear to be clearly larger than another set, the two sets may still have the same size/cardinal number. For example, the function $f : \mathbb{N} \mapsto \mathbb{Z}$ such that $f(n) = \begin{cases} \frac{n}{2} & \text{if } n \text{ is even} \\ -\frac{n+1}{2} & \text{if } n \text{ is odd} \end{cases}$ is bijective and hence $\mathbb{N} \sim \mathbb{Z}$. But clearly, $\mathbb{N} \subset \mathbb{Z}$ and $\mathbb{Z}$ appears to have twice as many elements as $\mathbb{N}$. What we need to understand here is that the arithmetic for the infinite cardinal numbers differs from the arithmetic of the finite cardinal numbers. The finite cardinal numbers are $0, 1, 2, \cdots$ and are the sizes of the finite sets (like $\mathbb{N}_{<0}$, $\mathbb{N}_{<1}$, $\cdots$). The infinite cardinal numbers start with the first new cardinal number which is the size $\#\mathbb{N}$ of the countable set $\mathbb{N}$ which we showed to be infinite. Other infinite cardinal numbers are the size $\#\mathbb{R}$ of the uncountable set $\mathbb{R}$ and the size $\#(0, 1]$ of the uncountable set $(0, 1]$. We have not said anything yet about how to or whether we can compare these infinite cardinals $\#\mathbb{N}$, $\#\mathbb{R}$ and $\#(0, 1]$.

> As it turns out, $\#\mathbb{N}$ is strictly smaller than $\#\mathbb{R}$ and $\#(0, 1]$ is equal to $\#\mathbb{R}$ and there are infinitely many cardinals after $\#\mathbb{R}$.

It is furthermore assumed that there is no other cardinal between $\#\mathbb{N}$ and $\#\mathbb{R}$ (this is known as the Continuum Hypothesis). We will meet these results and more in this chapter.

Definition 5.1.4 (Finite/transfinite cardinal numbers, $\mathfrak{a}$, $\mathfrak{c}$, Comparing cardinal numbers). *Let S, T and U be sets.*

- *If S is finite, we define its cardinal number $\#S$ to be its size $|S|$ as given in Definition 4.1.6 and say that $\#S$ is a finite cardinal number.*

- *If S is infinite, we denote its cardinal number by $\#S$ and say that $\#S$ is a transfinite cardinal number.*

- *We define the two transfinite cardinal numbers $\mathfrak{a} = \#\mathbb{N}$ and $\mathfrak{c} = \#\mathbb{R}$.*

- *We say that S and T have the same cardinal number and write $\#S = \#T$ iff $S \sim T$.*

- *We say that $\#S \leq \#T$ if and only if $S \preceq T$. In this case, we also say that $\#T \geq \#S$.*

- *We say that $\#S < \#T$ if and only if $\#S \leq \#T$ and $\#S \neq \#T$. In this case we also say that $\#T > \#S$.*

- *If $\bowtie, \bowtie \in \{=, \leq, <\}$ then we write $\#S \bowtie \#T \bowtie \#U$ to stand for $\#S \bowtie \#T$ and $\#T \bowtie \#U$. For example $\#S \leq \#T = \#U$ stands for $\#S \leq \#T$ and $\#T = \#U$.*

Example 5.1.5.   - *For each $n \in \mathbb{N}$, $\#\mathbb{N}_{<n} = |\mathbb{N}_{<n}|$.*

- *$\#\{a, b, c\} = 3$, $\#\{a, b, c\} = \#\mathbb{N}_{<3}$, $\#\mathbb{N}_{<6} \geq \#\{a, b, c\}$ and $\#\{a, b, c\} < \#\mathbb{N}_{<10}$.*

- *$\#\mathbb{N}^+ = \#\mathbb{N} = \#\mathbb{Z} = \#\mathbb{Q}$.*

- *$\#\mathbb{N} \leq \#\mathbb{R} = \#(0, 1]$.*

- *$\mathfrak{a} = \#\mathbb{N} < \#\mathbb{R} = \mathfrak{c}$ since clearly $\mathbb{N} \preceq \mathbb{R}$ and hence $\#\mathbb{N} \leq \#\mathbb{R}$ but $\#\mathbb{N} \neq \#\mathbb{R}$ (since otherwise $\mathbb{N} \sim \mathbb{R}$ and $\mathbb{R}$ would be countable, a contradiction).*

The following theorem whose proof is left as an exercise (see Exercise 5.2) gathers important properties about $=$, $\leq$ and $<$.

Theorem 5.1.6. *The following hold:*

1. *$\#S = 0$ iff $S = \emptyset$.*

2. *If $\#S = \#T$ then either S and T are both finite or are both infinite.*

3. *$=$ is an equivalence relation on cardinal numbers.*

4. $\leq$ *is reflexive and transitive on cardinal numbers.*

5. $\#S = \#T$ *iff* $(\#S \leq \#T$ *and* $\#T \leq \#S)$.

6. $\#S \leq \#T$ *iff* S *is equivalent to a subset of* T.

7. $\#S < \#T$ *iff* S *is equivalent to a subset of* T *and* T *is not equivalent to a subset of* S.

It is clear from Definition 5.1.4 that all infinitely countable sets have the same cardinal number, and that cardinal number is not finite. We called this cardinal number of the infinitely countable sets $\mathfrak{a}$. We named the cardinal number $\#\mathbb{R}$ as $\mathfrak{c}$. Definition 5.1.4 also establishes that $\mathfrak{a} = \#\mathbb{N} < \#\mathbb{R} = \mathfrak{c}$.

Now, for every finite cardinal number n, we have $n < \mathfrak{a}$. Hence we have the next theorem whose proof is left as an exercise (see Exercise 5.3).

Theorem 5.1.7. *If S and T are infinitely countable sets then $\#S = \#T = \mathfrak{a}$. Moreover, for any finite cardinal number n, we have $n < \mathfrak{a} < \mathfrak{c}$.*

In particular,

$$\#\emptyset < \#\{0\} < \#\{0,1\} < \#\{0,1,2\} < \cdots < \#\mathbb{N} = \#\mathbb{Z} < \#\mathbb{R}$$

That is:
$$0 < 1 < 2 < \cdots < \mathfrak{a} < \mathfrak{c}$$

What about $\#(0,1]$? By Theorem 4.1.17, $(0,1]$ is uncountable and hence $(0,1] \not\sim \mathbb{N}$ and $\#(0,1] \neq \mathfrak{a}$. Furthermore, $\#(0,1] \not\leq \mathfrak{a}$ because if we assume otherwise, then $(0,1] \preceq \mathbb{N}$ and hence there is an injection $f : (0,1] \mapsto \mathbb{N}$ which by Lemma 4.3.1.6 means (since $(0,1]$ is uncountable) that $\mathbb{N}$ is also uncountable, which is absurd. Now knowing that $\#(0,1] \not\leq \mathfrak{a}$ does not allow us to deduce that $\mathfrak{a} \leq \#(0,1]$ because we have not established yet that for any two cardinal numbers $\mathfrak{m}$ and $\mathfrak{n}$, either $\mathfrak{m} \leq \mathfrak{n}$ or $\mathfrak{n} \leq \mathfrak{m}$.

So, to summarise our knowledge about cardinal numbers so far:

Remark 5.1.8. *The following hold:*

- *We have finite cardinals: for every nonnegative integer n we have*
  $\#\{0,1,\cdots,n-1\} = n$.

- *We have one transfinite cardinal $\mathfrak{a} = \#\mathbb{N}$ such that for every nonnegative integer n we have*

$$n < \#\mathbb{N}_{\geq n} = \#\mathbb{N} = \#\mathbb{Z} = \#\mathbb{Q} = \mathfrak{a}.$$

- *By Lemma 4.2.23.5,*

$$\#\mathcal{P}\mathbb{N} = \#\mathcal{P}\mathbb{Z} = \#\mathcal{P}\mathbb{Q}.$$

  But we still don't know how $\#\mathcal{P}\mathbb{N}$ compares to $\#\mathbb{N}$ and how for any set S, $\#S$ compares to $\#\mathcal{P}S$.

- *We have also shown that $(0,1]$ and $\mathbb{R}$ are uncountable (see Theorems 4.1.17 and 4.1.18) and that we have the transfinite cardinal number $\mathfrak{c} = \#\mathbb{R}$ such that by Theorem 5.1.7, we conclude:*

$$0 < 1 < 2 < \cdots < \mathfrak{a} < \mathfrak{c} = \#\mathbb{R}$$

  But, although we know that $\#(0,1] \nleq \mathfrak{a}$ and $\#(0,1] \neq \mathfrak{a}$, we still don't know whether $\mathfrak{a} \leq \#(0,1]$ (and hence whether $\mathfrak{a} < \#(0,1]$) and we don't know how $\#(0,1]$ and $\mathfrak{c}$ compare.

Before proceeding with further properties of cardinal numbers that extend $\#\mathbb{N} = \mathfrak{a} < \mathfrak{c} = \#\mathbb{R}$ (where we will prove that $\mathfrak{a} < \#(0,1] = \mathfrak{c} = \#\mathcal{P}\mathbb{N} < \#\mathcal{P}\mathbb{R}$ as well as other properties), we give the next lemma that shows that all finite intervals are uncountable and have cardinality $\#(0,1]$.

Lemma 5.1.9. *All finite intervals are equivalent to each other. It follows that for any finite interval (a,b), where a and b are real numbers with $a < b$, we have $\#(0,1] = \#(a,b)$.*

Proof. This proof will be in stages.

- We begin by proving that $(0,1] \sim (0,1)$. We do this with the following correspondence:

  - For each $x \in (0,1]$, if $\frac{1}{2} < x \leq 1$, we let it correspond to $y = \frac{3}{2} - x$; then $\frac{1}{2} \leq y < 1$;
  - For each $x \in (0,1]$, if $\frac{1}{4} < x \leq \frac{1}{2}$, we let it correspond to $y = \frac{3}{4} - x$; then $\frac{1}{4} \leq y < \frac{1}{2}$;
  - For each $x \in (0,1]$, if $\frac{1}{8} < x \leq \frac{1}{4}$, we let it correspond to $y = \frac{3}{8} - x$; then $\frac{1}{8} \leq y < \frac{1}{4}$;
  - Etc.

  It is clear that to every $x \in (0,1]$, we hereby make correspond one and only one $y \in (0,1)$ and, conversely, to every $y \in (0,1)$ we hereby make to correspond exactly one $x \in (0,1]$.

- Next, by a similar proof we can show that $[0,1) \sim (0,1)$.

- Now, we can show that $[0, 1) \sim [0, 1]$ by letting 0 in each of these intervals correspond to 0 in the other and then use the correspondence between $(0, 1)$ and $(0, 1]$ shown above.

- We now show that if $a < b$ for real numbers a and b, then $(0, 1) \sim (a, b)$. For $x \in (0, 1)$ and $y \in (a, b)$, we let $y = (b - a)x + a$ correspond to x. This is clearly a one-to-one correspondence.

- It should be clear that the methods used above to prove $(0, 1) \sim (0, 1] \sim [0, 1) \sim [0, 1]$ can be used to prove for any real numbers a and b with $a < b$, $(a, b) \sim [a, b) \sim (a, b] \sim [a, b]$.

$\square$

Not only finite intervals are uncountable and of cardinality $\#(0, 1]$, but also the two bounded infinite intervals $(-\infty, a)$ and (a, ∞) are uncountable and of cardinality $\#(0, 1]$.

Lemma 5.1.10. *For a half-line $\{x : x > a\} = (a, \infty)$ or $\{x : x < a\} = (-\infty, a)$ for any real number a, we have $\#(a, \infty) = \#(0, 1]$ and $\#(-\infty, a) = \#(0, 1]$.*

Proof. If $a = 1$, we have easily that $(0, 1) \sim (1, \infty)$ by letting x, for $x \in (0, 1)$, correspond to $\frac{1}{x} \in (1, \infty)$. This shows that $\#(1, \infty) = \#(0, 1) = \#(0, 1]$.

Then, for $a \neq 1$, we have that $(1, \infty) \sim (a, \infty)$ by letting $y \in (a, \infty)$ correspond to $x - 1 + a$ for $x \in (1, \infty)$.

The second part of the lemma is proved by reversing the signs. $\square$

We even have the entire unbounded interval $(-\infty, \infty)$ is uncountable and of cardinality $\#(0, 1]$. We have already shown that $(-\infty, \infty)$ which is $\mathbb{R}$ is uncountable (see Theorem 4.1.18). But, how do we show that $\#\mathbb{R} = \#(0, 1]$? So far, since $(1, \infty) \subseteq \mathbb{R}$ and $\#(1, \infty) = \#(0, 1]$ then we have $\#(0, 1] \leq \#\mathbb{R}$. The next theorem shows that $\#(0, 1] = \#\mathbb{R}$.

Theorem 5.1.11. $\#\mathbb{R} = \#(0, 1]$ *and hence $\#\mathbb{R} = \#X$ for any X such that $I \subseteq X \subseteq \mathbb{R}$ for some subinterval I of $\mathbb{R}$.*

Proof. Bijections can be built between $\mathbb{R}$ and any of the finite intervals we have already proven to have cardinality $\mathfrak{c}$. For example, the function $\tan : (-\frac{\pi}{2}, \frac{\pi}{2}) \mapsto \mathbb{R}$ is a bijection. Also, $f : (0, 1) \mapsto \mathbb{R}$ such that $f(x) = \ln(\frac{1}{x} - 1)$ and $f' : (0, 1) \mapsto \mathbb{R}$ such that $f'(x) = \frac{2x - 1}{2x(1 - x)}$ are also bijections. Similarly, the function $g : \mathbb{R} \mapsto (0, 1)$ such that $g(x) = \frac{1}{1 + e^x}$ is bijective. However, this book has not covered the functions tan, ln or quadratic equations or enough calculus for you to prove that these functions are bijective. Instead, we will give another proof.

- There is $f : \mathcal{P}\mathbb{N} \mapsto [0,1]$ which is injective. The function f can be defined as $f(S) = 0.a_0 a_1 a_2 \cdots$ where for all $n \geq 0$, $a_n = \begin{cases} 1 & \text{if } n \in S \\ 0 & \text{otherwise} \end{cases}$

  Obviously f is injective. Hence $\#\mathcal{P}\mathbb{N} \leq \#[0,1]$.

- There is $g : \mathbb{R} \mapsto \mathcal{P}\mathbb{Q}$ which is injective. Define $g(x) = \{r \in \mathbb{Q} : r < x\}$. To show g injective, assume $x \neq x'$, say $x < x'$. Then, there is $r \in \mathbb{Q}$ such that $x < r < x'$ and hence $g(x) \neq g(x')$. Hence by Remark 5.1.8 and the above item, $\#\mathbb{R} \leq \#\mathcal{P}\mathbb{Q} = \#\mathcal{P}\mathbb{N} \leq \#[0,1]$.

- But $[0,1] \subseteq \mathbb{R}$ and hence $\#[0,1] \leq \#\mathbb{R}$. Now since $\#\mathbb{R} \leq \#[0,1]$ and $\#[0,1] \leq \#\mathbb{R}$, by Theorem 5.1.6.5, $\#[0,1] = \#\mathbb{R}$.

Finally, by Lemmas 5.1.9 and 5.1.10, $\#\mathbb{R}$ is $\#I$ for any I subinterval of $\mathbb{R}$. Since $I \subseteq X \subseteq \mathbb{R}$ then $\#I \leq \#X \leq \#\mathbb{R}$. Since $\#I = \#\mathbb{R}$ then $\#\mathbb{R} \leq \#X \leq \#\mathbb{R}$ and hence by Theorem 5.1.6.5, $\#\mathbb{R} = \#X$. □

Next, we show that $\#\mathbb{R} = \#\mathcal{P}\mathbb{N}$.

Theorem 5.1.12. $\mathbb{R} \sim \mathcal{P}\mathbb{N}$.

Proof. You could use parts of the proof of Theorem 5.1.11 where we showed that $\#\mathbb{R} \leq \#\mathcal{P}\mathbb{Q} = \#\mathcal{P}\mathbb{N} \leq \#[0,1]$ and $\#[0,1] \leq \#\mathbb{R}$, and hence $\#\mathbb{R} \leq \#\mathcal{P}\mathbb{N} \leq \#\mathbb{R}$, and so, $\#\mathbb{R} = \#\mathcal{P}\mathbb{N}$ and $\mathbb{R} \sim \mathcal{P}\mathbb{N}$. □

We can also prove another cardinal number is strictly greater than $\mathfrak{c}$.

Theorem 5.1.13. *Let $\mathbb{R}^{\mathbb{R}}$ be the set of real-valued functions of one argument of a real number. Let $\#\mathbb{R}^{\mathbb{R}} = \mathfrak{f}$. We have that $\mathfrak{c} < \mathfrak{f}$.*

Proof. First, let us prove that $\mathbb{R}^{\mathbb{R}}$ is not equivalent to the set $\mathbb{R}$. Suppose the set of real-valued functions of one argument $\mathbb{R}^{\mathbb{R}}$ were in one-to-one correspondence with $\mathbb{R}$. That is, $\mathbb{R}^{\mathbb{R}} \sim \mathbb{R}$. Then this correspondence would identify a real number that corresponds to each such function. Let the function corresponding to the real number r be f_r. Define the function $g : \mathbb{R} \mapsto \mathbb{R}$ by

$$g(x) = f_x(x) + 1.$$

But then g differs from each f_r in at least one value, the value at r, since

$$g(r) = f_r(r) + 1,$$

which is not $f_r(r)$. Hence, $g \in \mathbb{R}^{\mathbb{R}}$ but g does not correspond to any f_r which contradicts the meaning of a one-to-one correspondence. Therefore, $\mathbb{R}^{\mathbb{R}} \not\sim \mathbb{R}$.

Now, we will show that $\mathbb{R} \preceq \mathbb{R}^{\mathbb{R}}$. Let $h : \mathbb{R} \mapsto \mathbb{R}^{\mathbb{R}}$ such that $h(r) = c_r : \mathbb{R} \mapsto \mathbb{R}$ where $c_r(r') = r$ for every $r' \in \mathbb{R}$. Clearly, $h(r) = h(r')$ implies $r = r'$ and hence h is an injection.

Since $\mathbb{R} \preceq \mathbb{R}^{\mathbb{R}}$ and $\mathbb{R} \not\sim \mathbb{R}^{\mathbb{R}}$, we get $\#\mathbb{R} < \#\mathbb{R}^{\mathbb{R}}$. Hence, $\mathfrak{c} < \mathfrak{f}$. □

But we can go further than this. We can prove that there is no greatest cardinal number.

Theorem 5.1.14. *For every set S, $\#\mathcal{P}S > \#S$; i.e., the power set of S has a greater cardinal number than S.*

Remark 5.1.15. *For finite sets S this is clear, since if $\#S = n$, then $\#\mathcal{P}S = 2^n$. This result will now be proved for all sets, finite and infinite.*

Proof. Let S be a set, and let S' be the set which results from S if each element x of S is replaced by $\{x\}$, the subset of S whose only element is x. Since $\{x\} \subseteq S$, $S' \in \mathcal{P}S$ and $S' \sim S$, S is equivalent to a subset of $\mathcal{P}S$.

We need to show that $\mathcal{P}S$ is not equivalent to any subset of S. Suppose that there is a subset S_0 of S equivalent to $\mathcal{P}S$. Let ϕ be the function from S_0 to $\mathcal{P}S$ whose value for an element $x \in S_0$ is the set $p \in \mathcal{P}S$ which corresponds to it under the one-to-one correspondence between S_0 and $\mathcal{P}S$. Since every element of $\mathcal{P}S$ is a subset of S, it makes sense to ask whether a given element of S_0 is also an element of a given p. Let p' be the set consisting of those elements x of S_0 which are not elements of the corresponding $p = \phi(x)$; i.e.,

$$p' = \{x \in S_0 : x \notin p = \phi(x)\}.$$

Then p', is a (possibly empty) subset of S.

But this leads to a contradiction. For there is an element $x' \in S_0$ which would correspond to p' under the mapping ϕ; i.e., we would have $p' = \phi(x')$ for some x' in S_0. But then we would either have $x' \in p'$ or $x' \notin p'$, and either of these possibilities leads to a contradiction. The definition of p' excludes the possibility that $x' \in p'$. On the other hand, if $x' \notin p'$, then $x' \notin p' = \phi(x')$, so by the definition of p', we must have $x' \in p'$ after all. This contradiction shows that the assumption that there is a subset S_0 equivalent to $\mathcal{P}S$ must be false. □

Remark 5.1.16. *By Remark 5.1.8 and Theorems 5.1.11, 5.1.12, 5.1.13, we can deduce the following (in)equalities for every $n \in \mathbb{N}$:*

$$0 < 1 < 2 < \cdots < n < \cdots < \mathfrak{a} < \mathfrak{c} < \mathfrak{f}.$$

where

$$\mathfrak{a} = \#\mathbb{N} = \#\mathbb{N}_{\geq n} = \#\mathbb{Z} = \#\mathbb{Q}$$

and for I a subinterval of $\mathbb{R}$:

$$\mathfrak{c} = \#\mathbb{R} = \#I = \#\mathcal{P}\mathbb{N} = \#\mathcal{P}\mathbb{Z} = \#\mathcal{P}\mathbb{Q}$$

and

$$\#\mathbb{R}^{\mathbb{R}} = \mathfrak{f}.$$

Furthermore, since Theorem 5.1.14, for every set S, $\#\mathcal{P}S > \#S$, we deduce that

$$\mathfrak{a} < \mathfrak{c} = \#\mathbb{R} < \#\mathcal{P}\mathbb{R} < \#\mathcal{P}(\mathcal{P}\mathbb{R}) < \#\mathcal{P}(\mathcal{P}(\mathcal{P}\mathbb{R})) < \cdots$$

Hence, we have an infinite number of transfinite cardinals. All these transfinite cardinals represent uncountable sets with the only exception of the transfinite cardinal $\mathfrak{a}$ which is the only possible cardinal of a countable infinite set.

In fact, there are so many more uncountable sets than there are countable sets that any arbitrary set we pick up is most likely to be an uncountable set rather than a countable set. This is a very surprising fact considering that most of the sets we have seen so far are countable.

5.1.1 Exercises

Exercise 5.1. Prove Theorem 5.1.3.

Exercise 5.2. Prove Theorem 5.1.6.

Exercise 5.3. Prove Theorem 5.1.7

Exercise 5.4. In Lemma 5.1.9, we gave an explicit one-to-one correspondence between $(0, 1]$ and $(0, 1)$. Can you give another explicit one-to-one correspondence between these 2 intervals?
Similarly, give an explicit one-to-one correspondence between $[0, 1]$ and $(0, 1)$.

Exercise 5.5. Give another proof of Theorem 5.1.12.

Exercise 5.6. Give an explicit one-to-one correspondence between $[0, 1]$ and the irrationals on $[0, 1]$. I.e., an explicit one-to-one correspondence between $[0, 1]$ and $[0, 1] \setminus \mathbb{Q}$.

Exercise 5.7. Let $a, b \in \mathbb{R}$ such that $a < b$. Give an explicit one-to-one correspondence between $[-1, 1]$ and $[a, b]$.

Exercise 5.8. Let sets S, T and R such that R and $S \cup T$ are disjoint. Show that if $S \sim T$ then $S \cup R \sim T \cup R$.

Exercise 5.9. Let sets S, T, R and U be all mutually disjoint. Show that if $S \sim T$ and $R \sim U$ then $S \cup R \sim T \cup U$.

Exercise 5.10. Show that if $S \preceq \mathbb{N}$ then either S is finite or $\#S = \mathfrak{a}$.

5.2   Arithmetic of Cardinal Numbers

You are already familiar with the arithmetic of finite cardinal numbers since finite cardinal numbers are simply the natural numbers. The arithmetic of all cardinal numbers (including those which are not finite) is an extension of that for finite cardinal numbers. But, as we shall see, there are results for finite cardinal numbers which cannot be extended to the transfinite case.

Throughout this section, recall Theorem 5.1.6.1 which states that $\#S = 0$ iff $S = \emptyset$.

5.2.1   Addition of Cardinal Numbers

Let us begin with defining addition. In the finite case, it is easy to see that simply taking the union of sets with two given cardinal numbers does not give us the cardinal of the sum. For example, let us consider the two sets from Example 4.2.2, namely $S = \{1, 2, 3\}$ and $T = \{2, 3, 4, 5\}$. We clearly have $\#S = 3$ and $\#T = 4$. But $\#(S \cup T) = \#\{1, 2, 3, 4, 5\} = 5$, whereas $3 + 4 = 7 \neq 5$.

The problem here is that these sets S and T overlap; they have some elements in common. If we started with two disjoint sets, then the sum of the cardinal numbers would be the cardinal number of the union of the two sets.

Fortunately, it is easy given two sets to find two equivalent sets (sets with the same cardinal numbers) that are disjoint. In this case, we can take $S' = S \times \{1\}$ and $T' = T \times \{2\}$. Then $S' = \{(1, 1), (2, 1), (3, 1)\}$ and $T' = \{(2, 2), (3, 2), (4, 2), (5, 2)\}$, and these two sets are disjoint. Also, $S' \cup T' = \{(1, 1), (2, 1), (3, 1), (2, 2), (3, 2), (4, 2), (5, 2)\}$, and the cardinal number of this set is 7.

When we include the transfinite cardinal numbers, we have the following definition:

Definition 5.2.1 (Addition on cardinals). *If* $\mathfrak{m} = \#S$ *and* $\mathfrak{n} = \#T$, *then* $\mathfrak{m} + \mathfrak{n} = \#(S' \cup T')$, *where* S' *and* T' *are disjoint sets such that* $\#S' = \#S = \mathfrak{m}$ *and* $\#T' = \#T = \mathfrak{n}$.

For finite cardinal numbers, the addition defined here is the same as the familiar addition we all learned in school. But there are differences when we consider transfinite cardinal numbers as we will start seeing in Theorem 5.2.3.

For all cardinal numbers we have the following properties of addition:

Theorem 5.2.2. *For any cardinal numbers* $\mathfrak{m}, \mathfrak{n}, \mathfrak{p}$, *we have*

1. *(Well Definedness) If* $\mathfrak{m}_1 = \mathfrak{m}_2$ *and* $\mathfrak{n}_1 = \mathfrak{n}_2$. *then*

$$\mathfrak{m}_1 + \mathfrak{n}_1 = \mathfrak{m}_2 + \mathfrak{n}_2.$$

2. *(Identity)* $0 + \mathfrak{m} = \mathfrak{m} = \mathfrak{m} + 0$.

3. *(Commutativity)* $\mathfrak{m} + \mathfrak{n} = \mathfrak{n} + \mathfrak{m}$.

4. *(Associativity)* $\mathfrak{m} + (\mathfrak{n} + \mathfrak{p}) = (\mathfrak{m} + \mathfrak{n}) + \mathfrak{p}$.

5. *If* $\mathfrak{m}_1 \leq \mathfrak{m}_2$ *and* $\mathfrak{n}_1 \leq \mathfrak{n}_2$, *then* $\mathfrak{m}_1 + \mathfrak{m}_2 \leq \mathfrak{n}_1 + \mathfrak{n}_2$.

Proof. 1. Let S_1, S_2, T_1, and T_2 be disjoint sets such that $\#S_1 = \mathfrak{m}_1, \#S_2 = \mathfrak{m}_2$ and $\#T_1 = \mathfrak{n}_1$, and $\# T_2 = \mathfrak{n}_2$. Then by definition, $S_1 \sim S_2$ and $T_1 \sim T_2$. It is easy to show (see Exercise 5.9) that

$$S_1 \cup T_1 \sim S_2 \cup T_2,$$

and hence,

$$\#(S_1 \cup T_1) = \#(S_2 \cup T_2),$$

as desired.

2. By Definition 5.2.1 and Theorem 4.2.3.

3. By Definition 5.2.1 and Theorem 4.2.3.

4. By Definition 5.2.1 and Theorem 4.2.3.

5. Suppose the four cardinal numbers $\mathfrak{m}_1 = \#S_1, \mathfrak{m}_2 = \#S_2, \mathfrak{n}_1 = \#T_1$, and $\mathfrak{n}_2 = \#T_2$, where the sets S_1, S_2, T_1, and T_2 are all disjoint. By the hypotheses, $S_1 \preceq S_2$ and $T_1 \preceq T_2$ and there are injections $f : S_1 \mapsto S_2$ and $g : T_1 \mapsto T_2$. Now, let $h : S_1 \cup T_1 \mapsto S_2 \cup T_2$ such that $h(x) =$
$$\begin{cases} f(x) & \text{if } x \in S_1 \\ g(x) & \text{if } x \in T_1. \end{cases}$$
Since $S_1 \cap T_1 = \emptyset$ and f and g are injections, h is also an injective function and hence $S_1 \cup T_1 \preceq S_2 \cup T_2$. Therefore, $\#(S_1 \cup T_1) \leq \#(S_2 \cup T_2)$ and $\mathfrak{m}_1 + \mathfrak{m}_2 \leq \mathfrak{n}_1 + \mathfrak{n}_2$.

$\square$

So far, the properties of addition of cardinal numbers that we have seen are properties of all cardinal numbers, finite and transfinite. But there are some properties of transfinite cardinal numbers that are very different from those of finite cardinal numbers.

Theorem 5.2.3. *Recalling that $\mathfrak{a} = \#\mathbb{N}$ is the cardinal number of all countable sets, we have*

1. $\mathfrak{a} + \mathfrak{a} = \mathfrak{a}$.

2. *If n is a finite cardinal number, $n + \mathfrak{a} = \mathfrak{a} + n = \mathfrak{a}$,*

Proof.   1. Let E_0^+ be the set of even nonnegative integers and O^+ be the set of nonnegative odd integers. In other words,

$$E_0^+ = \{m \in \mathbb{N} : \text{there is } k \text{ such that } m = 2k\}$$
$$O^+ = \{m \in \mathbb{N} : \text{there is } k \text{ such that } m = 2k + 1\}.$$

The correspondence in which $2k$ corresponds to k shows that $\#E_0^+ = \mathfrak{a}$. The correspondence in which $2k + 1$ corresponds to k shows that $\#O^+ = \mathfrak{a}$. Also, E_0^+ and O^+ are disjoint, and $E_0^+ \cup O^+ = \mathbb{N}$, from which it follows that $\mathfrak{a} + \mathfrak{a} = \mathfrak{a}$.

2. Recall that $\mathbb{N}_{<n} = \{0, 1, \ldots n - 1\}$ and $\#\mathbb{N}_{<n} = n$. Let $\mathbb{N}_{\geq n} = \{n, n + 1, \ldots\} = \{m \in \mathbb{N} : m \geq n\}$ and $f : \mathbb{N} \mapsto \mathbb{N}_{\geq n}$ such that $f(m) = m + n$. Clearly f is a one-to-one correspondence and hence $\mathbb{N} \sim \mathbb{N}_{\geq n}$ and therefore, $\#\mathbb{N}_{\geq n} = \mathfrak{a}$. Also, $\mathbb{N}_{<n} \cup \mathbb{N}_{\geq n} = \mathbb{N}$ and $\mathbb{N}_{<n} \cap \mathbb{N}_{\geq n} = \emptyset$. Hence, $n + \mathfrak{a} = \mathfrak{a}$. $\square$

Theorem 5.2.4. *Recalling that $\mathfrak{c} = \#\mathbb{R}$, we have:*

1. $\mathfrak{c} + \mathfrak{c} = \mathfrak{c}$.

2. *If $\mathfrak{m}$ is a countable cardinal number (i.e., $\mathfrak{m} = \mathfrak{a}$ or $\mathfrak{m} = n$) then $\mathfrak{c} + \mathfrak{m} = \mathfrak{c}$. More precisely:*

   (a) $\mathfrak{c} + \mathfrak{a} = \mathfrak{c}$.

   (b) *If n is a finite cardinal number, $\mathfrak{c} + n = \mathfrak{c}$.*

Proof.   1. By Theorem 5.1.11, $\#\mathbb{R} = \#I$ for any I subinterval of $\mathbb{R}$. Since $[0, 1) \cup [1, 2] = [0, 2]$, $[0, 1) \cap [1, 2] = \emptyset$ and $\#[0, 1) = \#[1, 2] = \#[0, 2] = \mathfrak{c}$, then $\mathfrak{c} + \mathfrak{c} = \mathfrak{c}$.

2. For the last two parts of the theorem, we use Theorem 5.2.2 and 1. above to show that

$$\mathfrak{c} = \mathfrak{c} + 0 \leq \left\{ \begin{array}{c} \mathfrak{c} + n \\ \mathfrak{c} + \mathfrak{a} \end{array} \right\} \leq \mathfrak{c} + \mathfrak{c} = \mathfrak{c}.$$

$\square$

Theorem 5.2.5. *For every transfinite cardinal number* $\mathfrak{m}$, *we have* $\mathfrak{m} + \mathfrak{a} = \mathfrak{m}$.

Proof. Let S and T be disjoint sets with $\#S = \mathfrak{m}$ and $\#T = \mathfrak{a}$. Since S is an infinite set, by Lemma 4.3.1.5, it contains a subset R where $R \sim \mathbb{N}$. By Theorem 4.3.3, $R \cup T$ is also countable (actually, infinitely countable), and so $R \cup T \sim \mathbb{N}$. Since $\sim$ is an equivalence relation then $R \cup T \sim R$. Since $R \subseteq S$ then $S = (S \setminus R) \cup R$ and hence, $S \cup T = (S \setminus R) \cup (R \cup T)$. Since $R \cup T \sim R$ and $S \setminus R$ and $R \cup T$ are disjoint, by Exercise 5.8, $S \cup T = (S \setminus R) \cup (R \cup T) \sim (S \setminus R) \cup R = S$. Hence, $S \cup T \sim S$ and so, $\#S \cup T = \#S$. Since S and T are disjoint, $\#S \cup T = \#S + \#T$ and so, $\#S + \#T = \#S$ and $\mathfrak{m} + \mathfrak{a} = \mathfrak{m}$. $\square$

Corollary 5.2.6. *The cardinal number of the set of transcendental real numbers is* $\mathfrak{c}$.

Proof. Let S be the set of transcendental real numbers and A the set of algebraic real numbers. In Theorem 4.1.13 we showed that the set A is infinitely countable and hence $\#A = \mathfrak{a}$ (whether we include the complex algebraic numbers or not, the set will remain infinitely countable). We know that $\#\mathbb{R} = \mathfrak{c}$. Since S and A are disjoint, then $\#(S \cup A) = \#S + \#A = \#S + \mathfrak{a}$. Since $\#S$ is a transfinite cardinal then by Theorem 5.2.5, $\#S + \mathfrak{a} = \#S$. Hence, $\#(S \cup A) = \#S$. But, $S \cup A = \mathbb{R}$ and hence $\#(S \cup A) = \mathfrak{c}$. Therefore, $\#S = \mathfrak{c}$. $\square$

5.2.2 Product of Cardinal Numbers

Now let us consider products of cardinal numbers.

Definition 5.2.7. *The* product *of two cardinal numbers,* $\mathfrak{m}$ *and* $\mathfrak{n}$, *is defined as follows: if* S *and* T *are sets with* $\#S = \mathfrak{m}$ *and* $\#T = \mathfrak{n}$, *then the product* $\mathfrak{m}\mathfrak{n}$ *(or* $\mathfrak{m} \cdot \mathfrak{n}$*) is given by* $\#(S \times T)$.

We can easily see that this it correct for finite cardinal numbers; see Example 4.2.17. We are extending this to transfinite cardinal numbers.

Theorem 5.2.8. *The following hold:*

1. *(Well definedness)* Let S, S', T, and T', be disjoint sets such that $S \sim S'$ and $T \sim T'$. Then $S \times S' \sim T \times T'$. *(Thus, the product of two cardinal numbers does not depend on which sets represent the sets being mulltiplied.)*

2. *(Commutative Law)* $mn = nm$.

3. *(Associative Law)* $(mn)p = m(np)$.

4. *(Zero)* $0 \cdot m = m \cdot 0 = 0$.

5. *(Identity)* $1 \cdot m = m \cdot 1 = m$.

6. *(Distributive Law)* $m(n + p) = mn + mp$.

7. If $m_1 \leq n_1$ and $m_2 \leq n_2$, then $m_1 m_2 \leq n_1 n_2$.

Proof.   1. Let S, S', T, and T' be disjoint sets such that $S \sim S'$ and $T \sim T'$. Then to show that $S \times T \sim S' \times T'$, we proceed as follows: $(x, y) \in S \times T$ corresponds to $(x', y') \in (S' \times T')$ if $x \in S$ corresponds to $x' \in S'$ and $y \in T$ corresponds to $y' \in T'$. This correspondence is clearly bijective.

2. It is obvious that for any two sets S and T, $S \times T \sim T \times S$, by letting the pair (x, y) correspond to the pair (y, x).

3. Similarly, by letting $((x, y), z)$ correspond to $(x, (y, z))$, we get that $(S \times T) \times P \sim S \times (T \times P)$.

4. Let $\#S = m$. Since $\emptyset \times S = S \times \emptyset = \emptyset$ then $\#\emptyset \cdot \#S = \#S \cdot \#\emptyset = \#\emptyset$. By Theorem 5.1.6.1, $\#\emptyset = 0$ and hence, $0 \cdot \#S = \#S \cdot 0 = 0$. That is, $0 \cdot m = m \cdot 0 = 0$.

5. By letting (x, y) correspond to y, we get that $\{x\} \times S \sim S$.

6. Let R, S, and T be disjoint sets with $\#R = m$, $\#S = n$, and $\#T = p$. Then $R \times (S \cup T)$ consists of pairs (x, y), were $x \in R$ and $y \in S \cup T$. Now since S and T are disjoint, $n + p = \#(S \cup T)$. Also, $mn = \#(R \times S)$, and $mp = \#(R \times T)$. Thus,

$$
\begin{aligned}
mn + mp &= \#((R \times S) \cup (R \times T)) \\
&= \#(\{(x, y) : x \in R \text{ and } y \in S\} \\
&\quad \cup \{(x, y) : x \in R \text{ and } y \in T\}) \\
&= \#\{(x, y) : x \in R \text{ and } (y \in S \text{ or } y \in T)\} \\
&= m(n + p).
\end{aligned}
$$

7. Let S_1, S_2, T_1, and T_2 be mutually disjoint sets with $\mathfrak{m}_1 = \#S_1$, $\mathfrak{m}_2 = \#S_2$, $\mathfrak{n}_1 = \#T_1$, and $\mathfrak{n}_2 = \#T_2$. By hypothesis, there are injections $f_1 : S_1 \mapsto T_1$ and $f_2 : S_2 \mapsto T_2$. Now, we take $f : S_1 \times S_2 \mapsto T_1 \times T_2$ such that $f((x,y)) = (f_1(x), f_2(y))$. If $f((x,y)) = f((x',y'))$ then $(f_1(x), f_2(y)) = (f_1(x'), f_2(y'))$ and $f_1(x) = f_1(x')$ and $f_2(y) = f_2(y')$. Since f_1 and f_2 are injective, we get $x = x'$ and $y = y'$ and hence $(x,y) = (x',y')$. Hence f is injective and $\#(S_1 \times S_2) \leq \#(T_1 \times T_2)$. Therefore, $\mathfrak{m}_1\mathfrak{m}_2 \leq \mathfrak{n}_1\mathfrak{n}_2$.

$\square$

Now let us start looking at the multiplication table for transfinite cardinal numbers.

Theorem 5.2.9. *The following hold:*

1. $\mathfrak{a} \cdot \mathfrak{a} = \mathfrak{a}$.

2. *If $n \neq 0$ is finite, then $n \cdot \mathfrak{a} = \mathfrak{a}$.*

3. $\mathfrak{a} \cdot \mathfrak{c} = \mathfrak{c}$.

4. *If $n \neq 0$ is finite, then $n \cdot \mathfrak{c} = \mathfrak{c}$.*

5. $\mathfrak{c} \cdot \mathfrak{c} = \mathfrak{c}$.

Proof. 1. Let us use the fact that $\#\mathbb{N}^+ = \#\{1, 2, 3, \ldots\} = \mathfrak{a}$. Then $\mathfrak{a} \cdot \mathfrak{a} = \#\{(m, n) : m, n \in \mathbb{N}^+\}$. Let us arrange the pairs of elements of $\mathbb{N}^+$ the way we arranged the fractions in the proof of Theorem 4.1.10:

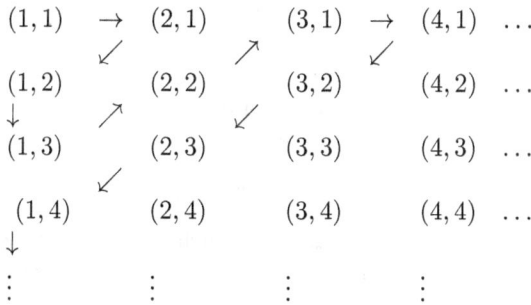

$$
\begin{array}{cccc}
(1,1) & \to \quad (2,1) & (3,1) & \to \quad (4,1) \quad \ldots \\
& \swarrow \qquad \nearrow & \swarrow & \\
(1,2) & (2,2) & (3,2) & (4,2) \quad \ldots \\
\downarrow \quad \nearrow & & \swarrow & \\
(1,3) & (2,3) & (3,3) & (4,3) \quad \ldots \\
& \swarrow & & \\
(1,4) & (2,4) & (3,4) & (4,4) \quad \ldots \\
\downarrow & & & \\
\vdots & \vdots & \vdots & \vdots
\end{array}
$$

As with the rational numbers in the proof of Theorem 4.1.10, this gives us a sequence of pairs that is clearly in one-to-one correspondence with $\mathbb{N}^+$.

2. By part 1 above and Theorem 5.2.8, we have

$$\mathfrak{a} = 1 \cdot \mathfrak{a} \leq n \cdot \mathfrak{a} \leq \mathfrak{a} \cdot \mathfrak{a} = \mathfrak{a}.$$

3. As in Part 1 above, we use the fact that $\mathfrak{a} = \#\mathbb{N}^+$. We also use the fact (see Theorem 5.1.11) that $\mathfrak{c} = \#[0,1) = \#[1,\infty)$. We then have

$$\mathfrak{a} \cdot \mathfrak{c} = \#(\mathbb{N}^+ \times [0,1)).$$

A typical element of $\mathbb{N}^+ \times [0,1)$ has the form (n,x), where $n \in \mathbb{N}+$ and $x \in [0,1)$. If we let (n,x) correspond to $n+x$, we have a one-to-one correspondence between $\mathbb{N}^+ \times [0,1)$ and the interval $[1,\infty)$, and the cardinal number of that interval is $\mathfrak{c}$. Hence, $\mathfrak{a} \cdot \mathfrak{c} = \mathfrak{c}$.

4. By part 3 and Theorem 5.2.8, we have

$$\mathfrak{c} = 1 \cdot \mathfrak{c} \leq n \cdot \mathfrak{c} \leq \mathfrak{a} \cdot \mathfrak{c} = \mathfrak{c}.$$

5. Recall that $\#(0,1] = \mathfrak{c}$. So, since $\#(0,1] \times (0,1] = \mathfrak{c} \cdot \mathfrak{c}$, it is sufficient to prove that $(0,1] \times (0,1]$ can be put into one-to-one correspondence with $(0,1]$. Let us represent the elements of $(0,1]$ as nonterminating decimals; i.e., decimals with infinitely many places none of which terminate. This means that whenever we have a decimal which does terminate, such as 0.284, we replace this by 0.283999..., where all the remaining digits are 9. (Note that this means that 1 is represented by 0.9999....) If we do this systematically, then we will have decimal fractions that are in one-to-one correspondence with the real numbers in $(0,1]$. Then we want to show that we can make pairs of real numbers (x,y) with $0 < x, y \leq 1$ correspond in a one-to-one manner with real numbers $z \in (0,1]$.

To do this, we consider a pair of nonterminating decimals (x,y) be given, and suppose, for example, that we have

$$x = 0.4\ 05\ 008\ 7\ 9\ \ldots$$

and
$$y = 0.002\ 4\ 2\ 005\ 7\ \ldots.$$

The nonterminating decimal z which will correspond to this pair will be obtained by alternating the parts ending in a nonzero digit; in this example,
$$z = 0.4\ 002\ 05\ 4\ 008\ 2\ 7\ 005\ 9\ 7\ldots.$$

On the other hand, from any such z we can get the pair (x,y) associated with it by marking off the groups of digits ending with a nonzero digit and arranging these groups in alternating order. This establishes the desired one-to-one correspondence.

$\square$

5.2.3   Exponentiation of Cardinal Numbers

Now let us define the *exponentiation* of cardinal numbers. We start with a different definition, one of an operation on sets.

Definition 5.2.10. *If S and T are sets, then S^T is the set of functions from T to S. That is:*

$$S^T = \{f : T \mapsto S : f \text{ is a function where for every } x \in T, f(x) \in S\}.$$

Note that this means that every element of T is an argument for every function $f \in S^T$. On the other hand, not every element of S is necessarily the value of such a function.

So, for example, $\mathbb{R}^\mathbb{R}$ would include the function $f(x) = x^2$, since any real number can be an argument of this function, but only nonnegative real numbers can be values of it. On the other hand, for the function $g(x) = x^3$, every real number is a value for some argument. But the function $h(x) = \sqrt{x}$ can take as arguments only real numbers that are nonnegative, so this function is in the set $\mathbb{R}^{\{x \in \mathbb{R} : x \geq 0\}}$.

To take another example, if we think of the elements of a sequence $s_1, s_2, s_3, \ldots$ as values of a function $j(1), j(2), j(3), \ldots$, each such function is in the set $\mathbb{R}^{\mathbb{N}^+}$.

We are now in a position to give the definition of exponentiation of cardinal numbers:

Definition 5.2.11. *If $\mathfrak{m}$ and $\mathfrak{n}$ are cardinal numbers and S and T are sets such that $\mathfrak{m} = \#S$ and $\mathfrak{n} = \#T$, then $\mathfrak{m}^\mathfrak{n} = \#(S^T)$.*

We need to prove that this definition is well-defined:

Theorem 5.2.12. *If $S \sim S'$ and $T \sim T'$, then $S^T \sim S'^{T'}$.*

Proof. Since $S \sim S'$ and $T \sim T'$, let the one-to-one correspondences (i.e., bijections) $\phi : S \mapsto S'$ and $\psi : T \mapsto T'$. By Definition 4.2.21, since the functions ϕ and ψ represent one-to-one correspondences, we have that each of these functions has an inverse: $\psi^{-1} : T' \mapsto T$ and $\phi^{-1} : S' \mapsto S$. By Lemma 4.2.23, the functions ϕ and ψ are injective and surjective. Now, define a one-to-one correspondence ρ between S^T and $S'^{T'}$ as follows: for $f \in S^T$, let $\rho(f) = g$ where $g \in S'^{T'}$ is defined by

$$g(x) = \phi(f(\psi^{-1}(x))) \text{ for } x \in T'.$$

Using the injectivity and surjectivity of ϕ and ψ^{-1}, we can show that ρ is a one-to-one correspondence between S^T and $S'^{T'}$ (see Exercise 5.12). □

Corollary 5.2.13. *Exponentiation of cardinal numbers is well defined: the value of* $\mathfrak{m}^{\mathfrak{n}}$ *does not depend on which sets we use with the cardinal numbers* $\mathfrak{m}$ *and* $\mathfrak{n}$.

Now we will prove some of the usual properties of exponentiation.

Theorem 5.2.14. *If* $\mathfrak{m}$ *is a cardinal number such that* $\mathfrak{m} \neq 0$, *then* $\mathfrak{m}^0 = 1$.

Proof. Let S such that $\#S = \mathfrak{m}$ where $\mathfrak{m} \neq 0$. By Theorem 5.1.6.1, $\#\emptyset = 0$. By definition, $\#(S^\emptyset) = (\#S)^{\#\emptyset} = \mathfrak{m}^0$. But, $S^\emptyset$ contains one element only (the empty function since its domain is $\emptyset$) and hence $\#(S^\emptyset) = 1$. Therefore, $\mathfrak{m}^0 = 1$. $\qquad\qquad\square$

Theorem 5.2.15. *If* $R, S,$ *and* T *are sets such that* S *and* T *are disjoint, then*

$$R^{S \cup T} \sim R^S \times R^T.$$

Proof. Elements of the set $R^{S \cup T}$ are functions $f : S \cup T \mapsto R$ which take arguments from $S \cup T$ and have values in R. Elements of $R^S \times R^T$ are pairs of functions (g, h) such that $g : S \mapsto R$ has arguments in S and values in R and $h : T \mapsto R$ has arguments in T and values in R. Now since by hypothesis S and T are disjoint, every x in $S \cup T$ is in either S or T but not both. Hence, given $f : S \cup T \mapsto R$, any argument of f is in either S or T but not both. Thus, given the f, we can get a function $g : S \mapsto R$ by applying f to arguments in S, and we can get $h : T \mapsto R$ by applying f to arguments in T. Thus, to f, we get a pair (g, h) of functions with $g : S \mapsto R$ and $h : T \mapsto R$, and so this pair (g, h) corresponds to f. Conversely, given the pair (g, h) of functions with $g : S \mapsto R$ and $h : T \mapsto R$, we can define f to be the function which takes arguments from $S \cup T$ by letting $f(x) = g(x)$ if $x \in S$ and $f(x) = h(x)$ if $x \in T$. This establishes the one-to-one correspondence of the theorem. In other words, the function $\Phi : R^{S \cup T} \mapsto R^S \times R^T$ defined for $f : S \cup T \mapsto R$ by: $\Phi(f) = (g, h)$ where $g : S \mapsto R$ and $h : T \mapsto R$ such that $g(x) = f(x)$ for every $x \in S$ and $h(x) = f(x)$ for every $x \in T$, is bijective. $\qquad\qquad\square$

Corollary 5.2.16. *If* $\mathfrak{m}, \mathfrak{n},$ *and* $\mathfrak{p}$ *are cardinal numbers, then*

$$\mathfrak{m}^{\mathfrak{n}+\mathfrak{p}} = \mathfrak{m}^{\mathfrak{n}} \cdot \mathfrak{m}^{\mathfrak{p}}.$$

Theorem 5.2.17. *For any sets* $R, S,$ *and* T, *we have* $(R \times S)^T \sim R^T \times S^T$.

Proof. If $f \in (R \times S)^T$, then arguments of f are in T and $f(x) = (u, v)$ where $u \in R$ and $v \in S$. Since u and v depend on x, we can write $u = f_1(x)$ and $v = f_2(x)$, and we have $f_1 \in R^T$ and $f_2 \in S^T$, and hence $(f_1, f_2) \in R^T \times S^T$.

The correspondence between f and (f_1, f_2) gives us the desired one-to-one correspondence. In other words, the function $\Phi : (R \times S)^T \mapsto R^T \times S^T$ defined for $f : T \mapsto R \times S$ by: $\Phi(f) = (f_1, f_2)$ where $f_1 : T \mapsto R$ and $f_2 : T \mapsto S$ such that $f(x) = (f_1(x), f_2(x))$ for every $x \in T$ is bijective. For a proof of this bijection, see Exercise 5.13. $\square$

Corollary 5.2.18. *If* $\mathfrak{m}, \mathfrak{n}$, *and* $\mathfrak{p}$ *are cardinal numbers, then*

$$(\mathfrak{mn})^{\mathfrak{p}} = \mathfrak{m}^{\mathfrak{p}} \mathfrak{n}^{\mathfrak{p}}.$$

Theorem 5.2.19. *For any sets* R, S, *and* T, *we have* $(R^S)^T \sim R^{S \times T}$.

Proof. Let $f \in (R^S)^T$. Then f is a function defined for $x \in T$ such that $f(x) \in R^S$, which means that $f(x)$ is a function $g_x \in R^S$ defined for $y \in S$ such that $g_x(y) \in R$. Now let $z \in S \times T$, so that $z = (y, x)$ for $y \in S$ and $x \in T$. Let $h(z) = g_x(y)$: we have $h(z) \in R$ for $z \in S \times T$, so $h \in R^{S \times T}$. Letting this h correspond to f gives us a one-to-one correspondence as desired. In other words, the function $\Phi : (R^S)^T \mapsto R^{S \times T}$ defined for $f : T \mapsto R^S$ by: $\Phi(f) = h : S \times T \mapsto R$ where for every $z = (y, x) \in S \times T$, $h(z) = f(x)(y)$ is bijective. For a proof of this bijection, see Exercise 5.14. $\square$

Corollary 5.2.20. *If* $\mathfrak{m}, \mathfrak{n}$, *and* $\mathfrak{p}$ *are cardinal numbers, then* $(\mathfrak{m}^{\mathfrak{n}})^{\mathfrak{p}} = \mathfrak{m}^{\mathfrak{np}}$.

We proved in Theorem 5.1.14 that $\#\mathcal{P}S > \#S$, and in Remark 5.1.15 we stated that for finite sets S, $\#\mathcal{P}S = 2^{\#S}$. We can now prove this statement for all sets.

Theorem 5.2.21. *If* S *is any set and* $\#S = \mathfrak{m}$, *then* $\#\mathcal{P}S = 2^{\mathfrak{m}}$.

Proof. Let S be a set with $\#S = \mathfrak{m}$, and recall that $\mathcal{P}S$ is the set of all subsets of S. For each subset T of S, let its *characteristic function* ξ_T be defined for $x \in S$ by $\xi_T(x) = 1$ if $x \in T$ and $\xi_T(x) = 0$ if $x \notin T$. Then $\xi_T \in \{0, 1\}^S$. Note that $\#\{0, 1\} = 2$. By letting each subset of S correspond to its characteristic function, we get a one-to-one correspondence between $\mathcal{P}S$ and $\{0, 1\}^S$. That is, the function $\Phi : \mathcal{P}S \mapsto \{0, 1\}^S$ such that $\Phi(T) = \xi_T$ for every $T \in \mathcal{P}S$, is bijective. For a proof of this bijection, see Exercise 5.15. Since by Definition 5.2.11, $\#\{0, 1\}^S = 2^{\mathfrak{m}}$ and since $\mathcal{P}S \sim \{0, 1\}^S$, then $\#\mathcal{P}S = 2^{\mathfrak{m}}$. $\square$

Corollary 5.2.22. $\mathfrak{c} = 2^{\mathfrak{a}}$.

Proof. By Theorem 5.1.12, $\mathbb{R} \sim \mathcal{P}\mathbb{N}$ and so, $\mathfrak{c} = \#\mathbb{R} = \#\mathcal{P}\mathbb{N}$. By Theorem 5.2.21, $\#\mathcal{P}\mathbb{N} = 2^{\#\mathbb{N}}$. Since $\#\mathbb{N} = \mathfrak{a}$, we get $\mathfrak{c} = 2^{\mathfrak{a}}$. $\square$

Theorem 5.2.23. *For any cardinals* $\mathfrak{m}, \mathfrak{n}$ *and* $\mathfrak{p}$, *if* $\mathfrak{m} \leq \mathfrak{n}$ *then* $\mathfrak{m}^{\mathfrak{p}} \leq \mathfrak{n}^{\mathfrak{p}}$ *and* $\mathfrak{p}^{\mathfrak{m}} \leq \mathfrak{p}^{\mathfrak{n}}$.

Proof. Let sets S, T and U such that $\#S = \mathfrak{m}$, $\#T = \mathfrak{n}$ and $\#U = \mathfrak{p}$. Since $\mathfrak{m} \leq \mathfrak{n}$, let $f : S \mapsto T$ be an injection.

1. Assume $\varphi : S^U \mapsto T^U$ such that $\varphi(g) = h$ where for any $x \in U$, $h(x) = f(g(x))$. We will show that φ is injective. Assume $\varphi(g) = \varphi(g')$ and let $x \in U$. Then, $\varphi(g)(x) = \varphi(g')(x)$ and so, $f(g(x)) = f(g'(x))$ and since f is injective we get $g(x) = g'(x)$. Hence, by function extensionality, $g = g'$. Therefore, φ is injective and $\mathfrak{m}^{\mathfrak{p}} \leq \mathfrak{n}^{\mathfrak{p}}$.

2. Let $\varphi' : U^S \mapsto U^T$ such that $\varphi'(g) = h$ where for any $x \in T$,

$$h(x) = \begin{cases} g(y) & \text{if } x = f(y) \\ p \in U \setminus f[S] & \text{if } x \notin f[S]. \end{cases}$$

Note here that in case $U \setminus f[S] \neq \emptyset$, we are choosing a particular element $p \in U \setminus f[S]$. Since f is injective, for any $x \in T$, $h(x)$ is unique and hence h is a function in U^T.

We will show that φ' is injective. Assume $\varphi'(g) = \varphi'(g')$ and let $y \in S$. Then, $g(y) = \varphi'(g)(f(y)) = \varphi'(g')(f(y)) = g'(y)$ and so, by function extensionality $g = g'$ and φ' is injective. Therefore, $\mathfrak{p}^{\mathfrak{m}} \leq \mathfrak{p}^{\mathfrak{n}}$.

$\square$

5.2.4   Exercises

Exercise 5.11. Let sets R, S, and T be such that S and T are disjoint and let $\Phi : R^{S \cup T} \mapsto R^S \times R^T$ defined by: $\Phi(f) = (g, h)$ where $g : S \mapsto R$ and $h : T \mapsto R$ such that $g(x) = f(x)$ for every $x \in S$ and $h(x) = f(x)$ for every $x \in T$. Show that Φ is a bijective function.

Exercise 5.12. Show that $\rho : S^T \mapsto S'^{T'}$ as defined in the proof of Theorem 5.2.12 is a bijection.

Exercise 5.13. Show that the function $\Phi : (R \times S)^T \mapsto R^T \times S^T$ defined in the proof of Theorem 5.2.17 is bijective.

Exercise 5.14. Show that the function $\Phi : (R^S)^T \mapsto R^{S \times T}$ defined in the proof of Theorem 5.2.19 where $\Phi(f) = h$ such that $h((y, x)) = f(x)(y)$ is bijective.

Exercise 5.15. Show that the function $\Phi : \mathcal{P}S \mapsto \{0, 1\}^S$ defined in the proof of Theorem 5.2.21 such that $\Phi(T) = \xi_T$ for every $T \in \mathcal{P}S$, is bijective.

Exercise 5.16. Let S be a set whose elements are also sets X such that $X \in S$ iff there is $Y \in S$ such that $\#X < \#Y$. Show that for all $X \in S$ we have $\#X < \#\bigcup S$ where $\bigcup S = \{x : x \in T \text{ for some } T \in S\}$.

Exercise 5.17. Show that $\mathfrak{a}^{\mathfrak{a}} = 2^{\mathfrak{a}}$.

Exercise 5.18. For each $n \in \mathbb{N}$, let $U_n = \{S \subseteq \mathbb{N} : S \sim \mathbb{N}_{<n}\}$.

1. Give $\#U_n$ for each $n \in \mathbb{N}$.

2. Give $\#U$ where $U = \{S \subseteq \mathbb{N} : S \text{ is finite}\}$.

3. Give $\#V$ where $V = \{S \subseteq \mathbb{N} : S \text{ is infinite}\}$.

Exercise 5.19. The *difference* of two cardinal numbers, $\mathfrak{m}$ and $\mathfrak{n}$, is defined as follows: if S and T are sets with $T \subseteq S$, $\#S = \mathfrak{m}$, $\#T = \mathfrak{n}$ and $\mathfrak{m} > \mathfrak{n}$ then $\mathfrak{m} - \mathfrak{n}$ is given by $\#(S \setminus T)$.
 Show the following:

1. This definition does not work if we remove the condition $\mathfrak{m} > \mathfrak{n}$.

2. If $S \sim S'$, $T \sim T'$, $T \subseteq S$, $T' \subseteq S'$ and $\#S > \#T$ then $\#(S \setminus T) = \#(S' \setminus T')$.

3. If $\mathfrak{m}, \mathfrak{n}, \mathfrak{p}$ are cardinal numbers such that $\mathfrak{n} > \mathfrak{p}$ then $\mathfrak{m}(\mathfrak{n}-\mathfrak{p}) = \mathfrak{m}\mathfrak{n}-\mathfrak{m}\mathfrak{p}$.

4. If $\mathfrak{m}, \mathfrak{n}, \mathfrak{p}$ are cardinal numbers such that $\mathfrak{n} > \mathfrak{p}$ and $\mathfrak{m} + \mathfrak{p} = \mathfrak{n}$ then $\mathfrak{m} = \mathfrak{n} - \mathfrak{p}$.

5.3 Paradoxes and Uncertainties of Set Theory

5.3.1 Paradoxes

Throughout this chapter and the previous, we have used the notation $\{x \in U : P(x)\}$ to indicate a set and we only deviated from this when the set U was clear from the context. Assuming such a U is crucial. We cannot create sets $\{x : P(x)\}$ where the elements x are not clearly assigned to a certain set. For, if we could do that, we could form the set

$$R = \{x : x \notin x\}.$$

But this leads to a contradiction. For suppose $R \in R$. Then R should have to satisfy $x \notin x$, so we would have $R \notin R$. On the other hand, if $R \notin R$, then R does satisfy $x \notin x$, so we would then have $R \in R$ again.

 The use of R here is for Russell: In 1902, Bertrand Russell found this contradiction in the system of Gottlob Frege, and so this contradiction is known as *Russell's Paradox*.

 There are other paradoxes in set theory and logic, and we will list some which can be understood with those who have read to this point.

1. *Cantor's paradox*. We have proved above (Theorem 5.1.14) that the cardinal number of the power set of a set is greater than the cardinal number of the set. But suppose there is a set of all sets? This set should have the greatest cardinal number that exists. However, the power set of the set of all sets has a greater cardinal number.

2. *Liar paradox*. This one has several forms. The simplest is the man who says, "I am lying." If he is lying, then he is telling the truth, whereas if he is telling the truth then he is lying.

3. *Richard's paradox*. Consider the set of all definable numerical functions, where "definable" means definable in a fixed language, such as mathematical English, with a fixed dictionary and grammar. Since the number of words in this language is finite, the number of expressions is infinitely countable, and since every definable function is defined by an expression in this language, the number of such functions is also infinitely countable. Let f_n be the nth such function in an enumeration of such functions. Then let

$$g(n) = f_n(n) + 1.$$

Then g differs from f_n for every n since $g(m) = f_m(m) + 1 \neq f_m(m)$, which contradicts the fact that g can clearly be defined.

4. *Berry paradox*. The number of natural numbers which can be named in English in less than a fixed number of syllables (or letters) is certainly finite, and hence there must be a least number which cannot be so named.

5. *Grelling paradox*. Among English adjectives there are some, such as "short", "polysyllabic", and "English" which apply to themselves. Let us call adjectives which apply to themselves *autological* and those which do not apply to themselves *heterological*. So examples of heretological adjectives are "long", "monosyllabic", and "green" are heterological. Then if "hetrological" is heterological, it applies to itself, and so it is autological, whereas if it is autological then it must apply to itself, and so it is heterological.

   Paradoxes like these were first published at the very end of the Nineteenth Century and the beginning of the Twentieth. At the beginning of Chapter 4, we mentioned that before Cantor, mathematicians rejected the idea of an infinite set as a completed whole. In fact when Cantor first began to publish his ideas on infinite sets, he encountered strong opposition to these ideas from some mathematicians. One particular such mathematician who clearly opposed the work of Cantor was Leopold Kronecker.[2]

If the paradoxes that directly involve set theory had first appeared even as little as a decade before they actually did, mathematicians might have abandoned set theory. But by the time they appeared, set theory had come to seem so important to mathematics, and especially to its foundations that instead of abandoning it mathematicians started trying to construct systems of axiomatic set theory that were consistent. These systems are beyond the scope of this book.

5.3.2 The Continuum Hypothesis, the Axiom of Choice and Incompleteness

Recall Remark 5.1.16 where we gathered the results so far on the cardinalities of sets. In particular for every $n \in \mathbb{N}$:

$$0 < 1 < 2 < \cdots < n < \cdots < \mathfrak{a} < \mathfrak{c} \text{ where } \mathfrak{a} = \#\mathbb{N} \text{ and } \mathfrak{c} = \#\mathbb{R}.$$

We also showed that $\#\mathbb{N}$ is the smallest infinite cardinal, that $\#\mathcal{P}\mathbb{N} = \#\mathbb{R}$ and that

$$\#\mathbb{N} \quad < \quad \#\mathbb{R} \quad < \quad \#\mathcal{P}\mathbb{R} \quad < \quad \#\mathcal{P}(\mathcal{P}\mathbb{R}) \quad < \quad \#\mathcal{P}(\mathcal{P}(\mathcal{P}\mathbb{R})) \quad < \quad \cdots$$

But, are there cardinals between $\#\mathbb{N}$ and $\#\mathbb{R}$? or between $\#\mathbb{R}$ and $\#\mathcal{P}\mathbb{R}$ or between $\#\mathcal{P}\mathbb{R}$ and $\#\mathcal{P}(\mathcal{P}\mathbb{R})$, etc.? In other words, do we have sets S, S', S'', S''', etc., such that:

$$\#\mathbb{N} < \#S < \#\mathbb{R} < \#S' < \#\mathcal{P}\mathbb{R} < \#S'' <$$
$$\#\mathcal{P}(\mathcal{P}\mathbb{R}) < \#S''' < \#\mathcal{P}(\mathcal{P}(\mathcal{P}\mathbb{R})) < \cdots$$

---

[2]Leopold Kronecker was a powerful professor in Berlin and had serious disagreements with Cantor with respect to his transfinite set theory. Much is written on this disagreement and on Cantor's reactions to Kronecker's hostilities and whether these hostilities could have been the cause of the bouts of depressions that Cantor suffered at various intervals in his life after his 39th Birthday. Kronecker strongly believed in the power of the natural numbers which according to him were created by God and all other numbers were created by men. Kronecker was opposed to the use of irrational numbers and infinite series made by Dedekind, Weierstrauss and certainly by Cantor in his proof of the uncountability of the real numbers. Kronecker further believed that Cantor's proof of a one-to-one correspondence between $[0, 1]$ and points in the 2-dimensional space was meaningless and without any hope of salvation (which by the way, was described by Cantor to his friend Dedekind with the words: "I see it, but I don't believe it"). Kronecker opposed the publication of that result in Crelle's journal but at Dedekind intervention, the work was eventually accepted in the journal. It is not clear what took its toll on Cantor and there are many different theories on the reasons for this. Irrespective of the reasons, Cantor did spend the last 30 years of his life suffering from occasional depression breakdowns and going in and out of the Halle Sanatorium where he died in 1918.

Unfortunately, we do not know for sure the answer to this question and whether there are such S, S', S'', etc. Let us concentrate on the first part which is $\#\mathbb{N} < \#S < \#\mathbb{R}$. Cantor already knew that $\mathfrak{a} < \mathfrak{c}$. He also thought he could prove that there is no cardinal number $\mathfrak{m}$ such that $\mathfrak{a} < \mathfrak{m} < \mathfrak{c}$. However, he was not able to complete this proof. In fact, it was his attempts at solving this hypothesis that led him to develop the subject of set theory. As a result, this has become a hypothesis that is discussed. It is called the *Continuum Hypothesis*.

Continuum Hypothesis
There is no set S such that $\#\mathbb{N} < \#S < \#\mathbb{R}$.

This Continuum Hypothesis has been influential in the domain of logic and set theory. So much so that it was listed in 1900 by the famous German mathematician David Hilbert as the first problem of 23 famous problems that needed to be settled. Hilbert was famous for believing that eventually we can find an axiomatisation of logic in which we can prove or disprove all of mathematics. Alas, Cantor's Continuum Hypothesis led to the discovery of Gödel's[3] Incompleteness Theorem in the 1930s:

Gödel's Incompleteness Theorem
There are statements that can be syntactically expressed in first order logic but that can neither be proven nor disproven. Moreover, in any axiomatic system of mathematics we can express statements that can neither be proven nor disproven in the system.

Gödel's Incompleteness Theorem implies that the axiomatisation that Hilbert wished cannot be achieved. Gödel furthermore showed in 1938 that the Continuum Hypothesis which can be expressed in the language of the most famous set theory (ZFC), cannot be disproved in ZFC as it does not lead to any contradictions if added to ZFC. Let us write CH for the Continuum Hypothesis. Then:

---

[3]Gödel, established major results that changed the course of mathematics. His personal life was troubled however since like Cantor, he suffered from episodes of mental illnesses and paranoia.

Gödel 1938:
If ZFC is consistent then

- ZFC + CH is consistent.

- ZFC cannot prove ¬CH is true.

- ZFC cannot prove CH is false.

In the 1963, Cohen proved that the Continuum Hypothesis cannot be proved in ZFC.

Cohen 1963:

- ZFC + ¬CH is consistent.

- ZFC cannot prove CH is true.

- ZFC cannot prove ¬CH is false.

It is still the case that the Continuum Hypothesis remains neither proven nor disproven. Göedel's and Cohen's results imply that there is no proof of the Continuum Hypothesis amongst the set of all the proofs that can be written in ZFC.

One can compare the status of the Continuum Hypothesis in set theory to the status of the *parallel postulate* in plane Geometry. The parallel postulate states that:

parallel postulate
Given a point p not on a line l, there is exactly one line l' which passes through p and which is parallel to l.

The parallel postulate cannot be proven or disproven from the other axioms of plane geometry. Nonetheless, we already have:

- A long standing model of Euclidean plane Geometry which in addition to the Euclidean axioms includes the parallel postulate, and hence, the parallel postulate holds in this Euclidean model.

- Non-Euclidean models of Geometry which contain all of the axioms of Euclidean Geometry excluding the parallel postulate, and the parallel postulate cannot be proven in these non-Euclidean models.

In fact, we can safely choose to work in a plane geometry where we either add the parallel postulate as an axiom or add the negation of the parallel postulate as an axiom.

Just like we can accept or reject the parallel postulate, we can accept or reject the Continuum Hypothesis. These two results of Gödel and Cohen on the Continuum Hypothesis imply that the Continuum Hypothesis is independent of other parts of mathematics and we can either accept it or reject it. It is consistent to either accept or to reject the Continuum Hypothesis. We may be discouraged by the lack of certainty with respect to the Continuum Hypothesis (no proof and no disproof), but no matter the status of the Continuum Hypothesis, the Incompleteness theorem of Gödel means that we will always find statements that cannot be proven or disproven and this is something we need to accept. Gödel proved the opposite of what Hilbert wished for. Hilbert wanted an axiomatic system of mathematics in which every statement can either be proved or disproved. Gödel's Incompleteness theorem states that no such axiomatic system of all of mathematics can exist and there will always be statements that cannot be proven nor disproven. The collection of these undecidable statements is actually much larger than the collection of the statements whose truth can be decided. This fact, and Gödel's incompleteness theorem can now explain why Cantor and Hilbert had such difficulties proving the Continuum Hypothesis.

But, the Continuum Hypothesis did not only influence Gödel and his result on Incompleteness in the 1930s. Much earlier, and in particular in 1900 when Hilbert put the Continuum Hypothesis as number 1 in the list of 23 problems that should be settled, he also mentioned the need to settle the related well-ordering theorem in set theory which states that every set can be well ordered. A German mathematician Ernst Zermelo began then to work on set theory with Hilbert and by 1902, shortly after Bertrand Russell first wrote to Gottlob Frege about Russell's paradox, Zermelo had already obtained results on cardinal numbers and had managed by 1904 to prove the well-ordering theorem. In his proof, he introduced the idea of making a choice of one member of each set of a set of sets. It soon became apparent that there was no way to avoid doubts about the possibility of making such a choice, especially if the set of sets was infinite, and so a new axiom was introduced to allow for this: the *Axiom of choice*, or AC.[4]

---

[4]Shortly afterwards, it became apparent that the well-ordering theorem and the axiom of choice are actually equivalent.

> *Axiom of Choice*
> Given a set S of sets T, there is a set C consisting of exactly
> one element of each set $T \in S$.

Since this axiom is not obvious, it has become common for mathematicians writing on subjects to which this axiom may be relevant to state whether or not they are assuming this axiom. It should also be mentioned that Cantor often made use of this axiom.[5] Zermelo however pressed on with the axiomatisation of set theory which he published in 1908.[6]

Of course, with any new axiom, there is always a question of whether or not it is consistent with the rest of the theory. In his book from 1940, [17], Kurt Gödel proved that the axiom of choice is consistent with the basic axioms of set theory (ZF). In the same book, Gödel also proved that the *Generalised Continuum Hypothesis* is consistent with ZF where the Generalissd Continuum Hypothesis, as its name implies, generalised the Continuum Hypothesis which asked whether there is an S such that

$$\#\mathbb{N} < \#S < \#(\mathcal{P}\mathbb{N})$$

to asking whether for any set T, there is a set U such that

$$\#T < \#U < \#(\mathcal{P}T).$$

Since $\mathfrak{c} = \#(\mathcal{P}\mathbb{N}) = 2^{\mathfrak{a}}$, this is equivalent to asking the question of whether for any cardinal number $\mathfrak{m}$, there is a cardinal number $\mathfrak{n}$ such that

$$\mathfrak{m} < \mathfrak{n} < 2^{\mathfrak{m}}.$$

5.3.3 Exercises

Exercise 5.20. Assume the axiom of choice and let S and T be sets such that $S \neq \emptyset$. Let $f : T \mapsto S$ be a surjective function. Give an injection from S to T.

Exercise 5.21. Let S and T be sets such that $S \neq \emptyset$. Let $f : S \mapsto T$ be an injective function. Give a surjection from T to S. Here, you do not need the axiom of choice.

---

[5] And so we have here. For example when we proved that every infinite set has an infinitely countable subset (see Lemma 4.3.1.5). This lemma cannot be proved without the axiom of choice.

[6] Around a decade and a half later, The axioms of Zermelo's set theory were independently improved by Frankel and Skolem which led to the ZF set theory. In 1932, Zermelo edited Cantor's collected works.

Exercise 5.22. Let us define a number system as follows: $\overline{0} = \{\}$, $\overline{1} = \{\overline{1}\}$, $\overline{2} = \{\overline{1}, \overline{2}\}$, $\overline{3} = \{\overline{1}, \overline{2}, \overline{3}\}$, $\cdots$, $\overline{n} = \{\overline{1}, \overline{2}, \overline{3}, \cdots, \overline{n}\}$. Show that for each $\overline{n}$, $\#\overline{n} = n$.

Are you happy with this definition? If not, can you give an alternative?

Chapter 6

Formal Logic

In this book, we have been presenting many proofs. These proofs use logic, although we have been using it informally. Now, we are going to introduce *formal logic*, and we will see how it can be used to analyse the informal proofs we have been presenting so far. The results of these analyses of proofs will show how formalising the logic can be useful.

We will begin with the logic of propositions.

6.1 Propositional Logic

6.1.1 Well Formed Formulas

Propositions can be either true or false. We will begin with this analysis of truth and falsehood of propositions.

Let A and B be propositions. We can put propositions together with what we call *connectives* as follows:

1. *Negation.* We use the notation $\neg A$ to denote "not A".

2. *Conjunction.* We use $A \wedge B$ to denote "A and B".

3. *Disjunction.* We use $A \vee B$ to denote "A or B".

4. *Implication.* We use $A \supset B$ to denote "A implies B".

5. *Equivalence.* We use $A \backsim B$ to denote "A if and only if B".

But what will count as formal propositions? We will define a class of *well formed formulas*, or *wffs* for short.

Definition 6.1.1. *We start with a finite or countably infinite set of* atomic *wffs, $A_0, A_1, A_3, \ldots$.*

Then we define wffs as follows:

1. *Every atomic wff is a wff.*

2. *If A is a wff, then so is $\neg A$.*

3. *If A and B are wffs, then so are $(A \wedge B), (A \vee B), (A \supset B)$, and $(A \backsim B)$.*

4. *Nothing is a wff unless its being so follows from clauses 1—3 above.*

When writing out wffs in practice, the parentheses are often omitted when omitting them does not result in ambiguity. We use the following

priority scheme for the connectives:
$$\dfrac{\overline{\dfrac{\neg}{\wedge \quad \vee}}}{\underline{\supset}} \\ \backsim$$

Example 6.1.2.   • $A \vee A \backsim A$ *stands for* $(A \vee A) \backsim A$.

   • $A \vee B \backsim B \vee A$ *stands for* $(A \vee B) \backsim (B \vee A)$.

   • $\neg(A \vee B) \backsim \neg A \wedge \neg B$ *stands for* $\neg(A \vee B) \backsim (\neg A \wedge \neg B)$.

6.1.2   Truth Tables

As we said above, propositions are either true or false. We will now turn to the truth values of compound propositions.

If A is a wff, then as we have seen, $\neg A$ is a wff, and its truth value is the opposite of the value of A. So if A is true, then $\neg A$ is false, and if A is false, then $\neg A$ is true. $\neg A$ is called the *negation* of A. We can represent the truth values of the negation of A with the following *truth table*:

A	$\neg A$
T	F
F	T

If A and B are wffs, then so is $A \wedge B$, and its truth value is true just when both A and B are true; otherwise it is false. This is called the *conjunction* of A and B. It's truth table is as follows:

A	B	$A \wedge B$
T	T	T
T	F	F
F	T	F
F	F	F

Putting these last two truth tables together, we get a truth table for $\neg(A \wedge \neg A)$:

A	$\neg A$	$A \wedge \neg A$	$\neg(A \wedge \neg A)$
T	F	F	T
F	T	F	T

This truth table has the value true for every line. This means that $\neg(A \wedge \neg A)$ is a *tautology*, which means that it is true no matter what the truth values of its parts are. On the other hand, if we drop the last column of the table, we see that the truth value of $A \wedge \neg A$ is always false no matter what the truth value of A is, so this formula is a *contradiction*.

If A and B are wffs, then so is $A \vee B$, the *disjunction* of A and B. The wff $A \vee B$ is true just when at least one of A and B is true. (This differs from ordinary English, since in ordinary English A or B may be false if both A and B are true. For example, consider the sentence: "I will eat the fish or the chicken." This often means "but not both," although it can mean "or both." If it does mean "but not both," it is called the *exclusive or*.") In mathematics, "or" *always* means "or both," so it is what we call the *inclusive or*. Its truth table is

A	B	$A \vee B$
T	T	T
T	F	F
F	T	F
F	F	F

Here is a truth table for the wff $\neg A \vee A$:

A	$\neg A$	$\neg A \vee A$
T	F	T
F	T	T

So this wff is also a tautology.

Now let us consider implication. We will fill in this truth table by degrees. First note, that it is important that if A is true and B is false, then $A \supset B$ must be false. This gives us the following:

A	B	$A \supset B$
T	T	
T	F	F
F	T	
F	F	

Now note that we always want $A \supset A$ to be true. This extends our table to the following:

A	B	$A \supset B$
T	T	T
T	F	F
F	T	
F	F	T

This leaves us with the line in which A is false and B is true. If we put F here, then we will have a table in which $A \supset B$ is true if and only if A and B have the same truth value, and this should be the table for $A \backsim B$, not $A \supset B$. So we will complete the table by putting a T in the last column of row 3:

A	B	$A \supset B$
T	T	T
T	F	F
F	T	T
F	F	T

This is the only possible truth table for $A \supset B$. If this is not the way we determine the truth of $A \supset B$, then we cannot have that the truth value of $A \supset B$ is determined only by the truth values of A and B.[1]

Note that it follows from this truth table that if A is false, then $A \supset B$ is true regardless of the truth of B. In this case, it is said to be *vacuously true*. Similarly, if B is true, then $A \supset B$ is always true, regardless of the truth value of A.

Here is a truth table for $A \supset (B \supset A)$:

A	B	$B \supset A$	$A \supset (B \supset A)$
T	F	T	T
T	F	T	T
F	T	F	T
F	F	T	T

So $A \supset (B \supset A)$ is a tautology.

Finally, we come to the truth table for $A \backsim B$:

A	B	$A \backsim B$
T	T	T
T	F	F
F	T	F
F	F	T

---

[1]There are philosophers of logic who do not think the truth value of $A \supset B$ should depend only on the truth values of A and B, and there are systems of formal logic which follow this idea. But we will not spend time on them in this book.

Note that $A \backsim B$ is true if and only if A and B have the same truth value.
Here is a truth table for $(A \supset B) \backsim (B \supset A)$:

A	B	$A \supset B$	$B \supset A$	$(A \supset B) \backsim (B \supset A)$
T	T	T	T	T
T	F	F	T	F
F	T	T	F	F
F	F	T	T	T

Note that this is not a tautology.

We call $B \supset A$ the *converse* of $A \supset B$. The converse of an implication is the implication in the other direction. *The converse of an implication is not equivalent to the original implication.*

6.1.3 Exercises

Exercise 6.1. For each of the following, state whether it is a wff and in that case, whether it is also an atomic wff:

1. A_{100}.

2. A_1.

3. $A_3 \vee A_5$.

4. $(A_3 \vee A_5)$.

5. $A_3 \vee A_5)$.

6. $(A_3 \vee A_5$.

7. $A_0 \wedge$.

8. $\neg A_0 \wedge A_1 \vee A_2$.

9. $\neg A_0 \wedge (A_1 \vee A_2)$.

10. $(\neg A_0 \wedge A_1) \vee A_2$.

11. $\neg A_0 \backsim (A_1 \vee A_2)$.

Give an example where all the wffs mentioned in this exercise are false and another example where all the wffs mentioned in this exercise are true.

Exercise 6.2. Give the truth tables for

- $(A \supset B) \wedge (B \supset A)$.

- $(A \supset B) \vee (B \supset A)$.

- $\neg(A \supset B) \wedge \neg(B \supset A)$.

For each of these formulas state whether it is a tautology or a contradiction or neither.

Exercise 6.3. Give the truth tables for

- $(A \supset B) \backsim (\neg B \supset \neg A)$.

- $\neg A \backsim (A \supset B)$.

- $B \backsim (A \supset B)$.

For each of these formulas state whether it is a tautology or a contradiction or neither.

6.2   Evaluating Arguments

An *argument* is a claim that a certain formula, say A, follows logically from premise $A_1, A_2, A_3, \ldots, A_n$. We can use truth tables to evaluate arguments.

First, some notation.

The argument that A follows from premises $A_1, A_2, A_3, \ldots, A_n$ is usually written

$$A_1, A_2, A_3, \ldots, A_n \vdash A.$$

The argument is *valid* if and only if every line of the truth table which makes $A_1, A_2, A_3, \ldots, A_n$ all true also makes A true. To show that an argument is *invalid* (i.e., not valid), it is enough to find a line of the truth table which makes all of $A_1, A_2, A_3, \ldots, A_n$ true but makes A false.

Let us consider some simple examples:

Example 6.2.1.
$$A \supset B, A \vdash B.$$

This says that if $A \supset B$ and A are true, then B is also true. Let us consider the truth table:

A	B	$A \supset B$
T	T	T
T	F	F
F	T	T
F	F	T

In this table, we have both $A \supset B$ and A both true only in the first line. In lines 3 and 4, A is false, so these lines do not affect the truth of the

*argument, and in line 2 $A \supset B$ is false so again, this line does not affect
the truth of the argument. So the truth table shows indeed that the argument
is valid. This argument is important historically, and it has a Latin name:*
modus ponens.

Example 6.2.2.

$$A \supset B, B \vdash A.$$

The truth table is

A	B	$A \supset B$
T	T	T
T	F	F
F	T	T
F	F	T

*The third line of the truth table makes $A \supset B$ and B true but A false. So this
argument is invalid. This argument also has a Latin name:* modus moron.

Let us now look at some results we can easily prove using truth tables:

Theorem 6.2.3. *Disjunction $(\vee)$ has the following properties:*

1. *(Idempotence) For any proposition A, $A \vee A \backsim A$ is a tautology.*

2. *(Commutativity) For any propositions A, and B, $A \vee B \backsim B \vee A$ is a
   tautology.*

3. *(Associativity) For any propositions A, B, and C, $A \vee (B \vee C) \backsim
   (A \vee B) \vee C$ is a tautology.*

Proof. The proofs are all truth tables.

1.

A	$A \vee A$	$A \vee A \backsim A$
T	T	T
F	F	T

2.

A	B	$A \vee B$	$B \vee A$	$A \vee B \backsim B \vee A$
T	T	T	T	T
T	F	T	T	T
F	T	T	T	T
F	F	F	F	T

3. Let Φ be $A \vee (B \vee C) \leftrightsquigarrow (A \vee B) \vee C$.

A	B	C	$B \vee C$	$A \vee (B \vee C)$	$A \vee B$	$(A \vee B) \vee C$	Φ
T	T	T	T	T	T	T	T
T	T	F	T	T	T	T	T
T	F	T	T	T	T	T	T
T	F	F	F	T	T	T	T
F	T	T	T	T	T	T	T
F	T	F	T	T	T	T	T
F	F	T	T	T	F	T	T
F	F	F	F	F	F	F	T

□

Theorem 6.2.4. *Conjunction ($\wedge$) has the following properties:*

1. *(Idempotence) For any proposition A, $A \wedge A \leftrightsquigarrow A$ is a tautology.*

2. *(Commutativity) For any propositions A and B, $A \wedge B \leftrightsquigarrow B \wedge A$ is a tautology.*

3. *(Associativity) For any propositions A, B, and C, $A \wedge (B \wedge C) \leftrightsquigarrow (A \wedge B) \wedge C$ is a tautology.*

The proof is left as an exercise (see Exercise 6.4).

Theorem 6.2.5. *Conjunction and Disjunction together have the following properties for propositions A, B, and C:*

1. *$A \wedge (B \vee C) \leftrightsquigarrow (A \wedge B) \vee (A \wedge C)$ is a tautology.*

2. *$A \vee (B \wedge C) \leftrightsquigarrow (A \vee B) \wedge (A \vee C)$ is a tautology.*

The proof is left as an exercise (see Exercise 6.5).

Example 6.2.6. *Recall that after Theorem 4.2.14 and before Corollary 4.2.15 we saw De Morgan's laws of logic:*

$$\neg(A \wedge B) \leftrightsquigarrow \neg A \vee \neg B, \qquad \neg(A \vee B) \leftrightsquigarrow \neg A \wedge \neg B.$$

We can establish these as tautologies using truth tables:

A	B	$\neg A$	$\neg B$	$A \wedge B$	$\neg(A \wedge B)$	$\neg A \vee \neg B$	$\neg(A \wedge B) \leftrightsquigarrow \neg A \vee \neg B$
T	T	F	F	T	F	F	T
T	F	F	T	F	T	T	T
F	T	T	F	F	T	T	T
F	F	T	T	F	T	T	T

and

A	B	$\neg A$	$\neg B$	$A \vee B$	$\neg(A \vee B)$	$\neg A \wedge \neg B$	$\neg(A \vee B) \leftrightarrow \neg A \wedge \neg B$
T	T	F	F	T	F	F	T
T	F	F	F	T	F	F	T
F	T	T	F	T	F	F	T
F	F	T	T	F	T	T	T

Now to compare this with the proof of Theorem 4.2.14 let A be $x \in T$ and B be $x \in R$. Then

$$
\begin{aligned}
x \in S \backslash (T \cap R) \;\; &\Leftrightarrow\;\; x \in S \text{ and } x \notin (T \cap R) \\
&\Leftrightarrow\;\; x \in S \text{ and not } x \in (T \cap R) \\
&\Leftrightarrow\;\; x \in S \text{ and not both } (x \in T) \text{ and } (x \in R) \\
&\Leftrightarrow\;\; x \in S \text{ and not both } A \text{ and } B \\
&\Leftrightarrow\;\; x \in S \text{ and } \neg(A \wedge B) \\
&\Leftrightarrow\;\; x \in S \wedge (\neg A \vee \neg B) \\
&\Leftrightarrow\;\; (x \in S \wedge \neg A) \vee (x \in S \wedge \neg B) \\
&\Leftrightarrow\;\; (x \in S \text{ and } x \notin T) \text{ or } (x \in S \text{ and } x \notin R) \\
&\Leftrightarrow\;\; x \in (S \backslash T) \text{ or } x \in (S \backslash R) \\
&\Leftrightarrow\;\; x \in (S \backslash T) \cup (S \backslash R)
\end{aligned}
$$

The equivalence of $x \in S \backslash (T \cup R)$ and $x \in (S \backslash T) \cap (S \backslash R)$ is similar.

Example 6.2.7. *Consider the proof of Lemma 4.1.2, Part 1. We need to prove $\emptyset \subseteq S$ for every set S, and for this we need to prove*

$$ x \in \emptyset \supset x \in S. $$

Now we know that $x \in \emptyset$ is always false, so we only have the lines for the truth table in which $x \in \emptyset$ is false. This leads to the following truth table:

$x \in \emptyset$	$x \in S$	$x \in \emptyset \supset x \in S$
F	T	T
F	F	T

So this is a tautology.

The above examples show how formal logic can analyse informal proofs.

6.2.1 Exercises

Exercise 6.4. Prove Theorem 6.2.4.

Exercise 6.5. Prove Theorem 6.2.5.

Exercise 6.6. Look again at what we did Example 6.2.6 for the first proof of Theorem 4.2.14, and repeat the same for the second proof where we showed $x \in S\backslash(T \cup R)$ and $x \in (S\backslash T) \cap (S\backslash R)$.

6.3 Inductive Definitions

Definition 6.1.1 is an *inductive definition*. This means that we have a property corresponding to the induction axiom for the natural numbers. Thus we have the following definition:

Definition 6.3.1. *If we have a property that we want to prove holds for all wffs, we can prove that it does by proving:*

1. *that it holds for all atomic wffs,*

2. *that if it holds for A it holds for $\neg A$, and*

3. *if it holds for A and B then it holds for $(A \wedge B)$, $(A \vee B)$, $(A \supset B)$, and $(A \backsim B)$.*

We call 1, the basic case and 2 and 3 the inductive steps in the induction.

This is another way of stating clause 4 of this definition.

Example 6.3.2. *We write $LP(A)$ to denote the number of left parentheses of wff A and $RP(A)$ to denote the number of right parentheses of wff A. We prove that in every wff, the number of left parentheses is equal to the number of right parentheses.*

1. *Basic case: For every atomic wff A_n, both $LP(A_n) = RP(A_n) = 0$.*

2. *Inductive case 1: Assume that the number of left parentheses of wff A is equal to the number of right parentheses of A. That is $LP(A) = RP(A)$. Then $LP(\neg A) = LP(A) = RP(A) = RP(\neg A)$.*

3. *Inductive case 2: Assume that for each of the wffs A and B, the number of its left parentheses is equal to the number of its right parentheses. Then the number of left parentheses of wff $\neg A$ is equal to the number of right parentheses of $\neg A$. That is $LP(A) = RP(A)$ and $LP(B) = RP(B)$. Then*

- $LP((A \wedge B)) = 1 + LP(A) + LP(B) = 1 + RP(A) + RP(B) = RP((A \wedge B))$.

- $LP((A \vee B)) = 1 + LP(A) + LP(B) = 1 + RP(A) + RP(B) = RP((A \vee B))$.

- $LP((A \supset B)) = 1 + LP(A) + LP(B) = 1 + RP(A) + RP(B) = RP((A \supset B))$.

- $LP((A \backsim B)) = 1 + LP(A) + LP(B) = 1 + RP(A) + RP(B) = RP((A \backsim B))$.

Hence, by induction, for every wff A, $LP(A) = RP(A)$.

Instead of doing inductions as described in Definition 6.3.1, some writers want to do induction on the number of symbols in the wffs in question. In a sense this amounts to the same thing as we suggest here, but what we suggest is more direct and shows more clearly what an inductive definition is like.

You might compare Definition 6.1.1 with the following definition of the natural numbers:

1. 0 is a natural number.

2. If n is a natural number, then $n + 1$ is a natural number.

3. Nothing is a natural number unless its being so follows from clauses 1–2 above.

The last clause of this definition is the usual axiom of mathematical induction. And one way to see that induction works is to note that if $P(x)$ is a proposition for every x, then $P(n) \supset P(n+1)$ for every n implies all of the following:

$$P(0) \supset P(1)$$
$$P(1) \supset P(2)$$
$$P(2) \supset P(3)$$
$$\vdots$$
$$P(n) \supset P(n+1)$$
$$\vdots$$

Since we have that $P(0)$ is true and all of these implications are true, it follows that $P(n)$ is true for every natural number n. Hence, to prove a property P holds for all the natural numbers, we first show it holds for 0, and then we show that if it holds for n then it will hold for $n + 1$, and then we deduce by induction that the property holds for every number in $\mathbb{N}$. That is:

1. We show $P(0)$.

2. We show that if $P(n)$ then $P(n + 1)$. I.e., $P(n) \supset P(n + 1)$.

Then, we deduce by induction that for every $n \in \mathbb{N}$ we have $P(n)$.

Now, let us look again at the induction axiom for the natural numbers that we gave in Chapter 4 and see how it fits with the above paragraph:

> **Induction axiom for $\mathbb{N}$:** If $S \subseteq \mathbb{N}$, $0 \in S$, and whenever $n \in S$ we also have $n + 1 \in S$, then $S = \mathbb{N}$.

If we want to prove a property P holds for $\mathbb{N}$, we let $S = \{n \in \mathbb{N} : P(n)\}$. Then, according to what we wrote above, if $P(0)$ holds and whenever $P(n)$ holds we deduce $P(n + 1)$ holds, then for every $n \in \mathbb{N}$ we have $P(n)$. Hence, $S = \mathbb{N}$.

Sometimes, in the inductive proof we may need more than $P(n)$ to prove $P(n + 1)$. We may need $P(k)$ for $k \leq n$ in order to prove $P(n + 1)$. And moreover, we may need to prove more base cases. Here, we use *Strong induction* which works as follows:

1. Base case: we show $P(0)$, $P(1)$, $\cdots$, $P(k)$.

2. Inductive step: we show that if $P(i)$ holds for all $i \leq n$, then $P(n + 1)$ holds.

Then, by strong induction, for every $n \in \mathbb{N}$ we have $P(n)$.

6.3.1 Exercises

Exercise 6.7. Show by induction that for every $n \in \mathbb{N}$,

$$\Sigma_{k=0}^{k=n} k \times k! = (n + 1)! - 1.$$

Exercise 6.8. The Fibonacci sequence is defined as follows for each $n \in \mathbb{N}$:

$$a_0 = 0, \qquad a_1 = 1, \qquad a_{n+2} = a_{n+1} + a_n.$$

Show that for every $n \in \mathbb{N}$,

$$a_{n+2} = 1 + \Sigma_{k=0}^{k=n} a_k.$$

Chapter 7

Study of Incommensurability

In Chapter 2, we saw how the ancient Greeks were thinking about geometric proofs, pebble proofs, statement proofs and the theory of odd/even numbers. We also saw there that the ancient Greeks were looking for a unit which divides evenly and exactly into both the side and diagonal of a square. They were looking for a unit and not an approximation.[1] This was in line with the Pythagorean idea that "number" is everything (they understood a number as a collection of units like I, II, III, etc., see Definition 2 of Book VII, Figure 7.1)

We saw in Chapter 2 that in their quest for the unit which divides evenly and exactly into both the side and diagonal of a square, the Greeks discovered that this unit does not exist (see Theorem 2.5.11 and Remark 2.5.12).

> **The incommensurability result**
> The unit which divides evenly and exactly into both the side and diagonal of a square could not be found. There is no common measure between the lengths of the side and the diagonal of a square.

During this quest, the Greeks found an irrational and then they found more irrationals. An irrational number can never be a whole number or the ratio of two whole numbers. The Greeks conception that "everything is a number"

---

[1]This does not coincide with our idea of measurement since for us, every measurement is an approximation.

started to crumble. Their notion of "number" became no longer enough. There was a need to treat other quantities which are not a discrete collection of units (like the naturals or the ratios of two naturals[2]), but which are continuous. The Greeks did not know how to handle these quantities.

> The main problem was that the Greeks treated mathematical objects as given and did not conceive of constructing them as we did for example when we constructed the positive rationals in Chapter 3 using equivalence classes.

The ancient Greeks juggled with two notions: their notion of "numbers" (as a multitude of units, see Definition 2 of Book VII, Figure 7.1) as well as the so-called *magnitudes* (which in addition to "numbers" include things like lines and areas and volumes, etc.). They developed arithmetic for their numbers, but treated their magnitudes geometrically. As we have seen in earlier chapters, in modern mathematical thinking we start from a basic set of mathematical objects which we use to construct new mathematical objects (e.g., like we constructed the integers and rationals from the naturals in Chapter 3). Starting in the 16th century, in order to construct magnitudes (like what we call nowadays the real numbers), approximations were used. Even though the Greeks have not thought of constructing new mathematical objects, they did introduce a procedure for approximating ratios that sheds light on the innovative thinking of the time. We discuss this procedure in the first section of this chapter. We then in the second section discuss the study on the theory of numbers after the discovery of the incommensurability of the side and diagonal of a square. Thereafter, we explain the "ruler and compass" constructions in Euclid's *Elements*.

7.1 Numbers *versus* Magnitudes and Anthyphairesis

The discovery of the incommensurability of the side and diagonal of a square, which is shown in Chapter 2, showed that the Pythagorean idea that "number" is everything would not work. It was necessary to treat quantities, or *magnitudes*, that were not numbers.

---

[2]The ancient Greeks never referred to fractions, at least not in their theoretical mathematics, but they could achieve the same effect by dividing their unit into smaller units to get another integer instead of a fraction.

> Today, we think of magnitudes as "real numbers," but this conception is not compatible with the ideas of the ancient Greeks.

Our conception of real number is based on a construction. But although the Greeks were willing to construct geometric figures, they regarded mathematical objects such as numbers and magnitudes (like lines, planes, etc.) as given. They did however differentiate between numbers, which are discrete, and various kinds of magnitudes, which are continuous. Figure 7.1 gives some definitions of Book VII of Euclid's Elements [20] where the Greeks set up their number system and related properties like *proportions* and *prime*, etc.[3]

Example 7.1.1. *The number 3 (a multitude of 3 units) measures the number 6 and is part of it (we also say that 6 is a multiple of 3), but 3 is parts of 7. Note that Definitions 3..5 are in preparation for ratio and proportion.*

- *Take for example 6 and 9 as well as 14 and 21. Then, we see that 6, 9, 14 and 21 are proportional (written as 6:9 = 14:21) because 6 is 2 parts of 3 parts of 9 and this is also the same for 14 and 21 (14 is 2 parts of 3 of 21).*

- *Another example is 6 and 12 as well as 14 and 28 (i.e., 6:12 = 14:28). Then, we see that 6 is half of 12 and 14 is half of 28.*

- *Similarly, 12:6 = 28:14.*

For numbers, the Greeks had a developed arithmetic. It was not the same as our arithmetic. Today we start with the natural numbers, including 0,[4] and construct rational numbers, real numbers, complex numbers, etc.

Magnitudes, on the other hand, the ancient Greeks treated geometrically. Furthermore, they differentiated between different kinds of magnitudes:

1. Length magnitudes: lengths of line segments.

2. Area magnitudes: areas of plane figures.

---

[3] For the entire Euclid Elements online, see http://aleph0.clarku.edu/~djoyce/elements/elements.html

[4] The ancient Greeks had no concept of 0. Conventional wisdom about 0 is that it was invented by Indian mathematicians and introduced into Europe by the Arabs. However, over a millennium before the flowering of ancient Greek mathematics, the Egyptians had used a symbol for 0 in the sense of a number minus itself, and they had used a concept equivalent to signed numbers. Apparently, they did not have the result that 0 multiplied by any number is 0.

DEFINITION 1 A *unit* is that by virtue of which each of the things that exist is called one.

DEFINITION 2 A *number* is a multitude composed of units.

DEFINITION 3 A number *is a part* of a number, the less of the greater, when it measures the greater;

DEFINITION 4 But *parts* of when it does not measure it.

DEFINITION 5 The greater number is a multiple of the less when it is measured by the less.

DEFINITION 11 A *prime number* is that which is measured by a unit alone.

DEFINITION 12 Numbers *relatively prime* are those which are measured by a unit alone as a common measure.

DEFINITION 13 A *composite number* is that which is measured by some number.

DEFINITION 14 Numbers *relatively composite* are those which are measured by some number as a common measure.

DEFINITION 15 A number is *said to multiply a number* when the latter is added as many times as there are units in the former.

DEFINITION 16 And, when two numbers having multiplied one another make some number, the number so produced be called plane, and its sides are the numbers which have multiplied one another.

DEFINITION 17 And, when three numbers having multiplied one another make some number, the number so produced be called solid, and its sides are the numbers which have multiplied one another.

DEFINITION 20 Numbers are proportional when the first is the same multiple, or the same part, or the same parts, of the second that the third is of the fourth.

DEFINITION 21 Similar plane and solid numbers are those which have their sides proportional.

DEFINITION 22 A *perfect number* is that which is equal to the sum its own parts.

Figure 7.1: Some DEFINITIONS of BOOK VII of Euclid's Elements

3. Volume magnitudes: volumes of solids.

4. Angle magnitudes: measurements of angles.

These kinds of magnitudes were distinct. Operations on magnitudes were guided by the following principles:

- **Addition and subtraction:** Magnitudes of the same kind could always be added, and a smaller one could be subtracted from a larger one of the same kind.

- **Multiplication of a magnitude by a number:** Any magnitude could be multiplied by a number, for multiplication is just repeated addition.

- **Multiplication of magnitudes:** Multiplication was a different matter. Two length magnitudes could be multiplied, but the result was not another length magnitude, but an area magnitude. Similarly, an area magnitude multiplied by a length magnitude was a volume magnitude.

Furthermore, their attitude towards areas and volumes was different from ours. We think of areas and volumes as the same kind of magnitudes as lengths, namely real numbers. The Greeks compared areas of various plane figures to areas of squares. Hence, the expression "squaring the circle" for finding the area of a circle from that of a square, an expression that survives to this day. It never occurred to them to compare the area of a plane figure to a rectangle one of whose sides is the unit, as we do now. And they compared volumes of various solids to volumes of cubes.

There is a construction of the kind the ancient Greeks used, a "ruler and compass" construction,[5] for multiplying lengths and getting another length. With a slight modification, it can be used to "divide" two lengths (i.e., find their ratio) and get another length. But this construction was totally unknown in ancient times. It was first published by René Descartes[6] in his *Geometrie*.[7] Thus, it was not the lack of the proper technical means that prevented the ancient Greeks from multiplying and dividing magnitudes (and getting another magnitude of the same kind), it was rather a matter of the way they looked on numbers and magnitudes. Descartes' construction was published at a time when it was generally accepted that numbers (including

---

[5]See Section 7.3.

[6]René Descartes, 1596–1650, was a French philosopher. He is probably known best for the saying "I think, therefore I am." He is also known for connecting geometry and algebra by introducing analytic geometry.

[7]Descartes originally published his *Geometrie* as a kind of appendix to his work *Discours de la Méthode*, which was published in 1637.

fractions) can be used to approximate magnitudes to any desired degree of accuracy. Although ancient Greek craftsmen must have taken measurements by approximating the desired quantities with numbers, it never occurred to their mathematicians to think this way in pure mathematics. The idea of using numbers with fractions to approximate continuous magnitudes developed in the Arab world during medieval times and only reached Europe in the 16th and 17th centuries C.E. (see [30]).

One consequence of the discovery that the side and diagonal of a square are incommensurable was a general study of incommensurability in general. Much of it was carried out at the Academy, founded in Athens by Plato[8] in the fourth century B.C.E. Although Plato himself was not a mathematician, he brought important mathematicians into his Academy whose entrance he engraved with: "Let no man destitute of geometry enter my door".

One of the objectives in this study of continuous magnitudes was to find something corresponding to the ratio, or quotient, of two numbers. For numbers, the Greeks had a procedure they called *anthyphairesis* (in Greek $\alpha\nu\theta\upsilon\phi\alpha\iota\rho\epsilon\sigma\iota\varsigma$) for approximating ratios. Anthyphairesis is composed of two Greek terms: $\upsilon\phi\alpha\iota\rho\epsilon\omega$ (meaning *subtract*) and $\alpha\nu\tau\iota$ (meaning *answer, alternating/reciprocal*) and hence $\alpha\nu\theta\upsilon\phi\alpha\iota\rho\epsilon\sigma\iota\varsigma$ may stand for *alternated/reciprocal subtraction*.

7.1.1 Anthyphairesis on Numbers

Euclid used anthyphairesis in his *Elements*, book VII to check whether two numbers are prime to one another or to find the greatest common divisor of two numbers (see propositions 1 and 2 of book VII).

> PROPOSITION 1. OF BOOK VII OF THE *Elements*
> Two unequal numbers being set out, and the less being continuously subtracted in turn from the greater, if the number which is left never measures the one before it until a unit is left, the original numbers will be prime to one another.

Example 7.1.2. *Take 17 and 3. Then:*
17-3 = 14, 14-3 =11, 11-3 = 8, 8-3 = 5, 5-3 =2, 3-2 =1.
Since the less being continuously subtracted in turn from the greater and the number which is left never measures the one before it until a unit is left,

---

[8]Plato, 427–347 B.C.E., was a major and influential Athenian philosopher, a disciple of Socrates (469–399 B.C.E.), another highly influential Athenian philosopher.

i.e., 14 never measures 17, 11 never measures 14, 8 never measures 11, 5 never measures 8, 2 never measures 5 and we are left with 1, then the original numbers 17 and 3 will be prime to one another.

Proof. Here is an adapted version of Euclid's proof of PROPOSITION 1: Assume a and b are the two numbers in question with $b < a$ and assume by contradiction that a and b are not prime to one another. Then there is a number $e > 1$ which measures both a and b. Now, repeated subtraction (anthyphairesis) is used to reach 1 as follows (for the sake of clarity, we assume there are only 3 steps).

- b is subtracted as many times (say q_0 times) from a until what is left (say r_1) is less than b. I.e., $a = q_0 b + r_1$ where $r_1 < b$.

- The same step is repeated for b and r_1 where r_1 is subtracted as many times (say q_1 times) from b until what is left (say r_2) is less than r_1. I.e., $b = q_1 r_1 + r_2$ where $r_2 < r_1$.

- The same step is repeated for r_i and r_{i+1} where r_{i+1} is subtracted as many times (say q_{i+1} times) from r_i until what is left (say r_{i+2}) is less than r_{i+1}. I.e., $r_i = q_{i+1} r_{i+1} + r_{i+2}$ where $r_{i+2} < r_{i+1}$.

- Again, the same step is repeated until as per hypothesis, the unit is left. I.e., $r_n = q_{n+1} r_{n+1} + 1$.

Since e measures a and b, then e measures $a - q_0 b = r_1$, hence e measures r_1. Similarly, e measures $b - q_1 r_1$ (hence r_2) and $r_i - q_{i+1} r_{i+1} = r_{i+2}$, etc. until we get e measures $r_n - q_{n+1} r_{n+1} = 1$. Hence $e > 1$ measures 1. Absurd. □

> PROPOSITION 2. OF BOOK VII OF THE *Elements*
> Given two numbers not prime to one another, to find their greatest common measure.

In his proof to PROPOSITION 2, Euclid also applies anthyphairesis to find the greatest common measure of two numbers. His method used there is similar to that used in textbooks of Algebra (see Heath [20, Vol. 2. page 299]). He proves that anthyphairesis applied to two relatively prime numbers leads to the unit, and applied to two non relatively prime numbers gives the greatest common divisor of these two numbers.

The next examples shows how anthyphairesis is used to find the greatest common divider of two numbers and also to show that two numbers are relatively prime to one another.

Example 7.1.3 (The greatest common divider of 136 and 6). *Applying anthyphairesis to 136 and 6 gives:*

	r_i	$=$	q_i	$\times$	r_{i+1}	$+$	$r_{i+2}.$
	136	$=$	22	$\times$	6	$+$	4.
$i=0$	r_0	$=$	q_0	$\times$	r_1	$+$	$r_2.$
	6	$=$	1	$\times$	4	$+$	2.
$i=1$	r_1	$=$	q_1	$\times$	r_2	$+$	$r_3.$
	4	$=$	2	$\times$	2	$+$	0.
$i=2$	r_2	$=$	q_2	$\times$	r_3	$+$	$r_4.$

Since the one before the final r in the series is not 1 ($r_3 = 2$), it is the greatest common divider (GCD) of 136 and 6.   □

Example 7.1.4 (The two numbers 125 and 12 are relatively prime to one another). *Applying anthyphairesis to 125 and 12 gives:*

	r_i	$=$	q_i	$\times$	r_{i+1}	$+$	$r_{i+2}.$
	125	$=$	10	$\times$	12	$+$	5.
$i=0$	r_0	$=$	q_0	$\times$	r_1	$+$	$r_2.$
	12	$=$	2	$\times$	5	$+$	2.
$i=1$	r_1	$=$	q_1	$\times$	r_2	$+$	$r_3.$
	5	$=$	2	$\times$	2	$+$	1.
$i=2$	r_2	$=$	q_2	$\times$	r_3	$+$	$r_4.$
	2	$=$	2	$\times$	1	$+$	0.
$i=3$	r_3	$=$	q_3	$\times$	r_4	$+$	$r_5.$

Since the one before the final r in the series is 1 ($r_4 = 1$), 125 and 12 are relatively prime to one another.   □

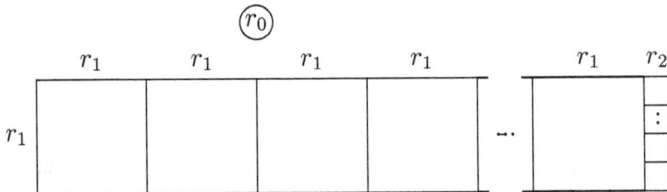

Figure 7.2: Anthyphairesis

Let us permit ourselves to use the notation r_0/r_1 (when $r_1 \neq 0$). As we saw above, to find the ratio of r_0 to r_1, subtract r_1 from r_0 as many times as possible and record that number of times, or, which amounts to the same thing, divide r_0 by r_1 and keep track of the quotient q_0 and remainder r_2. The Greeks would have done this geometrically, so they would have taken a rectangle with base r_0 and height r_1 and then cut off as many squares of side r_1 as possible, as indicated in Figure 7.2. Then the rectangle with base r_2 and height r_1 is left over, and it indicates that we need to find the ratio of r_1 to r_2, so we now cut off from this part squares of side r_2. To do this arithmetically, we divide r_1 by r_2 and keep track of the quotient q_1 and remainder r_3. Keep doing this until a 0 remainder is obtained.[9] This is described in the following definition:

Definition 7.1.5 (Continued fraction). *Let r_0 and $r_1 \neq 0$. Continue looking for q_{i-1}, r_{i+1} where $i \geq 1$ as follows:*

0. *Since $r_1 \neq 0$, let q_0 and $r_2 < r_1$ be the quotient resp. remainder of r_0/r_1. We write:*

$$r_0 = q_0 \times r_1 + r_2 \qquad or \qquad r_0/r_1 = q_0 + r_2/r_1.$$

1. *If $r_2 \neq 0$, let q_1 and $r_3 < r_2$ be the quotient resp. remainder of r_1/r_2. We write:*

$$r_1 = q_1 \times r_2 + r_3 \qquad or \qquad r_1/r_2 = q_1 + r_3/r_2.$$

:

:

i. *If $r_{i+1} \neq 0$, let q_i and $r_{i+2} < r_{i+1}$ be the quotient resp. remainder of r_i/r_{i+1}. We write:*

$$r_i = q_i \times r_{i+1} + r_{i+2} \qquad or \qquad r_i/r_{i+1} = q_i + r_{i+2}/r_{i+1}.$$

:

:

Once you reach q_{n-1} and r_{n+1} where $r_{n+1} = 0$ (and $n \geq 1$), stop and return $[q_0, q_1, \cdots, q_{n-1}]$. The sequence $[q_0, q_1, \cdots, q_{n-1}]$ is called the continued fraction representation of r_0/r_1. We say that the ratio $\frac{r_0}{r_1}$ is characterised by the sequence $[q_0, q_1, \cdots, q_{n-1}]$. Note that the ratio

$$\frac{r_0}{r_1} = q_0 + \frac{r_2}{r_1} = q_0 + \frac{1}{\frac{r_1}{r_2}} = q_0 + \frac{1}{q_1 + \frac{r_3}{r_2}} = q_0 + \frac{1}{q_1 + \frac{1}{\frac{r_2}{r_3}}} = \boxed{q_0} + \frac{1}{\boxed{q_1} + \frac{1}{\boxed{q_2} + \cdots}}$$

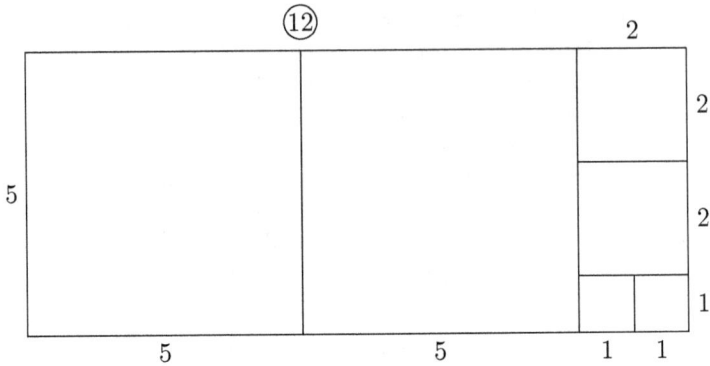

Figure 7.3: Ratio of 12 to 5

Let us see this with an example:

Example 7.1.6. *Find the ratio of 12 to 5.*

	r_i	$=$	q_i	$\times$	r_{i+1}	$+$	$r_{i+2}.$	$[q_0, \ldots, q_i]$
	12	$=$	2	$\times$	5	$+$	2.	$[2]$
$i=0$	r_0	$=$	q_0	$\times$	r_1	$+$	$r_2.$	$[q_0]$
	5	$=$	2	$\times$	2	$+$	1.	$[2,2]$
$i=1$	r_1	$=$	q_1	$\times$	r_2	$+$	$r_3.$	$[q_0, q_1]$
	2	$=$	2	$\times$	1	$+$	0.	$[[2,2,2]$
$i=2$	r_2	$=$	q_2	$\times$	r_3	$+$	$r_4.$	$[q_0, q_1, q_2]$

- *If 12 is divided by 5, the quotient is 2 and the remainder is 2. Save [2], and repeat for 5 and 2.*

- *Now, to find the ratio of 5 to 2, divide 5 by 2, and then the quotient is 2 and the remainder is 1. Save [2,2], and repeat for 2 and 1.*

- *What is left is the need to find the ratio of 2 to 1, so divide the divisor from the last step, 2, by the remainder from the last step, 1, to give a quotient of 2 and a remainder of 0. Save [2,2,2], and since remainder is 0, no more steps to be carried out.*

So this ratio is characterised by the sequence [2,2,2] (the quotients in order).

---

[9]This description is based on [13, Chapter 2].

We could also write this as the continued fraction:

$$\frac{12}{5} = 2 + \frac{2}{5} = 2 + \frac{1}{\frac{5}{2}} = 2 + \frac{1}{2 + \frac{1}{2}} = 2 + \frac{1}{2 + \frac{2}{1}} = \boxed{2} + \frac{1}{\boxed{2} + \frac{1}{\boxed{2}}}.$$

□

Figure 7.3 shows how the Greeks would have done this geometrically.

Example 7.1.7. *Find the ratio of 22 to 6. Let $r_0 = 22$ and $r_1 = 6$. Find q_{i-1}, r_{i+1} for $i \geq 1$ as follows:*

		r_i	=	q_i	×	r_{i+1}	+	r_{i+2}.	$[q_0, \ldots, q_i]$
		22	=	3	×	6	+	4.	$[3]$
$i=0$	r_0	=	q_0	×	r_1	+	r_2.		$[q_0]$
		6	=	1	×	4	+	2.	$[3, 1]$
$i=1$	r_1	=	q_1	×	r_2	+	r_3.		$[q_0, q_1]$
		4	=	2	×	2	+	0.	$[[3, 1, 2]$
$i=2$	r_2	=	q_2	×	r_3	+	r_4.		$[q_0, q_1, q_2]$

- *If 22 is divided by 6, the quotient is 3 and the remainder is 4. I.e., 22/6 gives $q_0 = 3$ and $r_2 = 4$.*

- *If 6 is divided by 4, the quotient is 1 and the remainder is 2. I.e., r_1/r_2 which is 6/4 gives $q_1 = 1$ and $r_3 = 2$.*

- *If 4 is now divided by 2, the quotient is 2 and the remainder is 0. I.e., r_2/r_3 which is 4/2 gives $q_2 = 2$ and $r_4 = 0$.*

Thus, this ratio is characterised by the sequence [3,1,2] (the quotients in order). We could also write this as the continued fraction:

$$\frac{22}{6} = 3 + \frac{4}{6} = 3 + \frac{1}{\frac{6}{4}} = 3 + \frac{1}{1 + \frac{2}{4}} = 3 + \frac{1}{1 + \frac{1}{\frac{4}{2}}} = \boxed{3} + \frac{1}{\boxed{1} + \frac{1}{\boxed{2}}}.$$

□

The Greeks would have used the diagram in Figure 7.4.

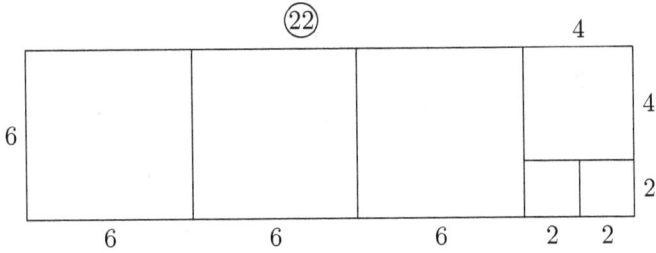

Figure 7.4: Ratio of 22 to 6

7.1.2 Anthyphairesis on Magnitudes

If the above procedure is applied to two "numbers", it will always terminate since we have seen, any two numbers are either prime to one another or have a common measure. But this is not the case for all magnitudes. After the Greeks discovered that not all magnitudes have a common measure, they called the magnitudes that have a common measure *commensurable* and those which do not *incommensurable* (see Definition 2.5.1). The first three propositions of the *Elements* book X, deal with common measures on magnitudes and we see there that Euclid applies again anthyphairesis (continuous subtraction):

> PROPOSITION 1 OF BOOK X OF THE *Elements*.
> Two unequal magnitudes being set out, if from the greater there be subtracted a magnitude greater than its half, and from that which is left a magnitude greater than its half, and if this process is repeated continuously, there will be left some magnitude which will be less than the lesser magnitude set out.

For example, take the line AB of length 11 and the line CD of length 3. If we take 6 from AB, we are left with the line EF of length 5. If we take again 3 from what is left (5), we get line GH of length 2. Clearly, 2 is lesser than the smallest 3 that we started with.

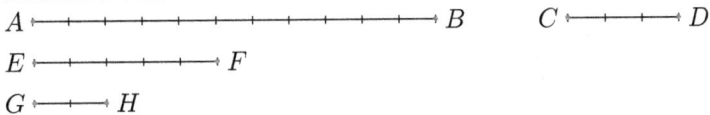

$A \longmapsto\!\longrightarrow B \qquad C \longmapsto\!\!\!\!\!\!\!\!\!\longrightarrow D$

$E \longmapsto\!\!\!\!\!\!\!\!\!\!\!\!\longrightarrow F$

$G \longmapsto\!\!\!\!\longrightarrow H$

> PROPOSITION 2 OF BOOK X OF THE *Elements*.
> If, when the less of two unequal magnitudes is continuously subtracted in turn from the greater, that which is left never measures the one before it, the magnitudes will be incommensurable.

You can see the connection of this Proposition 2. X to Proposition 1. VII given on page 174.[10] One deals with "numbers" while the other deals with Greek's quantities/magnitudes. A magnitude is a size that needs to be measured. For example the line AB we used above has measure 11 (which is a number). But, not all magnitudes have a number as their measure. For example there is no number that measures the line AB in the following triangle ABC whose size are of length 1 and whose hypotenuse is of length $\sqrt{2}$.

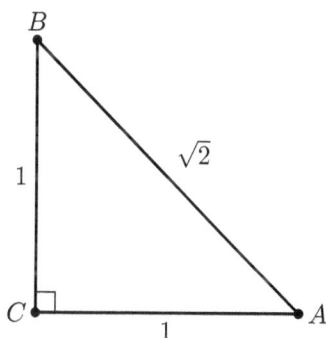

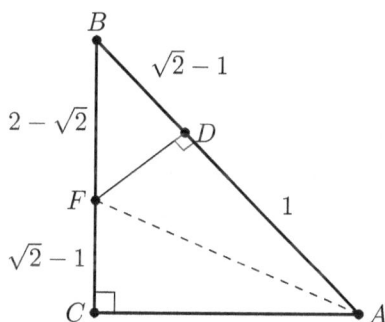

To see this, first let point D on the hypotenuse AB such that $AB = AD+DB$ where the length of AD is the length 1 of the sides. Let the bisector of angle $\angle BAC$ join line BC at point F. You can see that triangles ACF and ADF are similar (see Definition 7.2.3 and Exercise 7.7) and triangle BFD is isosceles. Hence, $CF = FD = DB$ and are all of length $\sqrt{2}-1$ and FB is of length $2 - \sqrt{2}$.

Now we can repeat the same process for the rectangular isosceles triangle BFD as we have done for ACB. That is, we let point E on the hypotenuse FB such that $FB = FE + EB$ where the length of FE is the length $\sqrt{2}-1$ of FD. Let the bisector of angle $\angle BFD$ join line BD at point G. You can see that triangles EFG and GFD are similar (again see Exercise 7.7) and triangle BEG is isosceles. Hence, $BE = EG = GD$ and are all of length $3 - 2\sqrt{2}$ and EF is of length $\sqrt{2} - 1$.

---

[10]Two unequal numbers being set out, and the less being continuously subtracted in turn from the greater, if the number which is left never measures the one before it until a unit is left, the original numbers will be prime to one another.

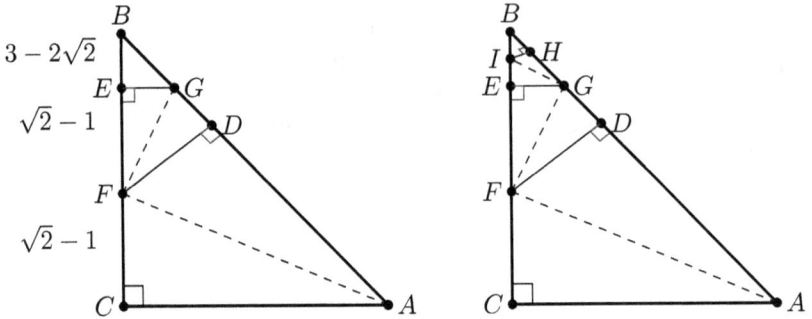

We repeat this process for the isosceles rectangular triangle BEG and for the new isosceles rectangular triangle BIH and so on. This process can be repeated infinitely.

In this repetition, you see that the less of two unequal magnitudes is continuously subtracted in turn from the greater, yet what is left never measures the one before it.

> **PROPOSITION 3 OF BOOK X OF THE** *Elements.*
> Given two commensurable magnitudes, to find their greatest common measure.

This says that no matter which two commensurable magnitudes we take, we can find their greatest common measure. For example, in Example 7.1.3 we found the greatest common divider of 136 and 6 to be 2 and in Example 7.1.4, we found that 125 and 12 are relatively prime to one another. Similarly, in Example 7.1.6 we found that 12 and 5 are relatively prime to one another and hence their greatest common factor is 1.

One can deduce by these propositions that two magnitudes are commensurable if and only if anthyphairesis terminates. On the other hand, if the anthyphairesis procedure of finding the ratio or GCD of two numbers is applied to incommensurable magnitudes, it will never terminate. In this case we never reach a remainder equal 0 and we continue generating equations of the form $r_i = q_i \times r_{i+1} + r_{i+2}$ or $r_i/r_{i+1} = q_i + r_{i+2}/r_{i+1}$. Here, our characterising sequence $[q_0, q_1, \ldots]$ will be infinite and our continued fraction representation will also be infinite:

$$\frac{r_0}{r_1} = q_0 + \cfrac{1}{q_1 + \cfrac{1}{q_2 + \cfrac{}{\ddots}}}.$$

We demonstrate this with the ratio of the incommensurable magnitude $\sqrt{2}$ to 1.

Example 7.1.8. *Find the ratio of $\sqrt{2}$ to 1. Here, division is not like long division with integers, so instead of dividing $\sqrt{2}$ by 1, we subtract 1 from $\sqrt{2}$ to get $\sqrt{2} - 1$. We now subtract the latter from 1 as many times as we can; we can do this twice, and we get $3 - 2\sqrt{2}$. We could continue this, but note that*

$$3 - 2\sqrt{2} = 1 - \sqrt{2} + \sqrt{2}(\sqrt{2} - 1) = -(\sqrt{2} - 1) + \sqrt{2}(\sqrt{2} - 1) = (\sqrt{2} - 1)^2,$$

so the ratio of $3 - 2\sqrt{2}$ to $1 - \sqrt{2}$ is the same as the ratio of $1 - \sqrt{2}$ to 1. Hence, no further steps will change the ratio involved, and the process will never terminate. Applying Definition 7.1.5, we get:

- *Let $r_0 = \sqrt{2}$ and $r_1 = 1$. Since $1 < \sqrt{2} < 2$ then $0 < \sqrt{2} - 1 < 1$. Let $q_0 = 1$ and $r_2 = \sqrt{2} - 1$. Note $0 < r_2 < r_1$. We have*

$$r_0 = q_0 \times r_1 + r_2 \qquad or \qquad \frac{r_0}{r_1} = q_0 + \frac{r_2}{r_1}.$$

$$\sqrt{2} = 1 + (\sqrt{2} - 1) \qquad or \qquad \frac{\sqrt{2}}{1} = 1 + \frac{\sqrt{2} - 1}{1}$$

- *Since $(\frac{1}{3} + 1)^2 < (\sqrt{2})^2 < (\frac{1}{2} + 1)^2$ then $\frac{1}{3} < \sqrt{2} - 1 < \frac{1}{2}$, $0 < 2(\sqrt{2} - 1) < 1$ and $3 - 2\sqrt{2} < \sqrt{2} - 1$. Let $r_3 = 3 - 2\sqrt{2}$ and $q_1 = 2$. Then $0 < r_3 < r_2$ and*

$$r_1 = q_1 \times r_2 + r_3 \qquad or \qquad \frac{r_1}{r_2} = q_1 + \frac{r_3}{r_2}.$$

$$1 = 2 \times (\sqrt{2} - 1) + (3 - 2\sqrt{2}) \qquad or \qquad \frac{1}{\sqrt{2} - 1} = 2 + \frac{3 - 2\sqrt{2}}{\sqrt{2} - 1}.$$

But $\dfrac{3 - 2\sqrt{2}}{\sqrt{2} - 1} = \dfrac{(\sqrt{2} - 1)^2}{\sqrt{2} - 1} = \sqrt{2} - 1 = \dfrac{1}{\dfrac{1}{\sqrt{2} - 1}}$. *Hence*

$$\frac{1}{\sqrt{2} - 1} = 2 + \frac{3 - 2\sqrt{2}}{\sqrt{2} - 1} = 2 + \frac{1}{\dfrac{1}{\sqrt{2} - 1}} = 2 + \cfrac{1}{2 + \cfrac{1}{\sqrt{2} - 1}}$$

$$= 2 + \cfrac{1}{2 + \cfrac{1}{2 + \cfrac{1}{\sqrt{2} - 1}}} = \ldots$$

So the ratio of $\sqrt{2}$ to 1 is characterised by the repeating (after 2) infinite sequence $[1, 2, 2, 2, ...]$ which we also write as $[1, \overline{2}]$. Written in continued fractions, this is:

$$\sqrt{2} = 1 + \cfrac{1}{2 + \cfrac{1}{2 + \cdots}}.$$

$\sqrt{2}$ is called a quadratic irrational because it is the solution to the quadratic equation $x^2 - 2 = 0$. Note that these continued fractions (especially the sequence characterising the ratio) provide an approximation to $\sqrt{2}$ as follows:

- $\sqrt{2} \approx 1$,

- $\sqrt{2} \approx 1 + \frac{1}{2} = 1.5$,

- $\sqrt{2} \approx 1 + \cfrac{1}{2 + \cfrac{1}{2}} = 1.4$,

- $\sqrt{2} \approx 1 + \cfrac{1}{2 + \cfrac{1}{2 + \cfrac{1}{2}}} = 1.417$,

- $\sqrt{2} \approx 1 + \cfrac{1}{2 + \cfrac{1}{2 + \cfrac{1}{2 + \cfrac{1}{2}}}} = 1.4139$ *etc.*   □

Figure 7.5 shows the diagram version.

The ancient Greeks would have constructed the length of $\sqrt{2}$ by a geometric construction. Today it is easier to draw the diagram by using an approximation for $\sqrt{2}$ that we can obtain from a calculator.

7.1.3   Problems with Anthyphairesis

The anthyphairesis method we described in the earlier sections was the first method used to define ratios of magnitudes. But a problem arose with this definition: certain obvious theorems could not be proved easily using it. One such obvious theorem is:

> *If the ratio of A to C is the same as the ratio of B to C, then $A = B$.*

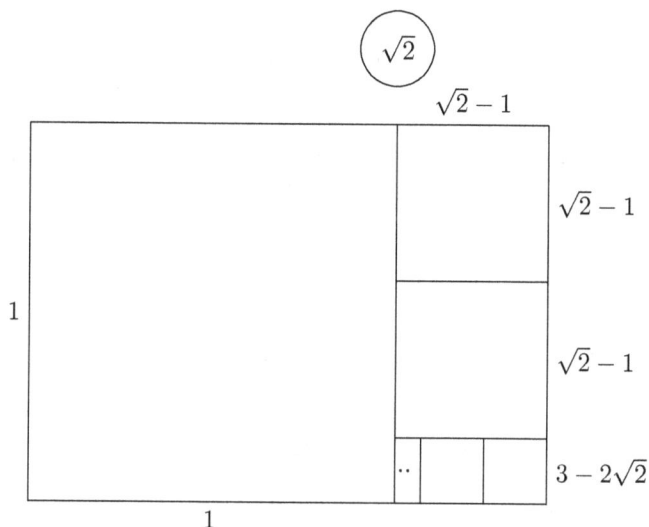

Figure 7.5: Ratio of $\sqrt{2}$ to 1

Although this is an obvious truth,[11] proving it turned out to be extremely difficult.[12] This experience is one of the reasons that mathematicians now insist on proofs of results that seem obvious. A solution to this problem was found by Eudoxus,[13] who found a way to define *proportion* (having the same ratio) for magnitudes instead of ratios of magnitudes.

Eudoxus builds his theory of proportions on the notion of *ratio* which as we have seen, relates the size of two magnitudes of the same kind. So, we say that a ratio exists between two quantities of the same kind if either can be multiplied to exceed the other.

So we speak of:

- The ratio of a number to a number since either can be multiplied to exceed the other.

- The ratio of an area to an area since again, either can be multiplied to exceed the other.

- The ratio of a volume to a volume where again, either can be multiplied to exceed the other.

---

[11]If we allow division, it says that if $\frac{A}{C} = \frac{B}{C}$, then $A = B$.
[12]See [26, p. 302].
[13]Eudoxus of Cnidus, 408–355 B.C.E., a leading mathematician of Plato's academy.

But we cannot speak of the ratio of an area to a volume. For example, a pyramid can be multiplied to exceed a certain square, a square cannot be multiplied to exceed a pyramid. The same can be said for a line and a circle or a line and a square, etc.

However, even though a ratio is a size relation between two magnitudes of the same kind, a ratio of one kind can still be compared to a ratio of another kind:

- A ratio between two numbers can be compared to the ratio between two lines.

- A ratio between two line can be compared to that between two areas.

This ratio comparison allowed Eudoxus to define the theory of proportions where one can make statements like: "the ratio between the area of square $ABCD$ and the area of square $EFGH$ is equal to the ratio of between the numbers 1 and 2 (in other words, the area of the square $EFGH$ is double that of square $ABCD$)". Despite the fact that the Greeks never used arithmetic or numerical representations for their geometric magnitudes, they still managed to calculate their areas and volumes and lengths, etc. They did this using Eudoxus' theory of proportions. This definition of proportion is the one given in Book V of Euclid's *Elements*. We will see how Euclid and Archimedes[14] made use of Eudoxus ideas in Chapter 8.[15]

DEFINITION 5 OF BOOK V OF THE *Elements*.
Magnitudes are said to be in the same ratio, the first to the second and the third to the fourth, when, if any equimultiples whatever are taken of the first and third, and any equimultiples whatever of the second and fourth, the former equimultiples alike exceed, are alike equal to, or alike fall short of, the latter equimultiples respectively taken in corresponding order.

---

[14] Archimedes, c. 287–212 B.C.E., mathematician who lived in Syracuse in Sicily and is credited with the anticipation of the integral calculus. Archimedes calculated areas bounded by curves (e.g. parabola) using ingenious techniques. He approximated π, developed the theory of centres of gravity. Despite the Greeks reluctance to use infinitesimals, and in spite of the lack of a notion of function in the work of the Greeks, Archimedes still managed to construct tangents to spirals.

[15] It may seem strange that Definitions 5 and 6 of Book V appear before the definitions we gave in Figure 7.1 which come from Chapter VII. It is generally assumed that historically, this is not the case, an that the Greeks first developed these definitions for numbers (as in Book VII), and then when they discovered incommensurability, they adapted them for more general magnitudes.

DEFINITION 6 OF BOOK V OF THE *Elements.*
Let magnitudes which have the same ratio be called proportional.

What this means is that x_1 is to x_2 as x_3 is to x_4 if and only if, for any
two numbers m and n, we have

$$mx_1 < nx_2 \text{ and } mx_3 < nx_4, \text{ or}$$

$$mx_1 = nx_2 \text{ and } mx_3 = nx_4, \text{ or}$$

$$mx_1 > nx_2 \text{ and } mx_3 > nx_4.$$

If we allow ourselves to think in terms of division, this is equivalent to

$$\frac{m}{n} < \frac{x_1}{x_2} \text{ and } \frac{m}{n} < \frac{x_3}{x_4} \text{ or}$$

$$\frac{m}{n} = \frac{x_1}{x_2} \text{ and } \frac{m}{n} = \frac{x_3}{x_4} \text{ or}$$

$$\frac{m}{n} > \frac{x_1}{x_2} \text{ and } \frac{m}{n} > \frac{x_3}{x_4}.$$

This says that two ratios of magnitudes of the same rational numbers are
less than both, are equal to both, or are greater than both, which seems
obviously true.

7.1.4 Exercises

Exercise 7.1. Draw on a piece of paper a rectangular triangle. Cut out this
rectangular triangle. Now, use scissors to cut this triangle into pieces that
when you put together will form a square such that no piece of the original
triangle is wasted so that the area of the square is exactly that of the original
rectangular triangle. Repeat this exercise with different sizes of rectangular
triangles.

Draw on a piece of paper a rectangle. Cut out this rectangle. Now, use
scissors to cut this rectangle into pieces that when you put together will
form a square such that no piece of the original rectangle is wasted so that
the area of the square is exactly that of the original rectangle. Repeat this
exercise with different sizes of rectangles.

Repeat the above using a parallelogram. Then repeat for a pentagon,
then a hexagon, then a septagon, then an octagon.

What about the circle? Can you cut a circle into pieces that will all be
used to form a square whose area is that of the original circle?

Exercise 7.2.   1. Use repeated subtraction (anthyphairesis) to show that 7 and 5 are relatively prime to one another.

2. Use repeated subtraction (anthyphairesis) to find the greatest common divisor of 212 and 24.

Exercise 7.3. Let a, b and $m \neq 1$ be numbers. Use repeated subtraction (anthyphairesis) to show the relationship between the GCD of a, b and the GCD of ma, mb, the characterising sequence $[q_0, \ldots, q_n]$ of the ratio a/b and that of ma/mb, and the remainders $r_1, \ldots, r_{i+1}$ of calculating the ratio a/b, and ma/mb.

Exercise 7.4. Use anthyphairesis to evaluate the following ratios of numbers. Show geometric diagrams.

1. The ratio of 15 to 4.

2. The ratio of 20 to 7.

3. The ratio of 15 to 10.

4. The ratio of 3 to 2.

5. The ratio of 7 to 2.

6. The ratio of 14 to 4.

Exercise 7.5.   1. Use anthyphairesis to evaluate the following ratios of magnitudes. Show the diagrams.

  (a) $\sqrt{3}$ to 1.
  (b) $\sqrt{5}$ to 1.
  (c) $\sqrt{7}$ to 1.

Exercise 7.6. Find n in each of the following cases:

1. $n = \cfrac{2}{1 + \cfrac{2}{1 + \cfrac{2}{1 + \ddots}}}.$

2. $n = \cfrac{6}{1 + \cfrac{6}{1 + \cfrac{6}{1 + \ddots}}}.$

3. $n = \dfrac{12}{1 + \dfrac{12}{1 + \dfrac{12}{1 + \cdots}}}$.

4. $n = \dfrac{20}{1 + \dfrac{20}{1 + \dfrac{20}{1 + \cdots}}}$.

7.2 Early Study of Incommensurability

After the discovery of the incommensurability of the side and diagonal of a square, there began a study of the phenomenon of incommensurability. The first person to undertake this study was Theodorus.[16] To understand his work, we need some results on Pythagorean triples which were introduced in Definition 2.5.2.

Theorem 7.2.1. *In a Pythagorean triple (a, b, c), if not all of a, b, and c are divisible by four, then either a or b must be divisible by 4, and it is the only one of a, b, and c which is so divisible.*

Proof. By Corollary 2.5.7, c cannot be a multiple of 4, for if it were, then so would be a and b, contrary to hypothesis. Furthermore, it is clear that at most one of a and b is a multiple of 4, for if both were, then a^2 and b^2 would be multiples of 16, and then c^2 would be a multiple of 16, from which it would follow that c is a multiple of 4.

So it remains to prove that at least one of a and b is a multiple of 4.

First, consider the case in which not all three terms are even. Then by Theorem 2.5.10, c and at least one leg, say b, are odd. Then b^2 and c^2 are each one more than a multiple of 8 by Corollary 2.4.3. Hence, $a^2 = c^2 - b^2$ is a multiple of 8. This means that $\frac{a^2}{4}$ is even, so it is a multiple of 4. Hence, a^2 is a multiple of 16, and so a is a multiple of 4.

In the remaining case, all three numbers are even (but not multiples of 4). Then by halving each number, we get a Pythagorean triple in which not all numbers are even, and the previous argument applies. □

---

[16]Theodorus of Cyrene, 465–398 B.C.E., was a philosopher and mathematician. He was a pupil of Protagoras and was also tutor to Plato. Our knowledge of him comes entirely from the work of Plato. According to Plato, Theodorus of Cyrene was the first to prove the irrationality of $\sqrt{3}$, $\sqrt{5}$, ..., $\sqrt{17}$. We already gave his spiral in Figure 2.2.

These results all concern odd and even, or divisibility by 2. We also have some results about divisibility by 3.

First note that we have

$$
\begin{array}{rcl}
(3n)^2 & = & 9n^2 & = & 3(3n^2) \\
(3n+1)^2 & = & 9n^2 + 6n + 1 & = & 3(3n^2 + 2n) + 1 \\
(3n+2)^2 & = & 9n^2 + 12n + 4 & = & 3(3n^2 + 4n + 1) + 1
\end{array}
$$

Hence, *The square of every positive integer is either a multiple of 3 or one more than a multiple of 3.*

Using this we can prove the following:

Theorem 7.2.2. *In a Pythagorean triple (a, b, c), if c is a multiple of 3, then so are a and b. Moreover, if not all three are multiples of 3, then exactly one of a and b is a multiple of 3.*

Proof. First, suppose neither a nor b is a multiple of 3. Then each of a^2 and b^2 is one more than a multiple of 3, and hence $a^2 + b^2$ is two more than a multiple of 3, which is not a perfect square. This proves that at least one of a and b is a multiple of 3. If both a and b are multiples of 3, then c is also a multiple of 3. If, on the other hand, only one of a and b is a multiple of 3, then c cannot be a multiple of 3.   □

Before proceeding, we give the definition of similar triangles:

Definition 7.2.3. *Two polygons are said to be similar if they have the same angles, and whose sides are all in the same proportion.*
For two triangles, we say they are similar if either they have the same angles, or their sides are all in the same proportion. Equivalently, it is enough that two sides have lengths in the same ratio, and the angles between these sides are equal.

Example 7.2.4. *The inner and outer polygons below are similar since they have the same angles and their sides are in the same proportion. The same applies to the remaining two adjacent polygons.*

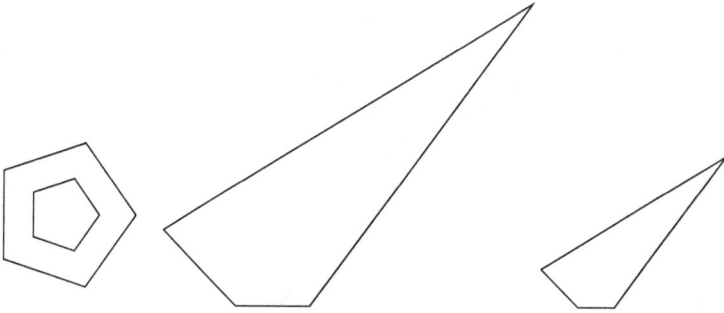

Similarly, the next two triangles AOC and COB where $BC = AC$ and OC is the bisector of $\angle ACB$, are similar:

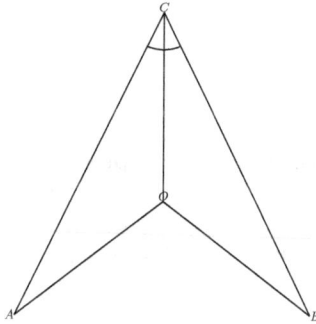

Furthermore, if we take the following 2 triangles whose sides are all in the same proportion, are similar:

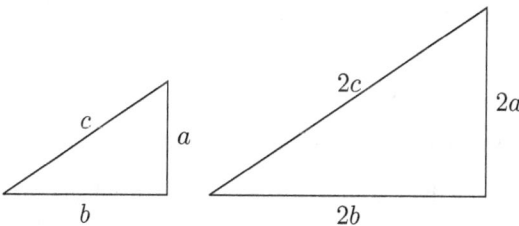

$\square$

We can now turn to the work of Theodorus as reconstructed by Knorr [26, Chapter VI, Section III].

Let us begin with $\sqrt{3}$.

Theorem 7.2.5. *The side of a square of area 3 units is incommensurable with the unit length.*

Proof. The square of area 3 units has a side of length $\sqrt{3}$. Consider a right triangle whose legs are of length 1 and $\sqrt{3}$. By the Pythagorean Theorem, the hypotenuse has length 2. Assuming that $\sqrt{3}$ is commensurable with the unit, there must be a similar [17] right triangle whose legs have lengths a and b and whose hypotenuse has length $2b$, where a and b are *positive integers*; see Figure 7.6.

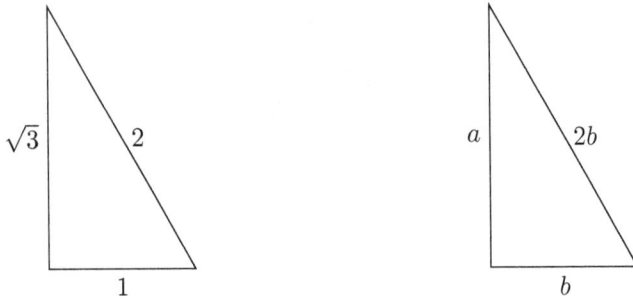

Figure 7.6: Diagram for the proof of Theorem 7.2.5

Now since the hypotenuse has length $2b$, it is even, and so by Theorem 2.5.5, both legs are even. This means that a and b are both even. Thus, we can get a smaller triangle of the same form whose linear dimensions are half of those of the triangle we started with. But then we have a triangle whose legs have lengths $a/2$ and $b/2$ and whose hypotenuse has length $2(b/2)$, or a', b' and $2b'$, where $a = a/2$ and $b' = b/2$. Assuming that a' and b' are positive integers, this is, again, a right triangle whose hypotneuse is even, and the above argument can be repeated. Clearly, we cannot indefinitely repeat this argument. Hence, there is no right triangle whose legs have lengths a and b and whose hypotenuse has length $2b$ where a and b are integers. It follows that $\sqrt{3}$ and 1 are incommensurable.   □

This theorem can be extended to any square whose area has the form $4n + 3$. See Exercises 7.8.1.

Theorem 7.2.6. *If a square has an area of 5 units, its side is incommensurable with the unit.*

Proof. Consider a right triangle whose legs have lengths $\sqrt{5}$ and 2. By the Pythagorean Theorem, the hypotenuse has length 3. If $\sqrt{5}$ and 1 are

---

[17]This means a right triangle whose angles are the same as the original one and whose sides are all in the same proportion.

commensurable, there must be a triangle whose legs have lengths a and $2b$ and whose hypotenuse has length $3b$, where a and b are positive integers; see Figure 7.7. Now b is either even or odd.

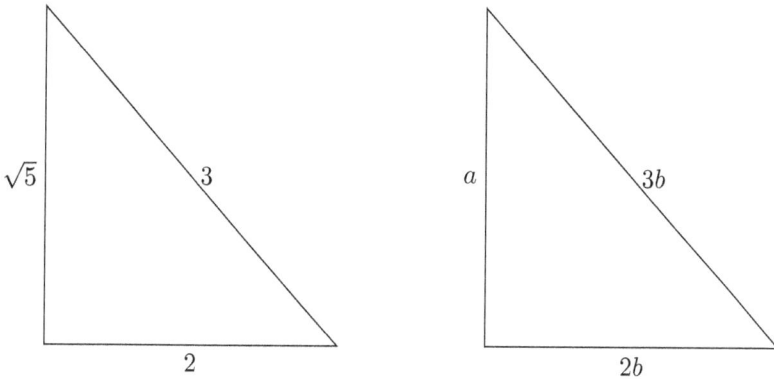

Figure 7.7: Diagram for the proof of Theorem 7.2.6

If b is even, then $3b$ is even, so the hypotenuse is even, and hence, by Theorem 2.5.5, both a and $2b$ are even. Then we can construct another triangle of the same form by halving each dimension. This cannot be repeated indefinitely, so there must be a triangle of this form in which b is odd. It follows that $3b$ is odd. Now we may assume that a and b have no common factors, since otherwise we can divide out these common factors, and we cannot keep doing this indefinitely. Now, since $3b$ is odd, it is not a multiple of 4. By Theorem 7.2.1, only one of a and $2b$ is a multiple of 4. If a is a multiple of 4, then it is even, so $3b$ must be the sum of two even squares and cannot be odd. Hence, a is not a multiple of 4, and so $2b$ must be a multiple of 4, from which it follows that b is even, contradicting its being odd. Hence, there is no right triangle whose legs have length a and $2b$, where a and b are positive integers, and $\sqrt{5}$ is incommensurable with 1. □

This theorem can be extended to any square whose area has the form $8n + 5$. See Exercises 7.8.2.

Theorem 7.2.7. *The side of a square of area 6 is incommensurable with the unit.*

Proof. Consider a right triangle whose legs are $2\sqrt{6}$ and 5, so that the hypotenuse is 7. If $\sqrt{6}$ is commensurable with 1, then there is a right triangle

whose legs are $2a$ and $5b$ and whose hypotenuse is $7b$, where a and b are
positive integers; see Figure 7.8.

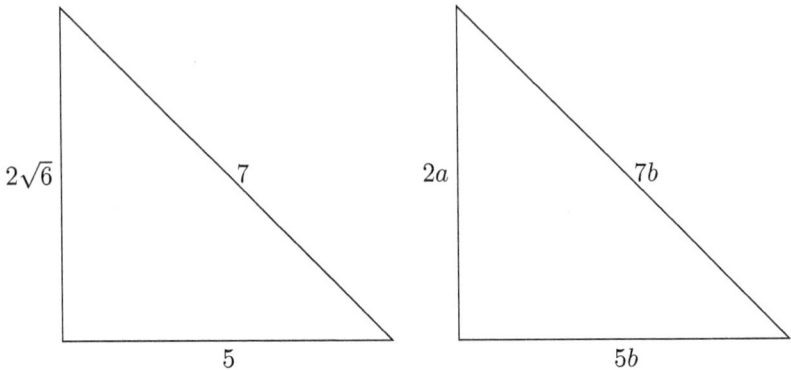

Figure 7.8: Diagram for the proof of Theorem 7.2.7

Now if a and b are both even, we could divide all the linear dimensions
by 2, and we cannot do this indefinitely. So we may assume that a and b
are not both even. Now suppose b is even, so that a must be odd. Then b
cannot be a multiple of 4, since then the hypotenuse would also be divisible
by 4, and so by Theorem 7.2.1, $2a$ would also be divisible by 4, contradicting
the oddness of a. Hence, b is not divisible by 4. Then neither are $7b$ (the
hypotenuse) and $5b$ divisible by 4, from which it follows by Theorem 7.2.1
that $2a$ is divisible by 4, again contradicting the oddness of a. Hence, b
cannot be even, so it must be odd. Then, as before, $7b$ and $5b$ are odd, and
hence not divisible by 4, so by Theorem 7.2.1, $2a$ is divisible by 4, and so a
is even. Say $a = 2c$. Now the Pythagorean condition implies that

$$(2a)^2 = (7b)^2 - (5b)^2,$$

and substituting $2c$ for a, this gives us

$$16c^2 = 49b^2 - 25b^2 = 24b^2.$$

This is equivalent to
$$2c^2 = 3b^2.$$

Now since b is odd, $3b^2$ is odd, so it cannot equal $2c^2$. It follows that there is
no right triangle whose legs are $2a$ and $5b$ and whose hypotenuse is $7b$ where
a and b are positive integers.   □

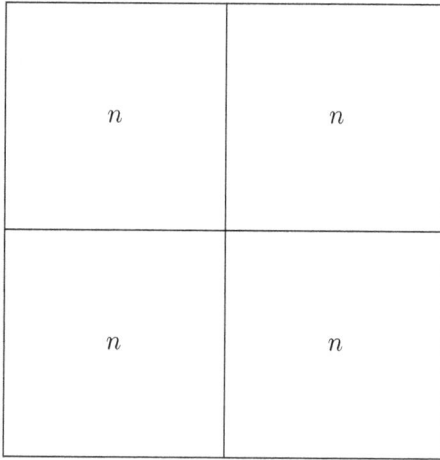

Figure 7.9: Squares of sides s and n

This theorem can be extended to any square whose area has the form $2(2n + 1)$, i.e., twice an odd number, or an even number which is not a multiple of 4. See Exercises 7.8.3.

Theorem 7.2.8. *The side of a square of area 4n is commensurable with the unit if and only if the side of a square of area n is commensurable with the unit.*

Proof. Let $s = 4n$. Then a square of area s can be made up of four squares of area n as in Figure 7.9, and the side of the square of area s is twice the size of the square of area n. It follows immediately that the side of the square of area n is incommensurable with the unit if and only if the side of the square of side $s = 4n$ is incommensurable with the unit.

In modern notation: $2\sqrt{n} = \frac{a}{b}$ where a and b are positive integers iff $\sqrt{n} = \frac{c}{d}$ where c and d are positive integers.

$\square$

Theorems 7.2.5, 7.2.6, 7.2.7, and 7.2.8, and the parts of Exercise 7.8 below take care of all numbers of the form $4n + 3$ (including 3), $8n + 5$ (including 5), $2(2n + 1)$, and $4n$. This leaves numbers of the form $8n + 1$, of which an example is 17. Let us now try to extend the above results to 17. Take a right triangle whose legs are $\sqrt{17}$ and 8. The hypotenuse is then 9. If $\sqrt{17}$ is commensurable with the unit, then there must be a right triangle whose legs are a and $8b$ and whose hypotenuse is $9b$, where a and b are positive integers;

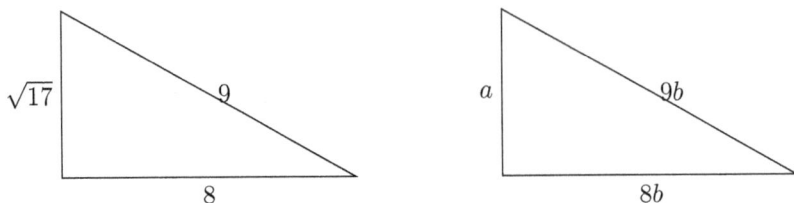

Figure 7.10: Diagram for the case of $\sqrt{17}$

see Figure 7.10. If b is even, then so is $9b$, the hypotenuse, so both legs must be even, a and b are even, and we can construct a new right triangle of the same kind by halving all the dimensions. Since we cannot continue halving indefinitely, we may assume that we have a triangle of this form in which b is odd. Then the hypotenuse, $9b$, is odd, but the leg $8b$ is even, and clearly a multiple of 4. Also, a must be odd. But there is no contradiction here.

Indeed, we should not expect to see a contradiction, since a square of area 9 has a side of 3, which is commensurable with the unit, and $9 = 8 + 1$ is of the form $8n + 1$. So we cannot expect this method to work here.

This is as far as Theodorus got in his analysis of incommensurability.

The next person to work on this problem was Theaetetus,[18] who is believed to have proved the following three theorems:[19]

Theorem 7.2.9. *If a square contains an integral number of unit areas, and if that number is a square integer, then the side of the square will be commensurable with the unit-line; but if that number is a non-square, then the side will be incommensurable with the unit.*

This implies that the square root of a positive integer is a rational number if and only if the positive integer is a perfect square.

Theorem 7.2.10. *If a cube contains n unit volumes, for n a cubic integer, then the side of the cube is commensurable with the unit line; but if n is a non-cubic integer, the side will be incommensurable with the unit.*

This implies that the cube root of a positive integer is rational if and only if the positive integer is a perfect cube.

---

[18]Theaetetus of Athens, 417–369 B. C. E., studied with Theodorus and later taught at Heraclea in southern Italy. Plato made him the subject ot two dialogues.

[19]See [26, Chapter VII Section II].

Theorem 7.2.11. *Two lines are commensurable in length if and only if the squares constructed on them have the ratio which a square integer has to a square integer.*

This is part of Proposition 9 of Book X of Euclid's *Elements*.

Before we give a proof for Theorem 7.2.9, it is interesting to take a look at the dialogue from Plato's *Theaetetus* which expresses Theaetetus's theorem that the square of a non perfect square is irrational and the cube of a non perfect cube is irrational. This dialogue is taken from [23], 148–149 and is given in Figure 7.11.

Here is now a reconstruction by Knorr [26, Chapter VII Section III] of a proof that Theaetetus might have given of Theorem 7.2.9:

Proof. Let a be the side of the larger square, let b the unit length, and let c be the non-square number of unit areas. Then by hypothesis, a^2 is to b^2 as c is to 1. If a and b are commensurable, then there are numbers d and e in the same ratio as a to b, and we can assume that d and e have no common factors (since otherwise we could divide out the common factors and get new numbers with no common factors in the same ratio). Since a is to b as d is to e, which we can write $a : b$ as $d : e$, it follows that $d^2 : e^2 = a^2 : b^2 = c : 1$. It follows that e^2 divides evenly into itself and d^2. But d and e have no common factors, and hence d^2 and e^2 have no common factors, so since e^2 divides evenly into d^2, we must have $e^2 = 1$, and hence $d^2 = c$. But c is a non-square number, so this is a contradiction. Hence, a and b are incommensurable. $\square$

There is another, more algebraic way to prove this. The proof depends on a result, attributed to Euclid (but not found in the *Elements*). Recall that a prime number is a number which can be divided evenly only by itself and 1.

Theorem 7.2.12 (Unique Factorisation into Primes). *Any positive integer greater than 1 can be factored into prime numbers in a way that is unique except for the order of the factors.*

Thus, for example, 12 can be factored into $2 \cdot 2 \cdot 3$ or $2^2 \cdot 3$. The only other possible ways to factor 12 into primes are $2 \cdot 3 \cdot 2$ and $3 \cdot 2 \cdot 2$, and the only difference between these different ways of factoring 12 is the order of the factors.

On the other hand, the only prime factorisation of 9 is $3 \cdot 3$.

Proof. The proof depends on the fact that if a prime number p divides a product ab, then p divides a or p divides b.

Theaet. Theodorus. was writing out for us something about roots, such as the roots of three or five, showing that they are incommensurable by the unit: he selected other examples up to seventeen; there he stopped. Now as there are innumerable roots, the notion occurred to us of attempting to include them all under one name or class.

Soc. And did you find such a class?

Theaet. I think that we did; but I should like to have your opinion.

Soc. Let me hear.

Theaet. We divided all numbers into two classes: those which are made up of equal factors multiplying into one another, which we compared to square figures and called square or equilateral numbers;—that was one class.

Soc. Very good.

Theaet. The intermediate numbers, such as three and five, and every other number which is made up of unequal factors, either of a greater multiplied by a less, or of a less multiplied by a greater, and when regarded as a figure, is contained in unequal sides;—all these we compared to oblong figures, and called them oblong numbers.

Soc. Capital; and what followed?

Theaet. The lines, or sides, which have for their squares the equilateral plane numbers, were called by us lengths or magnitudes; and the lines which are the roots of (or whose squares are equal to) the oblong numbers, were called powers or roots; the reason of this latter name being, that they are commensurable with the former [i. e. with the so-called lengths or magnitudes] not in linear measurement, but in the value of the superficial content of their squares; and the same about solids.

Soc. Excellent, my boys; I think that you fully justify the praises of Theodorus, and that he will not be found guilty of false witness.

Theaet. But I am unable, Socrates, to give you a similar answer about knowledge, which is what you appear to want; and therefore Theodorus is a deceiver after all.

Soc. Well, but if some one were to praise you for running, and to say that he never met your equal among boys, and afterwards you were beaten in a race by a grown-up man, who was a great runner would the praise be any the less true?

Theaet. Certainly not.

Soc. And is the discovery of the nature of knowledge so small a matter, as I just now said? Is it not one which would task the powers of men perfect in every way?

Theaet. By heaven, they should be the top of all perfection!

Soc. Well, then, be of good cheer; do not say that Theodorus was mistaken about you, but do your best to ascertain the true nature of knowledge, as well as of other things.

Figure 7.11: Theaetetus Dialogue

Now suppose there were two prime factorisations of a number, say $p_1 p_2 ... p_n$ and $q_1 q_2 ... q_m$. Then since p_1 divides $p_1 p_2 ... p_n$, it must divide $q_1 q_2 ... q_m$. Hence, it must divide one of the q_i. But if one prime divides another, the two primes are equal. Thus, $p_1 = q_i$ for some i. If that factor is removed from both sides, we are left with $p_2 p_3 ... p_n = q_1' q_2' ... q_{m'}'$, where on the right we have the original product with q_i removed. If this procedure is repeated n times, the conclusion is that the two original produces $p_1 p_2 ... p_n$ and $q_1 q_2 ... q_m$ are identical except for the order of the primes. □

It is easy to see that *a number is a perfect square if and only if each prime factor of its prime factorisation occurs an even number of times.*

Theorem 7.2.13. *A positive integer has a rational square root if and only if it is a perfect square.*

Proof. It is clear that if a positive integer is a perfect square then it has a rational square root, namely the positive integer which is its square root, so we need only prove the converse. Suppose k is a positive integer with a rational square root, say $\frac{p}{q}$. Then p and q are positive integers, and

$$\frac{p^2}{q^2} = k,$$

or

$$p^2 = kq^2.$$

Now consider the prime factorisations of the two sides. Since p^2 is a perfect square, the number of times each prime number occurs in its prime factorisation is even. Similarly, the number of times each prime number occurs in q^2 is even. Now since the number of times each prime factor occurs in kq^2 is the same as the number of times it occurs in p^2, the number of times it occurs in k is the number of times it occurs in p^2 minus the number of times it occurs in q^2. But this means that the number of times each prime number occurs in the prime factorisation of k is the difference of two even numbers and is therefore even. It follows that k is a perfect square. □

Note that this proof is a sequence of statements, each of which is supported by a reason.

Corollary 7.2.14. *If a positive integer has a rational square root, that square root is a positive integer.*

7.2.1 Exercises

Exercise 7.7. Take the triangle below built just after Proposition 2. X on Page 180. Show the similarity of the following pairs of triangles and give the measures of all their angles:

1. ACF and AFD.

2. BFD and BCA.

3. EFG and GFD and ACF.

4. BEG and BCA.

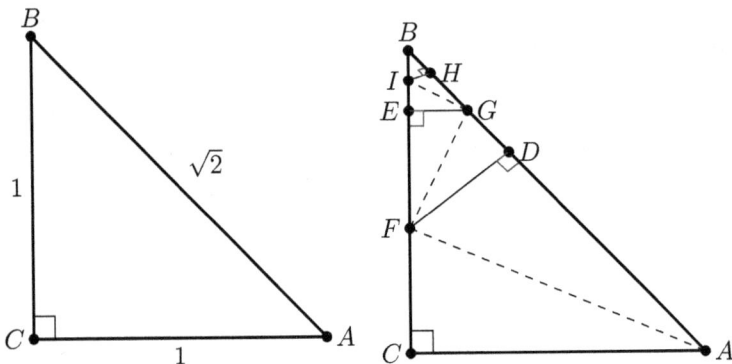

Exercise 7.8. The following series of exercises constitutes the rest of Knorr's reconstruction of the study of incommensurability by Theodorus.

1. (Extends Theorem 7.2.5.) Let p be a number of the form $4n + 3$. Note that $\frac{p-1}{2} = 2n + 1$ and $\frac{p+1}{2} = 2n + 2$. Now consider a right triangle whose legs have lengths $\sqrt{p}$ and $\frac{p-1}{2}$, which will mean that the hypotenuse has length $\frac{p+1}{2}$. Use this triangle to prove the irrationality of $\sqrt{p}$. Note that p can be any of the numbers 3, 7, 11, 15, or 19. (Hint: Assuming commensurability, the triangle will be a right triangle whose legs have lengths a and $(2n + 1)b$ and whose hypotenuse has length $(2n + 2)b$.)

2. (Extends Theorem 7.2.6.) Let q be a number of the form $8n + 5$. Note that $\frac{q-1}{2} = 4n + 2$ and $\frac{q+1}{2} = 4n + 3$. Consider a right triangle whose legs have lengths $\sqrt{q}$ and $\frac{q-1}{2}$ and and whose hypotenuse therefore has length $\frac{q+1}{2}$. Use this triangle to prove that $\sqrt{q}$ is irrational. Note that q can be 5, 13, 21, and 29. (Hint: Assuming commensurability, this

is a right triangle whose legs have lengths a and $(4n + 2)b$ and whose hypotenuse has length $(4n + 3)b$.)

3. (Extends Theorem 7.2.7.) Let r be a number of the form $2(2n+1)$ (i.e., an even number which is not a multiple of 4.) Consider a right triangle whose legs have lengths $2\sqrt{r}$ and $r - 1$ and whose hypotenuse therefore has length $r + 1$. Use this right triangle to prove that $\sqrt{r}$ is irrational. Note that r can be 6, 12, 20, etc. (Hint: Assuming commensurability, the right triangle has the form of one whose legs have lengths $2a$ and $(r - 1)b$ and whose hypotenuse has length $(r + 1)b$.)

Exercise 7.9. Extend the proof of Theorem 7.2.13 and Corollary 7.2.14 to kth roots for any positive integer k. That is, prove that $\sqrt[k]{n}$ is a rational number if and only if n is the kth power of an integer.

7.3 Ruler and Compass Constructions in Euclid's *Elements*

Euclid's *Elements* was written as a textbook of the elementary parts of mathematics, and it was used as a textbook into the nineteenth century.

The structure of the book is deductive, beginning with definitions, postulates, and common notions, all given at the beginning of Book I. These definitions, postulates and common notions are given in Figures 7.12, 7.13 and 7.14. Definitions, define concepts whereas both the postulates and the common notions are axioms (i.e. rules) we take for granted and use even though they are not proven (common notions are concerned with magnitudes of various kinds).

Now look at the Postulates. The first two of them allow for the drawing of a straight line through any two points. We may think of this as lining up a straightedge, or ruler, with the two points and then using that to draw the line. The third postulate allows us to take a compass, put its point at any (geometric) point, put the pencil end at the other end of the radius, and then draw the circle. (It is assumed here that once the circle is drawn and the compass lifted from the paper, it collapses again, so a compass cannot be used to transfer a length.) These methods of drawing lines and circles lead to *ruler and compass* constructions, which have played an important role in the history of mathematics.

Postulate 5 is less obvious than the others. It is equivalent to the *Parallel Postulate* which states that *given a line and a point not on that line, there is exactly one line through the point parallel to the original line (in the same plane)*. In the early 19th century, it was discovered that there are

DEFINITION 1 A *point* is that which has no part.

DEFINITION 2 A *line* is breadthless length.

DEFINITION 3 The ends of a line are points.

DEFINITION 4 A *straight line* is a line which lies evenly with the points on itself.

DEFINITION 5 A *surface* is that which has length and breadth only.

DEFINITION 6 The edges of a surface are lines.

DEFINITION 7 A *plane surface* is a surface which lies evenly with the straight lines on itself.

DEFINITION 8 A *plane angle* is the inclination to one another of two lines in a plane which meet one another and do not lie in a straight line.

DEFINITION 9 And when the lines containing the angle are straight, the angle is called *rectilinear*.

DEFINITION 10 When a straight line standing on a straight line makes the adjacent angles equal to one another, each of the equal angles is *right*, and the straight line standing on the other is called a *perpendicular* to that on which it stands.

DEFINITION 11 An *obtuse angle* is an angle greater than a right angle.

DEFINITION 12 An *acute angle* is an angle less than a right angle.

DEFINITION 13 A *boundary* is that which is an extremity of anything.

DEFINITION 14 A *figure* is that which is contained by any boundary or boundaries.

DEFINITION 15 A *circle* is a plane figure contained by one line such that all the straight lines falling upon it from one point among those lying within the figure equal one another.

DEFINITION 16 And the point is called the *centre* of the circle.

DEFINITION 17 A *diameter* of the circle is any straight line drawn through the centre and terminated in both directions by the circumference of the circle, and such a straight line also bisects the circle.

DEFINITION 18 A *semicircle* is the figure contained by the diameter and the circumference cut off by it. And the centre of the semicircle is the same as that of the circle.

DEFINITION 19 *Rectilinear figures* are those which are contained by straight lines, *trilateral* figures being those contained by three, *quadrilateral* those contained by four, and *multilateral* those contained by more than four straight lines.

DEFINITION 20 Of trilateral figures, an *equilateral triangle* is that which has its three sides equal, an *isosceles triangle* that which has two of its sides alone equal, and a *scalene triangle* that which has its three sides unequal.

DEFINITION 21 Further, of trilateral figures, a *right-angled triangle* is that which has a right angle, an *obtuse-angled triangle* that which has an *obtuse angle*, and an acute-angled triangle that which has its three angles acute.

DEFINITION 22 Of quadrilateral figures, a *square* is that which is both equilateral and right-angled; an *oblong* that which is right-angled but not equilateral; a *rhombus* that which is equilateral but not right-angled; and a *rhomboid* that which has its opposite sides and angles equal to one another but is neither equilateral nor right-angled. And let quadrilaterals other than these be called *trapezia*.

DEFINITION 23 *Parallel* straight lines are straight lines which, being in the same plane and being produced indefinitely in both directions, do not meet one another in either direction.

Figure 7.12: DEFINITIONS of BOOK 1 of Euclid's Elements

Let the following be postulated:

POSTULATE 1 *To draw a straight line from any point to any point.*

POSTULATE 2 *To produce a finite straight line continuously in a straight line.*

POSTULATE 3 *To describe a circle with any centre and radius.*

POSTULATE 4 *That all right angles equal one another.*

POSTULATE 5 *That, if a straight line falling on two straight lines makes the interior angles on the same side less than two right angles, the two straight lines, if produced indefinitely, meet on that side on which are the angles less than the two right angles.*

Figure 7.13: POSTULATES of BOOK 1 of Euclid's Elements

COMMON NOTION 1 *Things which equal the same thing also equal one another.*

COMMON NOTION 2 *If equals are added to equals, then the wholes are equal.*

COMMON NOTION 3 *If equals are subtracted from equals, then the remainders are equal.*

COMMON NOTION 4 *Things which coincide with one another equal one another.*

COMMON NOTION 5 *The whole is greater than the part.*

Figure 7.14: COMMON NOTIONS of BOOK 1 of Euclid's Elements

perfectly good theories of geometry in which this postulate is violated, and *non-Euclidean geometry* came to flourish.

The Common Notions are properties of equality.

Some of the other books have additional definitions, but there are no other postulates or common notions.

As an example of the style of the proofs in the *Elements*, look at the first proposition of Book I and its proof given in Figure 7.15. Euclid's *Elements* was considered the model of deductive mathematical theories for over two millennia after it was originally written (about 300 B.C.E.). We have seen a reference to "ruler and compass constructions" above. These are constructions based on the first three Postulates. Postulates 1 and 2 are considered to allow the use of a rule or straightedge to construct a line segment that

PROPOSITION 1. OF BOOK I OF THE *Elements*

To construct an equilateral triangle on a given finite straight line.

Let AB be the given finite straight line.

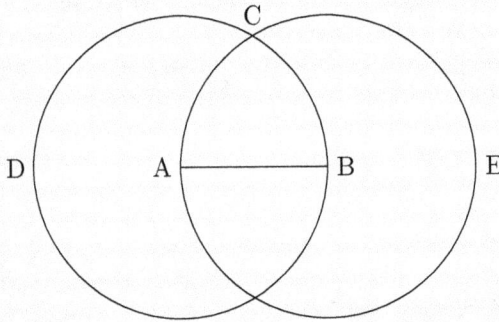

It is required to construct an equilateral triangle on the straight line AB.

Describe the circle BCD with centre A and radius AB. Again describe the circle ACE with centre B and radius BA. [Post.3]

Join the straight lines CA and CB from the point C at which the circles cut one another to the points A and B.   [Post.1]

Now, since the point A is the centre of the circle CDB, therefore AC equals AB. Again, since the point B is the centre of the circle CAE, therefore BC equals BA.   [I.Def.15]

But AC was proved equal to AB, therefore each of the straight lines AC and BC equals AB.

And things which equal the same thing also equal one another, therefore AC also equals BC.   [C.N.1]

Therefore the three straight lines AC, AB, and BC equal one another.

Therefore the triangle ABC is equilateral, and it has been constructed on the given finite straight line AB.   [I.Def.20]

Q.E.F.

Figure 7.15: Proposition 1 of Book I and its proof

passes through any two points. It does not permit the use of the ruler to make a measurement. Postulate 3 allows the use of a compass to draw a circle by placing the point at the centre of the circle and the end with the pencil at one of the points on the circle. However, the compass is assumed to collapse after the construction, so that it cannot be used for measurement. The construction in Proposition 1 of Book 1 of the *Elements* given above is an example of a ruler and compass construction. We will be discussing Euclid's *Elements* further in Section 10.1

7.3.1 Exercises

Exercise 7.10. Give ruler and compass constructions of the following:

1. The bisector of an angle; i.e., the line that goes through the point of the angle and divides the original angle into two equal angles.

2. The perpendicular bisector of a line segment; i.e., the line at right angles to a given line segment that passes through its midpoint.

3. Given a line and a point not on the line, the line that passes through the point and is perpendicular to the original line.

Chapter 8

Limits: the Basics and the Background

The calculus is based on limits. Therefore, in order to obtain a proper foundation for the calculus, we need a proper foundation for a theory of limits.

In your calculus course, you learned to evaluate limits. However, how carefully did you analyse the concept?

In this chapter, we will take up this subject. We will start by looking at the basic properties we would expect for limits of functions and sequences. In this process, we revisit again Zeno's dichotomy and discuss where Zeno's conclusion is false. We move on to explain that adding infinitely many numbers need not return an infinite and then we analyse the way the ancient Greeks dealt with limits especially through the so-called *method of exhaustion*. In particular, we will look at two theorems from the ancients: Archimedes theorem giving the area of a circle, and a theorem of Euclid which says that the areas of circles are to each other as the squares of their radii. Both theorems were proved by a method that relied on Eudoxus theory of proportions which was a geometric theory designed to overcome the difficulties expected because of the discovery of the incommensurability.

As we have already seen, the Greeks kept numbers separate from other quantities (especially the irrationals) and although their arithmetic could not handle the irrationals, their geometrical methods were extended to handle infinity and the irrationals. We saw in Chapter 7 that Anthyphairesis was the first method used to define ratios of magnitudes but that there was a problem with that definition which was solved by Eudoxus theory of ratio and proportion which became an important tool in the geometrical methods of the Greeks.

8.1   Limits of Functions: Basic Properties

If we go back to the example of Section 1.3, we see that

$$f(x) = \frac{x^2 - 4}{x - 2}$$

and

$$g(x) = x + 2$$

have exactly the same values as long as $x \neq 2$. Indeed, f is not defined if $x = 2$, so f and g have the same values wherever both are defined. But we are used to saying that

$$\lim_{x \to 2} f(x) = g(2) = 4.$$

By this, we mean that as x gets close to 2, $f(x)$ gets close to 4. More precisely, we can make $f(x)$ as close to 4 as we want by making x close enough to 2.

We can also see this by comparing the graphs. See Figures 8.1 and 8.2.

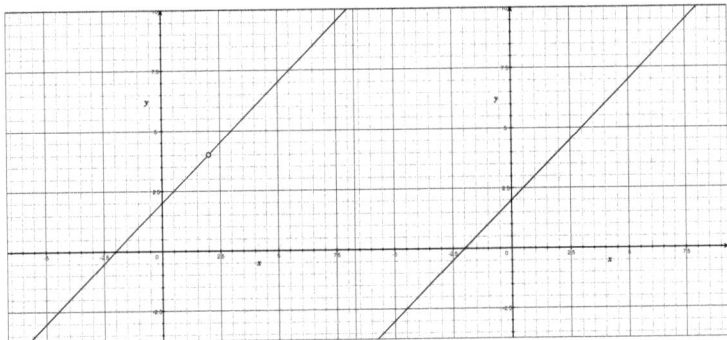

Figure 8.1: Graph of f Figure 8.2: Graph of g

More generally, the idea is that

$$\lim_{x \to c} f(x) = l$$

means that when x is close to c but not equal to c, then $f(x)$ is close to l, and, in particular, we can make $f(x)$ as close to l as we please by taking x close enough to c.

There are other versions of limits of functions. One of these ideas is *one-sided limits*:

- Here,
$$\lim_{x \to c^+} f(x) = l$$
means that if x is close to c and greater than c, then $f(x)$ is close to l.

- Whereas
$$\lim_{x \to c^-} f(x) = l$$
means that if x is close to c and less than c, then $f(x)$ is close to l.

Again with this one-sided version of limit, we can make $f(x)$ as close to l as we want by making x close enough to c (but not equal to c).

Another version occurs when the values of the function are not close to a value, but become arbitrarily large, positively or negatively. This occurs, for example, when a function has a vertical asymptote. For example, consider the function
$$h_1(x) = \frac{1}{x}.$$
The graph of this function is shown in Figure 8.3. This function has a vertical asymptote at $y = 0$, and we say that

$$\lim_{x \to 0^-} h_1(x) = -\infty \qquad \text{and} \qquad \lim_{x \to 0^+} h_1(x) = \infty.$$

This means that if x is close to 0 but positive, then the value of $f(x)$ is very large, but if x is close to 0 but negative, then the value of $h_1(x)$ is very negative. If, on the other hand, we consider the function

$$h_2(x) = \frac{1}{x^2},$$

whose graph is shown in Figure 8.4, we see that

$$\lim_{x \to 0} h_2(x) = \infty.$$

This means that if x is close to 0 but not equal to 0, then the value of $h_2(x)$ is very large positively. Note that ∞ and $-\infty$ are not values that a variable

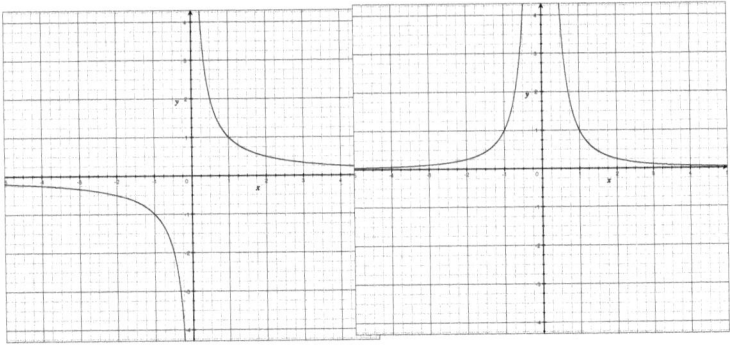

Figure 8.3: Graph of $h_1(x) = \frac{1}{x}$ Figure 8.4: Graph of $h_2(x) = \frac{1}{x^2}$

x can take or that can be the value of a function. They are signs that we use as a shorthand for values that are very large or very large negatively.

Note that both of these functions h_1 and h_2 have the x-axis as a horizontal asymptote. We express this by the notion of a limit as $x \to \infty$ or $x \to -\infty$, and we write

$$\lim_{x \to \infty} h_i(x) = 0 \quad \text{and} \quad \lim_{x \to -\infty} h_i(x) = 0,$$

where $i = 1$ or $i = 2$.

This limit notation can also be used to indicate that a function does not have a horizontal asymptote. If we consider the function g given above, we write

$$\lim_{x \to \infty} g(x) = \infty \quad \text{and} \quad \lim_{x \to -\infty} g(x) = -\infty.$$

It is useful at this stage to list the properties that we expect limits of functions to have. In what follows, remember Definition 4.1.16.

Definition 8.1.1. *The properties of limits of functions that we expect to be true are the following:*

LF1.   − *If c is a value or one of the symbols ∞ and $-\infty$, then*

$$\lim_{x \to c} x = c.$$

− *If c is a value then*

$$\lim_{x \to c^-} x = c \quad and \quad \lim_{x \to c^+} x = c.$$

LF2.   – *If k is any value and $f(x) = k$ for all x except possibly for $x = c$ in an open interval that includes c, then*

$$\lim_{x \to c} f(x) = k \quad and \quad \lim_{x \to c^-} f(x) = k \quad and \quad \lim_{x \to c^+} f(x) = k.$$

   – *If $f(x) = k$ for all values of x, then, in addition,*

$$\lim_{x \to \infty} f(x) = \lim_{x \to -\infty} f(x) = k.$$

LF3.   – *If c is any value or else ∞ or $-\infty$, and if*

$$\lim_{x \to c} f(x) = l \ and \ \lim_{x \to c} g(x) = m,$$

then

$$\lim_{x \to c} [f(x) + g(x)] = l + m.$$

   – *If c is a value and if*

$$\lim_{x \to c^-} f(x) = l \ and \ \lim_{x \to c^-} g(x) = m,$$

then

$$\lim_{x \to c^-} [f(x) + g(x)] = l + m.$$

   – *If c is any value and if*

$$\lim_{x \to c^+} f(x) = l \ and \ \lim_{x \to c^+} g(x) = m,$$

then

$$\lim_{x \to c^+} [f(x) + g(x)] = l + m.$$

LF4.   – *If c is any value or ∞ or $-\infty$, and if*

$$\lim_{x \to c} f(x) = l \ and \ \lim_{x \to c} g(x) = m,$$

then

$$\lim_{x \to c} [f(x)g(x)] = lm.$$

   – *If c is any value and if*

$$\lim_{x \to c^-} f(x) = l \ and \ \lim_{x \to c^-} g(x) = m,$$

then

$$\lim_{x \to c^-} [f(x)g(x)] = lm.$$

– *If c is any value and if*

$$\lim_{x \to c+} f(x) = l \ and \ \lim_{x \to c+} g(x) = m,$$

then

$$\lim_{x \to c+} [f(x)g(x)] = lm.$$

LF5.   – *If c is any value or else ∞ or $-\infty$, and if*

$$\lim_{x \to c} f(x) = l \neq 0,$$

then

$$\lim_{x \to c} \frac{1}{f(x)} = \frac{1}{l}.$$

– *If c is any value and if*

$$\lim_{x \to c-} f(x) = l \neq 0,$$

then

$$\lim_{x \to c-} \frac{1}{f(x)} = \frac{1}{l}.$$

– *If c is any value and if*

$$\lim_{x \to c+} f(x) = l \neq 0,$$

then

$$\lim_{x \to c+} \frac{1}{f(x)} = \frac{1}{l}.$$

LF6.   – *If c is any value or is ∞ or $-\infty$, and if*

$$\lim_{x \to c} f(x) = l,$$

and if $l \geq 0$ whenever n is even, then

$$\lim_{x \to c} \sqrt[n]{f(x)} = \sqrt[n]{l}.$$

– *If c is any value and if*

$$\lim_{x \to c-} f(x) = l,$$

and if $l \geq 0$ whenever n is even, then

$$\lim_{x \to c-} \sqrt[n]{f(x)} = \sqrt[n]{l}.$$

- *If c is any value and if*

$$\lim_{x \to c^+} f(x) = l,$$

and if $l \geq 0$ whenever n is even, then

$$\lim_{x \to c^+} \sqrt[n]{f(x)} = \sqrt[n]{l}.$$

LF7.   – *If $g(x) \leq f(x) \leq h(x)$ for all x except possibly for $x = c$ in some open interval which contains c, and if*

$$\lim_{x \to c} g(x) = \lim_{x \to c} h(x) = l,$$

then

$$\lim_{x \to c} f(x) = l.$$

- *If $g(x) \leq f(x) \leq h(x)$ for all x in an interval whose right endpoint is c, and if*

$$\lim_{x \to c^-} g(x) = \lim_{x \to c^-} h(x) = l,$$

then

$$\lim_{x \to c^-} f(x) = l.$$

- *If $g(x) \leq f(x) \leq h(x)$ for all x in an interval whose left end-point is c, and if*

$$\lim_{x \to c^+} g(x) = \lim_{x \to c^+} h(x) = l,$$

then

$$\lim_{x \to c^+} f(x) = l.$$

LF8.   – *If $f(x) \leq g(x)$ for all x except possibly for $x = c$ in some open interval which contains c, or if c is ∞ and if $f(x) \leq g(x)$ for all x greater than some value, or if c is $-\infty$ and $f(x) \leq g(x)$ for all x less than some negative value, then*

$$\lim_{x \to c} f(x) \leq \lim_{x \to c} g(x).$$

- *If $f(x) \leq g(x)$ for all x in an interval whose right end-point is c, then*

$$\lim_{x \to c^-} f(x) \leq \lim_{x \to c^-} g(x).$$

    – If $f(x) \le g(x)$ for all x in an interval whose left end-point is c, then

$$\lim_{x \to c^+} f(x) \le \lim_{x \to c^+} g(x).$$

LF9. If l is a value, then

$$\lim_{x \to c} f(x) = l \quad \text{if and only if} \quad \lim_{x \to c^-} f(x) = \lim_{x \to c^+} f(x) = l.$$

LF10.   – If c is a value or else ∞ or $-\infty$, and if

$$\lim_{x \to c} f(x) = \infty \quad \text{or} \quad \lim_{x \to c} f(x) = -\infty,$$

then

$$\lim_{x \to c} \frac{1}{f(x)} = 0.$$

– If c is a value, and if

$$\lim_{x \to c^-} f(x) = \infty \quad \text{or} \quad \lim_{x \to c^-} f(x) = -\infty,$$

then

$$\lim_{x \to c^-} \frac{1}{f(x)} = 0.$$

– If c is a value, and if

$$\lim_{x \to c^+} f(x) = \infty \quad \text{or} \quad \lim_{x \to c^+} f(x) = -\infty,$$

then

$$\lim_{x \to c^+} \frac{1}{f(x)} = 0.$$

LF11.   – If c is a value, and if

$$\lim_{x \to c} f(x) = 0,$$

and if, for all x in an open interval containing c, $f(x) \ge 0$ [respectively $f(x) \le 0$], then

$$\lim_{x \to c} \frac{1}{f(x)} = \infty \quad \text{[respectively } -\infty \text{]}.$$

– *If*
$$\lim_{x \to \infty} f(x) = 0,$$

and if $f(x) \geq 0$ *[respectively $f(x) \leq 0$] for all x greater than a certain value, then*

$$\lim_{x \to \infty} \frac{1}{f(x)} = \infty \quad \textit{[respectively } -\infty \textit{]}.$$

– *If*
$$\lim_{x \to -\infty} f(x) = 0,$$

and if $f(x) \geq 0$ *[respectively $f(x) \leq 0$] for all x less than a certain value, then*

$$\lim_{x \to -\infty} \frac{1}{f(x)} = \infty \quad \textit{[respectively } -\infty \textit{]}.$$

– *If c is a value and if*
$$\lim_{x \to c^-} f(x) = 0,$$

and if, for all x in an interval whose right end-point is c, $f(x) \geq 0$ *[respectively $f(x) \leq 0$], then*

$$\lim_{x \to c^-} \frac{1}{f(x)} = \infty \quad \textit{[respectively } -\infty \textit{]}.$$

– *If c is a value and if*
$$\lim_{x \to c^+} f(x) = 0,$$

and if, for all x in an interval whose left end-point is c, $f(x) \geq 0$ *[respectively $f(x) \leq 0$], then*

$$\lim_{x \to c^+} \frac{1}{f(x)} = \infty \quad \textit{[respectively } -\infty \textit{]}.$$

LF12. *If c is a value or ∞ or $-\infty$, and if*
$$\lim_{x \to c} f(x) = l \quad \textit{and} \quad \lim_{x \to c} g(x) = m,$$

then

1.
$$\lim_{x \to c} [f(x) - g(x)] = l - m.$$

2. *If $m \neq 0$, then*
$$\lim_{x \to c} \frac{f(x)}{g(x)} = \frac{l}{m}.$$

3. *If n is any positive integer,*
$$\lim_{x \to c} [f(x)]^n = l^n.$$

4. *If p and q are positive integers, then*
$$\lim_{x \to c} [f(x)]^{\frac{p}{q}} = l^{\frac{p}{q}}.$$

LF13. *If c is a value and n is a positive integer, then*
$$\lim_{x \to c} x^n = c^n.$$

LF14. *If c is a value, a is a constant, and n is a positive integer, then*
$$\lim_{x \to c} ax^n = ac^n.$$

LF15. *If c is a value and $p(x)$ is a polynomial, then*
$$\lim_{x \to c} p(x) = p(c),$$

LF16. *If c is a value and $r(x)$ is any rational function of the form $p(x)/q(x)$ and $q(x) \neq 0$, then*
$$\lim_{x \to c} r(x) = r(c).$$

LF17.
$$\lim_{x \to 0} \frac{\sin x}{x} = 1.$$

LF18.
$$\lim_{x \to 0} \frac{\tan x}{x} = 1.$$

LF19.
$$\lim_{x \to 0} \frac{1 - \cos x}{x} = 0.$$

LF20.
$$\lim_{x \to 0} \frac{1 - \cos x}{x^2} = \frac{1}{2}.$$

These properties are not all independent. Some can be proved from others as we see for example in Exercises 8.1 and 8.3.

8.1.1 Exercises

Exercise 8.1. Prove that LF12 can be derived from the properties given in Definition 8.1.1 that are listed before LF12.

Exercise 8.2. Prove all the parts of LF12 for

$$\lim_{x \to c^-}$$

and

$$\lim_{x \to c^+}$$

for c a real number.

Exercise 8.3. Prove that each of LF13, LF14, LF15 and LF16 can be derived from the properties given in Definition 8.1.1 that are listed before the property in question.

Exercise 8.4. Prove that LF17 can be derived from the properties given in Definition 8.1.1 that are listed before LF17.

[Hint: Consider Figure 8.5: Here, $B = (1,0)$, $OB = OC = 1$, $OD = \cos x$, $CD = \sin x$, $AB = \tan x$, and the arc CB is equal to the angle x (in radians), Consider the inequalities that can be shown from this diagram and use LF7.]

Exercise 8.5. Prove that LF18, LF19 and LF20 can be derived from the properties given in Definition 8.1.1 that are listed before the property in question.

8.2 Limits of Sequences: Basic Properties

8.2.1 Revisiting Zeno's Dichotomy

Let us consider the paradox of Zeno called *dichotomy* of Section 2.2. The argument is that a moving object would have to complete half its course before it reaches the end, but once it has done that it has to reach the half of the rest of its course, etc. Based on this, Zeno concluded that the moving object could never reach the end of its course.

Let us suppose that a moving object is moving along a course whose length is 1 meter and that it is moving at 1 meter per second. It must first reach the halfway point, $\frac{1}{2}$ meter from its starting point. The time it will take to get there will be $\frac{1}{2}$ second. Let us represent the time involved (in seconds) by a_1, so $a_1 = \frac{1}{2}$. Let the total time $t_1 = a_1$.

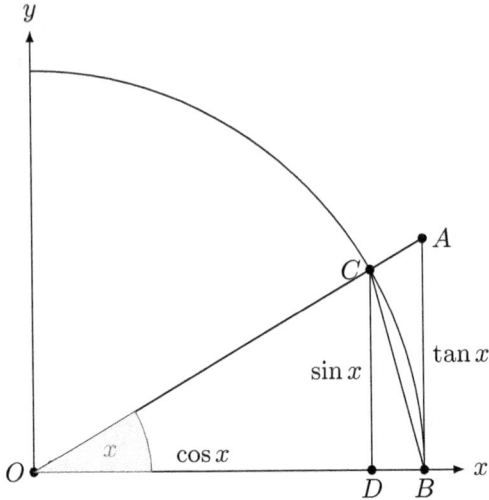

Figure 8.5: Diagram for $(\sin x)/x$

Now, from this halfway point, the object must move halfway from there to the end, which is $\frac{1}{4}$ meters from its new starting point. The total time it takes to get there from the beginning is $\frac{3}{4} = \frac{1}{2} + \frac{1}{4}$ seconds. Let $a_2 = \frac{1}{4}$ and $t_2 = \frac{3}{4} = a_1 + a_2$.

We clearly have the following infinite sequences:

$$a_1, a_2, a_3, \ldots = \frac{1}{2}, \frac{1}{4}, \frac{1}{8}, \cdots$$

$$t_1, t_2, \ldots = \frac{1}{2}, \frac{3}{4}, \frac{7}{8}, \ldots \text{ where each } t_n = a_1 + a_2 + \cdots + a_n$$

Zeno concluded that since we have an infinite sequence, the total time which is the sum of this infinite sequence must be infinite and hence we can never reach our destination.

We know that this is not the case, we can reach our destination in a finite time. So, where did Zeno get it wrong?

First, note that

$$t_n = \frac{2^n - 1}{2^n}.$$

and hence

$$t_n = 1 - \frac{1}{2^n}.$$

It follows that each term of this sequence is less than 1, so the sequence of times never exceeds 1 second. This is why Zeno's conclusion is false.

On the other hand, the term $\frac{1}{2^n}$ gets smaller as n gets bigger. By taking a large enough value of n, we can make $\frac{1}{2^n}$ smaller than any small value we choose. This leads us to say that

$$\lim_{n \to \infty} t_n = 1.$$

Furthermore, adding infinitely many numbers need not return an infinite, and can instead return a finite number. For example, it is easy to see that

$$a_n = \frac{1}{2^n}.$$

Hence,

$$\Sigma_{n=1}^{\infty} a_n = \Sigma_{n=1}^{\infty} \frac{1}{2^n}$$

and

$$2\Sigma_{n=1}^{\infty} a_n = 2a_1 + 2\Sigma_{n=2}^{\infty} \frac{1}{2^n} = 1 + \Sigma_{n=2}^{\infty} \frac{1}{2^{n-1}} = 1 + \Sigma_{n=1}^{\infty} \frac{1}{2^n} = 1 + \Sigma_{n=1}^{\infty} a_n.$$

Therefore,

$$\Sigma_{n=1}^{\infty} a_n = 1.$$

In fact,

$$lim_{n \to \infty} t_n = \Sigma_{n=1}^{\infty} a_n = 1.$$

8.2.2 Basic Properties of Limits of Sequences

Some sequences do not have limits. There are three ways a sequence can fail to have a limit:

1. The terms get arbitrarily large. E.g., the sequence $1, 2, 3, \ldots, n, \ldots$ has this property. We say that this sequence *diverges to infinity*, and we write

$$\lim_{n \to \infty} n = \infty.$$

2. The terms become negative with an arbitrarily large absolute value. For example, the sequence $-1, -2, -3, \ldots, -n, \ldots$ has this property. We say that this sequence *diverges to negative infinity*, and we write

$$\lim_{n \to \infty} (-n) = -\infty.$$

3. The terms do not get arbitrarily large, but they do not get close to a particular value. For example, consider the sequence $1, -1, 1, -1, \ldots$, where the general term is $a_n = (-1)^{n+1}$. For this sequence, we simply say that it *diverges*, and we say that

$$\lim_{n \to \infty} (-1)^{n+1}$$

does not exist.

Example 8.2.1. *The following are examples of properties of limits of sequences.*

LS1. *For any constant value k,*

$$\lim_{n \to \infty} k = k.$$

LS2. *If*

$$\lim_{n \to \infty} a_n = a \qquad and \qquad \lim_{n \to \infty} b_n = b,$$

then

$$\lim_{n \to \infty} (a_n + b_n) = a + b.$$

LS3. *If*

$$\lim_{n \to \infty} a_n = a \qquad and \qquad \lim_{n \to \infty} b_n = b,$$

then

$$\lim_{n \to \infty} (a_n b_n) = ab.$$

LS4. *If*

$$\lim_{n \to \infty} a_n = a \neq 0,$$

then

$$\lim_{n \to \infty} \frac{1}{a_n} = \frac{1}{a}.$$

LS5. *If $a_n \leq 0$ for all $n > N$ for some N, and if*

$$\lim_{n \to \infty} a_n = a,$$

then $a \leq 0$.

LS6.

$$\lim_{n \to \infty} a_n = 0$$

if and only if

$$\lim_{n \to \infty} |a_n| = 0.$$

LS7. Suppose

$$\lim_{n \to \infty} a_n = 0.$$

Let b_n be defined for $n > k$ (for some $k > 0$), and let there be a g, independent of n, such that

$$|b_n| \le g \qquad \text{for } n > k.$$

Then

$$\lim_{n \to \infty} (a_n b_n) = 0.$$

LS8. If a_n is defined for all $n > N$ for some N, and if

$$\lim_{n \to \infty} a_n$$

exists, then there is a g, independent of n, such that

$$|a_n| \le g \qquad \text{for all } n > N.$$

LS9. If $a_n \le b_n \le c_n$ for all $n > N$ for some N, and if

$$\lim_{n \to \infty} a_n = \lim_{n \to \infty} c_n = l,$$

then

$$\lim_{n \to \infty} b_n$$

exists and $\lim_{n \to \infty} b_n = l$.

LS10. If $|x| < 1$, then

$$\lim_{n \to \infty} x^n = 0.$$

LS11.

$$\lim_{n \to \infty} n = \infty.$$

LS12. *If*

$$\lim_{n \to \infty} a_n = \pm\infty \qquad and \lim_{n \to \infty} b_n = b,$$

or if

$$\lim_{n \to \infty} a_n = a \qquad and \lim_{n \to \infty} b_n = \pm\infty,$$

or if

$$\lim_{n \to \infty} a_n = \pm\infty \qquad and \lim_{n \to \infty} b_n = \pm\infty,$$

then

$$\lim_{n \to \infty} (a_n + b_n) = \pm\infty.$$

LS13. *If*

$$\lim_{n \to \infty} a_n = \pm\infty \qquad and \lim_{n \to \infty} b_n = b > 0,$$

or if

$$\lim_{n \to \infty} a_n = a > 0 \qquad and \lim_{n \to \infty} b_n = \pm\infty,$$

then

$$\lim_{n \to \infty} (a_n b_n) = \pm\infty.$$

LS14. *If*

$$\lim_{n \to \infty} a_n = \pm\infty \qquad and \lim_{n \to \infty} b_n = \pm\infty,$$

then

$$\lim_{n \to \infty} (a_n b_n) = \infty.$$

LS15. *If*

$$\lim_{n \to \infty} a_n = \pm\infty,$$

then

$$\lim_{n \to \infty} \frac{1}{a_n} = 0.$$

LS16. *If*
$$\lim_{n \to \infty} a_n = \infty$$
and if, for all $n > N$ for some N, we have $a_n \le b_n$, then
$$\lim_{n \to \infty} b_n = \infty.$$

LS17.
$$\lim_{n \to \infty} a_n = \pm\infty$$
if and only if
$$\lim_{n \to \infty} -a_n = \mp\infty$$

LS18. *If $k > 0$ and if, for each $1 \le i \le k$,*
$$\lim_{n \to \infty} a_{i,n} = a_i,$$
then
$$\lim_{n \to \infty} \left(\sum_{i=1}^{k} a_{i,n} \right) = \sum_{i=1}^{k} a_i.$$

LS19. *If k is any constant value and if*
$$\lim_{n \to \infty} a_n = a,$$
then
$$\lim_{n \to \infty} k a_n = ka.$$

LS20. *If $\lim_{n \to \infty} a_n = a$ and $\lim_{n \to \infty} b_n = b$, then*
$$\lim_{n \to \infty} (a_n - b_n) = a - b.$$

LS21
$$\lim_{n \to \infty} -n = -\infty.$$

LS22. *If $\lim_{n \to \infty} a_n = \pm\infty$ and $\lim_{n \to \infty} b_n = b$, then*
$$\lim_{n \to \infty} (a_n - b_n) = \pm\infty.$$

LS23. *If $\lim_{n \to \infty} a_n = a$ and $\lim_{n \to \infty} b_n = \pm\infty$, then*
$$\lim_{n \to \infty} (a_n - b_n) = \mp\infty.$$

LS24. *If* $\lim_{n \to \infty} a_n = \pm\infty$ *and* $\lim_{n \to \infty} b_n = \mp\infty$, *then*

$$\lim_{n \to \infty} (a_n - b_n) = \pm\infty.$$

LS25. *If* $k > 0$ *and for each* i, $0 < i < k$, $\lim_{n \to \infty} a_{i,n} = a_i$, *then*

$$\lim_{n \to \infty} \left(\prod_{i=1}^{k} a_{i,n} \right) = \prod_{i=1}^{k} a_i.$$

LS26. *If* $\lim_{n \to \infty} a_n = a$ *and if* k *is a positive integer, then*

$$\lim_{n \to \infty} a_n^k = a^k.$$

LS27. *If* $\lim_{n \to \infty} a_n = \pm\infty$ *and* $\lim_{n \to \infty} b_n = b < 0$, *or* $\lim_{n \to \infty} a_n = a < 0$ *and* $\lim_{n \to \infty} b_n = \pm\infty$, *then*

$$\lim_{n \to \infty} (a_n b_n) = \mp\infty.$$

LS28. *If* $\lim_{n \to \infty} a_n = \pm\infty$ *and* $\lim_{n \to \infty} b_n = \mp\infty$, *then*

$$\lim_{n \to \infty} (a_n b_n) = -\infty.$$

LS29. *If* $\lim_{n \to \infty} a_n = \pm\infty$ *and* $\lim_{n \to \infty} b_n = b > 0$, *then*

$$\lim_{n \to \infty} \frac{a_n}{b_n} = \pm\infty.$$

LS30. *If* $\lim_{n \to \infty} a_n = \pm\infty$ *and* $\lim_{n \to \infty} b_n = b < 0$, *then*

$$\lim_{n \to \infty} \frac{a_n}{b_n} = \mp\infty.$$

LS31. *If* $\lim_{n \to \infty} a_n = a$ *and* $\lim_{n \to \infty} b_n = b$ *and if, for all* $n > N$ *for some* N *we have* $a_n \le b_n$, *then* $a \le b$.

LS32. *If* $\lim_{n \to \infty} a_n = 0$, *where* $a_n \ge 0$ *for all* n, *and if, for all* $n > N$ *for some* N, $|b_n| \le a_n$, *then* $\lim_{n \to \infty} b_n = 0$.

LS33. *If* $a_n \le b_n \le c_n$ *for all* $n > N$ *for some* N, *and if*

$$\lim_{n \to \infty} a_n = \lim_{n \to \infty} c_n = l,$$

then

$$\lim_{n \to \infty} b_n = l.$$

LS34.

$$\lim_{n \to \infty} a_n = a$$

if and only if

$$\lim_{n \to \infty} (a_n - a) = 0.$$

These properties are not all independent. Some can be proved from others as we see for example in Exercises 8.6, 8.7 and 8.9.

8.2.3 Exercises

Exercise 8.6. Prove that LS18 can be derived from the properties given in Example 8.2.1 that occur before LS18.

Exercise 8.7. Prove that each of LS19 and LS20 can be derived from the properties given in Example 8.2.1 that occur before the respective property.

Exercise 8.8. Prove using the properties given in Example 8.2.1 that if

$$\lim_{n\to\infty} a_n = a, \quad \text{and} \quad \lim_{n\to\infty} b_n = b \neq 0,$$

then

$$\lim_{n\to\infty} \frac{a_n}{b_n} = \frac{a}{b}.$$

Exercise 8.9. Prove that each of LS21...LS34 can be derived from the properties given in Example 8.2.1 that occur before the respective property.

8.3 Ancient Greek Limits: the Method of Exhaustion

Definition 8.1.1 and Example 8.2.1 cannot constitute a theory of limits for a number of reasons. First, each has a lot of axioms, which can lead one to wonder if they are all truly independent of each other and we have seen in the exercises that they are not. Furthermore, one might wonder if they are, in fact, consistent. Finally, these axioms are incomplete: it is not possible to use these rules to prove that the sequence $1, -1, 1, -1, \ldots$, where the nth term is $(1)^{n+1}$, diverges. It seems desirable to construct a simpler theory.

Our goal here is to develop a theory of limits that would satisfy any sophist. We will approach this goal by looking at the way the ancient Greeks dealt with limits. In particular, we will look at two theorems from the ancients: Archimedes theorem giving the area of a circle, and a theorem of Euclid which says that the areas of circles are to each other as the squares of their radii. Both theorems were proved by a method due to Eudoxus.

8.3.1   On the Area of Regular Polygons

For both Archimedes' theorem and Euclid's theorem, we need a general formula for the area of a regular polygon (i.e., a polygon where all angles are equal and all sides are equal, for example, an equilateral triangle, or a square). To see how this formula is developed, let us look at a square of side s in a new way (see Figure 8.6). Instead of finding the area simply by taking

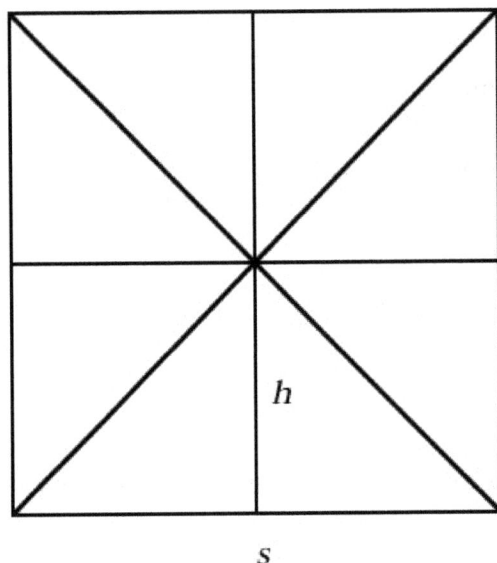

Figure 8.6: The area of a square

s^2, let us start with the bottom of the four triangles obtained by taking the diagonals. The altitude of each triangle is $h = \frac{1}{2}s$, so the area of the triangle is given by $\frac{1}{2}hs$. If we add the areas of all four triangles, we get

$$A = \frac{1}{2}h(4s) = \frac{1}{2}hp,$$

where $p = 4s$ is the perimeter. If we substitute the value of h and p in terms of s into this formula, we get s^2, which is what we expect.

Now let us consider a regular octagon (see Figure 8.7). If we divide it into triangles the same way, we get eight triangles, each of whose areas is $\frac{1}{2}hs$. If we take all eight triangles and note that here $p = 8s$, we get for the

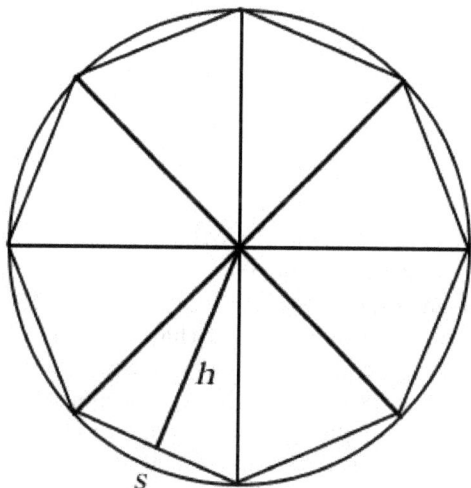

Figure 8.7: The area of a regular octagon/circle

area

$$A = \frac{1}{2}h(8s) = \frac{1}{2}hp.$$

This should be enough to establish the result that:

> The area of any regular polygon is one-half the altitude to a side times the perimeter, or $\frac{1}{2}hp$.

Now what about a circle? If we take regular polygons and keep increasing the number of sides, the perimeter will approach the circumference C and the altitude will approach the radius r (see Figure 8.7). This suggests that the formula for the area of a circle should be

$$A = \frac{1}{2}rC.$$

And since π is defined to be the ratio of the circumference of a circle to its diameter, or, what amounts to the same thing, the ratio of the circumference to twice its radius, we have

$$\pi = \frac{C}{2r},$$

from which our familiar formula $C = 2\pi r$ follows. If we now substitute this into the above formula for the area of a circle, we get

$$A = \frac{1}{2}r(2\pi r) = \pi r^2,$$

the formula we all learned in school.

This must have seemed obvious to the ancient Greeks from an early period in the history of their geometry. But how could they prove it? At one time some of them argued that a circle is a regular polygon with infinitely many sides, but they eventually decided that this kind of reasoning is not immune to attacks by sophists. For just because regular polygons with an increasing number of sides seems to be approaching a circle, we are not automatically justified in deducing this formula for the area of a circle. For evidence like this can be misleading. Consider the following example (see Figure 8.8): The length of the stepped line is clearly $2s$ no matter how many

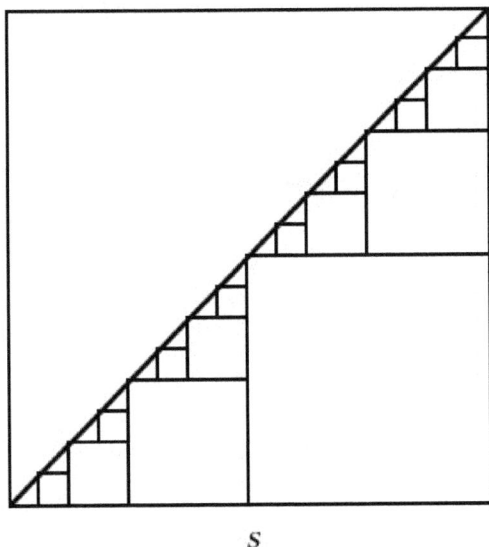

s

Figure 8.8: The length of the stepped line

steps there are. But as the number of steps increases, the stepped line seems to approach the diagonal, and the length of the diagonal is $\sqrt{2}\, s \neq 2s$.

8.3.2 Euclid on Areas of Circles and Squares

Although it must have been obvious to the ancient Greeks that the area of a circle is, in our terms, given by the formula

$$A = \frac{1}{2} r C,$$

where r is the radius and C is the circumference, it was a long time before a proof was given of this fact. And before that proof was given, Euclid proved that the areas of circles have the same proportion as the squares on their diameters. This is Proposition 2 of Book XII of Euclid's *Elements*. Its proof uses Proposition 1 of Book XII, which states that *Similar polygons inscribed in circles are to one another as the squares on the diameters of the circles.*

PROPOSITION 1 OF BOOK XII OF THE *Elements*.
Similar polygons inscribed in circles are to one another as the squares on the diameters of the circles.

Euclid's proof of Proposition 1 of Book XII involves a number of concepts and results that we have not seen and which are not directly relevant, so there is no point in reproducing his proof here. Instead, we can see that it is true using modern ideas. Let us start with the following fact related to similar polygons. Here, you need to recall Definition 7.2.3 where we defined similar figures to be those which have the same shape and said that in similar polygons the corresponding angles are equal and the corresponding sides all have the same proportion. For these similar polygons, we have the following properties:

The areas of similar polygons are proportional to all the following:

- The squares of their altitudes.

- The squares of their perimeters.

- The squares of any of their linear parts.

To see this, start with two similar *regular* polygons. If the altitudes and perimeters of two similar regular polygons are, respectively, h_1 and p_1 and h_2 and p_2, then since the polygons are similar, we have

$$\frac{p_1}{p_2} = \frac{h_1}{h_2}.$$

Then since

$$A_1 = \frac{1}{2}h_1 p_1 \quad \text{and} \quad A_2 = \frac{1}{2}h_2 p_2,$$

we have that

$$\frac{A_1}{A_2} = \frac{h_1 p_1}{h_2 p_2} = \frac{h_1}{h_2}\frac{p_1}{p_2} = \frac{h_1}{h_2}\frac{h_1}{h_2} = \frac{h_1^2}{h_2^2}.$$

This shows that the areas of similar regular polygons are proportional to the squares of their altitudes. Similarly, they are proportional to the squares of their perimeters. Indeed, they are proportional to the squares of any of their linear parts.[1] In fact, this is true of all similar polygons, regular or not.

Now we move to the proof of Proposition 1 of Book XII of Euclid.

Proof. Since we know that the areas of similar polygons are proportional to the squares of any of their linear parts, it remains to show that the diameters of the circles in which two similar polygons are inscribed are in the same proportion to any of their linear parts.

Let two similar polygons $ACDEFG$ and $A'C'D'E'F'G'$ inscribed respectively in the circles whose diameters are AB and $A'B'$. Join AF, BG, $A'F'$ and $B'G'$.

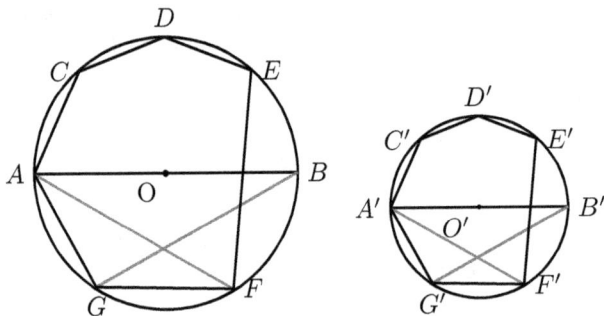

Recall Exercise 2.5 where we showed that $\angle AGB = \angle A'G'B' = 90°$ and also Exercise 2.6 where we showed that $\angle ABG = \angle AFG = 2\angle AOG$ and $\angle A'B'G' = \angle A'F'G' = 2\angle A'O'G'$.

Since $ACDEFG$ and $A'C'D'E'F'G'$ are similar, then the triangles AGF and $A'G'F'$ are similar and hence $\angle GFA = \angle G'F'A'$.

But, $\angle GFA = \angle GBA$ and $\angle G'F'A' = \angle G'B'A'$ and $\angle AGB = \angle A'G'B' = 90°$. Hence, the two triangles AGB and $A'G'B'$ are similar and hence $\frac{AB}{A'B'} = \frac{AG}{A'G'}$. Therefore, the diameters AB and $A'B'$ of the circles in

---

[1]For examples, see Exercise 8.10.

which the two similar polygons $ACDEFG$ and $A'C'D'E'F'G'$ are inscribed are in the same proportion to their linear parts AG and $A'G'$ and we are done. □

Now let us proceed to Book XII Proposition 2:

PROPOSITION 2 OF BOOK XII OF THE *Elements*.
Circles are to one another as the squares on the diameters.

Euclid starts his proof as follows:

Let $ABCD$, $EFGH$ be circles, and BD, FH their diameters; I say that, as the circle $ABCD$ is to the circle $EFGH$, so is the square on BD to the square on FH.

For, if the square on BD is not to the square on FH as the circle $ABCD$ is to the circle $EFGH$, then, as the square on BD is to the square on FH, so will the circle $ABCD$ be either to some less area than the circle $EFGH$ or to a greater.

Euclid's strategy is to prove his result by contradiction. In fact, it will be a double proof by contradiction. He will first assume that it will be in the ratio to a smaller area S, derive a contradiction from that, then assume that it will be in the ratio to a larger area S, and then derive a contradiction from that area as well. As a result, the only possibility left will be the result stated in the proposition.

First, let it be in that ratio to a less area S.

Let the square $EFGH$ be inscribed in the circle $EFGH$; then the inscribed square is greater than half of the circle $EFGH$, in as much as, if through the points E, F, G, H we draw tangents to the circle, the square $EFGH$ is half the square circumscribed about the circle, and the circle is less than the circumscribed square;

hence the inscribed square $EFGH$ is greater than half of the circle $EFGH$.

Note that Euclid feels the need to give a proof that the area of the inscribed square is more than half the area of the circle. He shows this by noting that the circumscribed square, which includes area outside the circle, has twice the area of the inscribed square (see Exercise 2.7).

Let the circumferences EF, FG, GH, HE be bisected at the points K, L, M, N, and let EK, KF, FL, LG, GM, MH, HN, NE be jointed;

therefore each of the triangles EKF, FLG, GMH, HNE is also greater than the half of the segment of the circle about it, inasmuch as, if through the points K, L, M, N we draw tangents to the circle and complete the parallelograms on the straight lines EF, FG, GH, HE, each of the triangles EKF, FLG, GMH, HNE will be half of the parallelogram about it, while the segment about it is less than the parallelogram; hence each of the triangles EKF, FLG, GMH, HNE is greater than the half of the segment of the circle about it (see Exercise 2.8).

Note how carefully Euclid proves that each time the circumference is bisected (in effect inscribing a new regular polygon with twice the number of sides as the previous one), we use up more than half the area inside the circle but outside the previous polygon.

> Thus, by bisecting the remaining circumferences
> and joining straight lines, and by doing this con-
> tinually, we shall leave some segments of the circle
> which will be less than the excess by which the cir-
> cle $EFGH$ exceeds the area S.
>
> For it was proved in the first theorem of the tenth
> book that, if two unequal magnitudes be set out,
> and if from the greater there be subtracted a mag-
> nitude greater than the half, and from that which
> left a greater than the half, and if this be done con-
> tinually, there will be left some magnitude which
> will be less than the lesser magnitude set out.

Euclid needs to give a proof that if the number of sides of the inscribed polygon is doubled over and over again, eventually the area inside the circle but outside the polygon will be less than the difference between the area of the circle and S. Without this proof, a sophist could find a place to attack his argument.

In our terms, the argument goes as follows: let a and b be two magnitudes with $b > a$. Let an amount greater $c_1 > \frac{1}{2}b$ be subtracted from b to leave $b_1 = b - c_1$, and then let $c_2 > \frac{1}{2}b_1$ be subtracted from b_1 to leave $b_2 = b_1 - c_2$, etc. At each step, starting with b_i, we let $c_{i+1} > \frac{1}{2}b_i$ and then let $b_{i+1} = b_i - c_{i+1}$. Eventually, we will reach a value b_n which is less than a.

The proof of this proposition depends on the fact that, given two magnitudes a and b with $b > a$, there is a number n such that $na > b$. From this, the proof proceeds (in modern algebraic notation) as follows: Let $b = b_0$. Let $c_1 > (1/2)b_0$, $b_1 = b_0 - c_1$, $c_2 > (1/2)b_1$, etc., so that $c_k > (1/2)b_{k-1}$ and $b_k = b_{k-1} - c_k$, etc., and keep going until $k = n - 1$. Then since $a \leq (1/2)na$ but $c_1 > (1/2)b_0$, and $na > b$, $(n-1)a > b_1$. Similarly, $(n-2)a > b_2$, and so on until we get to $a > b_{n-1}$, which is the desired conclusion.

> Let segments be left such as described, and let the segments of the circle $EFGH$ on EK, KF, FL, LG, GM, MH, HN, NE be less than the excess by which the circle $EFGH$ exceeds the area S.
>
> Therefore the remainder, the polygon $EKFLGMHN$, is greater than the area S.
>
> Let there be inscribed, also, in the circle $ABCD$ the polygon $AOBPCQDR$ similar to the polygon $EKFLGMHN$;
> therefore, as the square on BD is to the square on FH, so is the polygon $AOBPCQDR$ to the polygon $EKFLGMHN$.   [XII.1]

If we use the notation $\bigcirc A$ for the area of circle $ABCD$, $\bigcirc E$ for the area of circle $EFGH$, PA for the area of the polygon $AOBPCQDR$ and PE for the area of the polygon $EKFLGMHN$, this says, in our terms,

$$\frac{BD^2}{FH^2} = \frac{PA}{PE}.$$

> But as the square on BD is to the square on FH, so also is the circle $ABCD$ to the area S;
> therefore also, as the circle $ABCD$ is to the area S, so is the polygon $AOBPCQDR$ to the polygon $EKFLGMHN$;   [V.11]

This says that

$$\frac{BD^2}{FH^2} = \frac{\bigcirc A}{S},$$

and therefore

$$\frac{\bigcirc A}{S} = \frac{PA}{PE}.$$

> But the circle $ABCD$ is greater than the polygon inscribed in it;
> therefore, the area S is also greater than the polygon $EKFLGMHN$.

This says that $\bigcirc A > PA$ and therefore $S > PE$.

But it is also less;

which is impossible.

Therefore, as the square on BD is to the square on FH, so is not the circle $ABCD$ to any area less than the circle $EFGH$.

Similarly we can prove that neither is the circle $EFGH$ to any area less than the circle $ABCD$ as the square on FH is to the square on BD.

I say next that neither is the circle $ABCE$ to any area greater than the circle $EFGH$ as the square on BD is to the square on FH.

For, if possible, let it be in that ratio to a greater area S.

Therefore, inversely, as the square on FH is to the square on DB, so is the area S to the circle $ABCD$.

But, as the area S is to the circle $ABCD$, so is the circle $EFGH$ to some area less than the circle $ABCD$;

therefore also, as the square on FH is to the square on BD, so is the circle $EFGH$ to some area less than the circle $ABCD$:   [v.11]

which was proved impossible.

Therefore, as the square on BD is to the square on FH, so is not the circle $ABCD$ to any area greater than the circle $EFGH$.

And it was proved that neither is it in that ratio to any area less than the circle $EFGH$;

therefore, as the square on BD is to the square on FH, so is the circle $ABCD$ to the circle $EFGH$.

Therefore etc.   Q.E.D.

LEMMA.

I say that, the area S being greater than the circle $EFGH$, as the area S is to the circle $ABCD$, so is the circle $EFGH$ to some area less than the circle $ABCD$.

For let it be contrived that, as the area S is to the circle $ABCD$, so is the circle $EFGH$ to the area T.

I say that the area T is less than the circle $ABCD$.

For since, as the area S is to the circle $ABCD$, so is the cirdle $EFGH$ to the area T.   [v.16]

But the area is greater than the circle $EFGH$;

therefore the circle $ABCD$ is also greater than the area T.

Hence, as the area S is to the circle $ABCD$, so is the circle $EFGH$ to some area less than the circle $ABCD$.   Q.E.D.

To put this argument into modern algebraic notation, let the given circles have areas a and b respectively, and let the corresponding ratio of the squares of their diameters be k. Let the areas of the polygons inscribed in the circle with area a have areas $a_1, a_2, \ldots$. Let the polygons inscribed in the other circle have areas $b_1, b_2, \ldots$. We have

$$0 < a_1 < a_2 < \ldots < a_n < \ldots < a,$$

and

$$0 < b_1 < b_2 < \ldots < b_n < \ldots < b.$$

Furthermore, for each n, we have

$$k = \frac{a_n}{b_n},$$

So that

$$\frac{a_n}{k} = b_n.$$

In addition, we have for each n that

$$(a - a_{n+1}) < \frac{1}{2}(a - a_n), \qquad\qquad (b - b_{n+1}) < \frac{1}{2}(b - b_n).$$

We want to prove

$$k = \frac{a}{b}.$$

Now if $k \neq \frac{a}{b}$, then $k = \frac{a}{S}$, where $S < b$ or $S > b$.
I. Suppose $S < b$. Choose N so that

$$b - b_N < b - S.$$

The number N represents the number of times the number of sides of the inscribed polygon was doubled. Then

$$S < b_N.$$

But

$$S = \frac{a}{k} > \frac{a_N}{k} = b_N,$$

a contradiction.

II. Suppose $S > b$. This is similar to case I with a and b reversed.
It follows that

$$k = \frac{a}{b}.$$

Remark 8.3.1. *This proof depends on our being able to find an N so that $a - a_N < a - S$ or $b - b_N < b - S$. This, in turn, depends on*

$$(a - a_{n+1}) < \frac{1}{2}(a - a_n), \qquad\qquad (b - b_{n+1}) < \frac{1}{2}(b - b_n).$$

From this we get easily

$$(a - a_{n+1}) < \left(\frac{1}{2}\right)^n (a - a_1), \qquad\qquad (b - b_{n+1}) < \left(\frac{1}{2}\right)^n (b - b_1).$$

In other words, $(a - a_n)$ and $(b - b_n)$ can be made as small as we please. Another way to say this is

1. *For each $\epsilon > 0$, there is an N such that for all $n > N$, $|a - a_n| < \epsilon$.*

2. *For each $\epsilon > 0$, there is an N such that for all $n > N$, $|b - b_n| < \epsilon$.*

Shorter Proof. Let $\epsilon > 0$ be given. Choose N so that for all $n > N$,

$$|b - b_n| < \frac{1}{|k|}\epsilon.$$

Then

$$|kb - a_n| = |kb - kb_n| = |k||b - b_n| < \epsilon.$$

Hence, kb is the limit of the sequence $\{a_n\}$, and since a sequence can have only one limit, $kb = a$.

8.3.3 Archimedes' Measurement of a Circle

Getting back to the area of a circle, the result in question was finally proved by Archimedes in a book called (in English) "Measurement of a Circle," which can be found in [19].

PROPOSITION 1 OF ARCHIMEDES'S BOOK
"MEASUREMENT OF A CIRCLE".

The area of any circle is equal to a right-angled triangle in which one of the sides about the right triangle is equal to the radius, and the other to the circumference of the circle.

Let $ABCD$ be the given circle, K the triangle described.

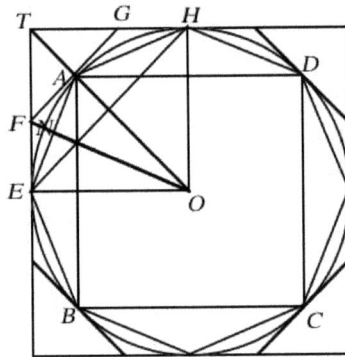

Then, if the circle is not equal to K, it must be either greater or less.

I. If possible, let the circle be greater than K.

Inscribe a square $ABCD$, bisect the arcs AB, BC, CD, DA, and then bisect (if necessary) the halves, and so on, until the sides of the inscribed polygon whose angular points are the points of division subtend segments whose sum is less than the excess of the circle over K.

At this point, Archimedes is relying on the construction of Euclid as quoted above, and he is not repeating all the details.

Thus the area of the polygon is greater than K.

Let AE be any side of it, and ON the perpendicular on AE from the centre O.

Then ON is less than the radius of the circle and therefore less than one of the sides about the right angle of the triangle K. Also, the perimeter of the polygon is less than the circumference of the circle, i.e. less than the other side about the right angle in K.

Therefore the area of the polygon is less than K; which is inconsistent with the hypothesis.

Thus the area of the circle is not greater than K.

II. If possible, let the circle be less than K.

Circumscribe a square, and let two adjacent sides, touching the circle in E, H, meet in T. Bisect the arcs between adjacent points of contact and draw tangents at the points of bisection. Let A be the middle point of the arc EH and FAG the tangent at A.

Then the angle TAG is a right angle.

Therefore

$$TG \; > \; GA$$
$$> \; GH.$$

It follows that the triangle FTG is greater than half the area $TEAH$.

Similarly, if the arc AH be bisected and the tangent at the point of bisection be drawn, it will cut off from the area GAH more than one-half.

Thus, by continuing the process, we shall ultimately arrive at a circumscribed polygon such that the spaces intercepted between it and the circle are together less than the excess of K over the area of the circle.

Thus the area of the polygon will be less than K.

Now, since the perpendicular from O on any side of the polygon is equal to the radius of the circle, while the perimeter of the polygon is greater than the circumference of the circle, it follows that the area of the polygon is greater than the triangle K; which is impossible.

Therefore the area of the circle is not less than K.

Since the area of the circle is neither greater nor less than K, it is equal to it.

Let us now put this proof into modern algebraic notation. We want to prove that

$$A = \frac{1}{2}rC,$$

where A is the area of the circle and C its circumference.

Let $K = \frac{1}{2}rC$ (the area of the triangle). If $A \neq K$, then we must have $A > K$ or $A < K$.

I. Suppose $A > K$. Inscribe a square, and let its side be s_1, the altitude to the side be h_1, and its perimeter be p_1. The area of the square is

$$a_1 = \frac{1}{2}h_1 p_1.$$

Now, double the number of sides of the inscribed polygon, and keep doubling it. For polygon n, the side is s_n, the altitude to the side is h_n, and the perimeter is p_n; hence the area is

$$a_n = \frac{1}{2}h_n p_n.$$

From the geometry of the situation, we have that

$$h_1 < h_2 < \ldots < h_n < \ldots r,$$

$$p_1 < p_2 < \ldots < p_n < \ldots < C,$$

and

$$a_1 < a_2 < \ldots < a_n < \ldots < A.$$

Now choose N so that

$$A - a_N < A - \frac{1}{2}rC.$$

It follows that

$$\frac{1}{2}rC < a_N.$$

But since $h_N < r$, $p_N < C$, and $a_N = \frac{1}{2}h_N p_N$, we have

$$a_N < \frac{1}{2}rC,$$

a contradiction.

II. Suppose, on the contrary, that $A < K$. Circumscribe a square, and let its perimeter be P_1; then the area is

$$A_1 = \frac{1}{2}rP_1.$$

Now double the number of sides of the circumscribed figure, and keep doing it. If, for the nth polygon, the perimeter is P_n, then the area is

$$A_n = \frac{1}{2} r P_n.$$

From the geometry, we have

$$C < \ldots < P_n < \ldots < P_2 < P_1$$

and

$$A < \ldots A_n < \ldots < A_2 < A_1.$$

Choose N so that

$$A_N - A < \frac{1}{2} r C - A.$$

Then

$$A_N < \frac{1}{2} r C.$$

But $C < P_N$ and $A_N = \frac{1}{2} r P_N$, so

$$\frac{1}{2} r C < A_N,$$

another contradiction.

It follows that $A = K = \frac{1}{2} r C$.

Summary of argument. We are given that

$$h_1 < h_2 < \ldots < h_n < \ldots < r,$$

$$p_1 < p_2 < \ldots < p_n < \ldots < C < \ldots P_n < \ldots P_2 < P_1,$$

and

$$\tfrac{1}{2} h_1 p_1 < \tfrac{1}{2} h_2 p_2 < \ldots < \tfrac{1}{2} h_n p_n < \quad \ldots < A <$$
$$\ldots < \tfrac{1}{2} r P_n < \ldots < \tfrac{1}{2} r P_2 < \tfrac{1}{2} r P_1.$$

We want to prove that

$$A = \frac{1}{2} r C.$$

The inequalities we are given imply that for every n,

$$\frac{1}{2} h_n p_n < \frac{1}{2} r C < \frac{1}{2} r P_n.$$

Furthermore, if $A \neq \frac{1}{2} r C$, then:

1. there is an N such that

$$A - \frac{1}{2}h_N p_N < \left| A - \frac{1}{2}rC \right|;$$

2. there is an M such that

$$\frac{1}{2}rP_M - A < \left| A - \frac{1}{2}rC \right|.$$

I. Suppose $A > \frac{1}{2}rC$. Then

$$\left| A - \frac{1}{2}rC \right| = A - \frac{1}{2}rC.$$

Hence, $\frac{1}{2}rC < \frac{1}{2}h_N p_N$, a contradiction.
 II. Suppose $A < \frac{1}{2}rC$. Then

$$\left| A - \frac{1}{2}rC \right| = \frac{1}{2}rC - A.$$

Hence, $\frac{1}{2}rP_M < \frac{1}{2}rC$, another contradiction.
 It follows that $A = \frac{1}{2}rC$.

Shorter summary of argument Give $\left| A - \frac{1}{2}rC \right|$ a name, say ϵ. Then we have:

1. For each $\epsilon > 0$, there is N such that

$$A - \frac{1}{2}h_N p_N < \epsilon.$$

2. For each $\epsilon > 0$, there is N such that

$$\frac{1}{2}rP_N - A < \epsilon.$$

Now let $\epsilon > 0$ be given. Then there are N_1 and N_2 such that

1. if $n > N_1$,

$$A - \frac{1}{2}h_n p_n < \frac{\epsilon}{2},$$

2. if $n > N_2$,

$$\frac{1}{2}rP_n - A < \frac{\epsilon}{2},$$

Hence, if N is the maximum of N_1 and N_2, then

$$\frac{1}{2}rP_N - \frac{1}{2}h_N p_N < \frac{\epsilon}{2} + \frac{\epsilon}{2} = \epsilon.$$

This is true for any $\epsilon > 0$. Furthermore, by the inequalities,

$$\frac{1}{2}h_N p_N < A < \frac{1}{2}rP_N$$

and

$$\frac{1}{2}h_N p_N < \frac{1}{2}rC < \frac{1}{2}rP_N.$$

It follows almost immediately that

$$\left| A - \frac{1}{2}rC \right| < \epsilon$$

for every $\epsilon > 0$, and hence $A = \frac{1}{2}rC$.

Note that the last part of this argument has essentially the following form: given that for every n, we have $A_n < a < B_n$ and $A_n < b < B_n$, and that for every $\epsilon > 0$, there is an N such that for all $n > N$, $B_n - A_n < \epsilon$, to prove $a = b$. To give a complete proof is to use the original proof of Archimedes: Suppose $a \neq b$. Then $a < b$ or $a > b$. If $a < b$, let $\epsilon = b - a$, and let $n > N$; we have $A_n < a < b < B_n$, and hence $\epsilon = b - a < B_n - A_n$, and this contradicts $B_n - A_n < \epsilon$. The case of $a > b$ is similar.

8.3.4 Exercises

Exercise 8.10. Let an equilateral triangle ABC whose side length is l. Draw the altitude AD. Then, draw the square $ADEF$. Give the all the angles of the triangles ABD and ADC, the length of AD, and the areas of the square $ADEF$ and of the equilateral triangle ABC.

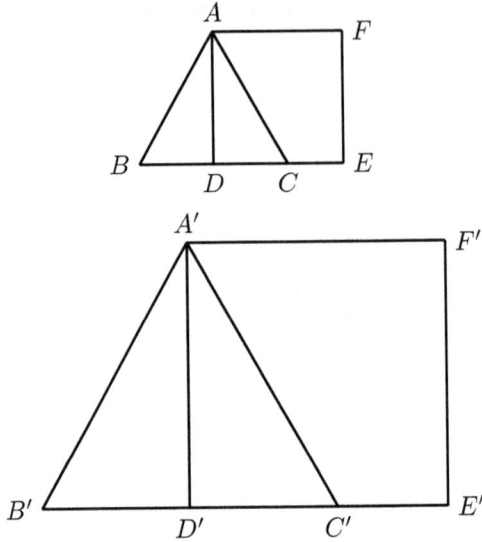

Repeat all the above but for an equilateral triangle $A'B'C'$ whose side length is $2l$, the drawn square is $A'D'E'F'$ and the altitude is $A'D'$.

Show that ABC is similar to $A'B'C'$ and that $ADEF$ is similar to $A'D'E'F'$.

Give the ratio of the area of the triangle ABC to the area of $A'B'C'$ and of the the area of the square $ADEF$ and the area of $A'D'E'F'$.

Chapter 9

The Theory of Limits

We start this chapter by using the method of exhaustion to suggest a definition of the limit of a sequence which we will use to prove the properties of limits of sequences discussed in the earlier chapter. We then demonstrate how to construct proofs in the theory of limits starting with scratch work that goes backwards. We then do for limits of functions what we have already done in this chapter for limits of sequences. All this is done without specifying that we are dealing with real numbers. This corresponds to the historical order of the development of calculus and analysis in European mathematics. The ancient Greeks had distinguished between numbers and magnitudes, and Descartes had shown that both could be considered together as quantities. But a rigorous theory of these quantities had not been developed, and many questions about them had not been answered. Most of the elementary calculus was developed before the real number system was defined. In fact, it had not even been settled whether or not there were infinitely small quantities; for example, for an indication that Cauchy **thought** that there were infinitely small quantities, see [27].

9.1 Limits of Sequences: Theory

In what follows, we will distinguish a sequence from its general term by writing $\{a_n\}$ for the sequence and a_n for its general term.

The method of exhaustion we saw in Section 8.3 suggests the following definition of the limit of a sequence:

Definition 9.1.1. *A sequence* $\{a_n\}$ *approaches* a limit a *as* $n \to \infty$,

$$\lim_{n \to \infty} a_n = a,$$

if and only if, for every $\epsilon > 0$, there is a natural number N such that for each $n > N$,

$$|a_n - a| < \epsilon.$$

Example 9.1.2. *Note that with this definition, we can prove that the sequence $\{(-1)^{n+1}\}$ diverges. For let ϵ be chosen so that $0 < \epsilon < \frac{1}{2}$. Then, no matter what value l we choose as a limit and no matter what value of N we have, there are (infinitely many) values of $n > N$ for which $|a_n - l| > \epsilon$. Hence, the sequence has no limit.*

For divergence to one of the infinities, the definitions are as follows:

Definition 9.1.3. *A sequence $\{a_n\}$ is said to* diverge to ∞,

$$\lim_{n \to \infty} a_n = \infty,$$

if, for every $M > 0$, there is an N such that for each $n > N$,

$$a_n > M.$$

Definition 9.1.4. *A sequence $\{a_n\}$ is said to* diverge to $-\infty$,

$$\lim_{n \to \infty} a_n = -\infty,$$

if, for every $M < 0$, there is an N such that for each $n > N$,

$$a_n < M.$$

We should now prove from these definitions all the properties of limits of sequences given in Example 8.2.1. Here are some examples:

LS1. If k is any real number.

$$\lim_{n \to \infty} k = k.$$

Proof. Let $\epsilon > 0$ be given. Let $N = 1$. Then, for all $n > N$, $a_n = k$, and so

$$|a_n - k| = |k - k| = 0 < \epsilon.$$

$\square$

LS2. If $\lim_{n \to \infty} a_n = a$ and $\lim_{n \to \infty} b_n = b$, then

$$\lim_{n \to \infty} (a_n + b_n) = a + b.$$

Proof. By the hypotheses,

1. For every $\epsilon > 0$, there is an N such that, for $n > N$, $|a_n - a| < \epsilon$.

2. For every $\epsilon > 0$, there is an N such that, for $n > N$, $|b_n - b| < \epsilon$.

Let $\epsilon > 0$ be given. Then there are N_1 and N_2 such that

1. for $n > N_1$, $|a_n - a| < \frac{\epsilon}{2}$, and

2. for $n > N_2$, $|b_n - b| < \frac{\epsilon}{2}$.

Let N be the maximum of N_1 and N_2. Then for $n > N$,

$$
\begin{aligned}
|(a_n + b_n) - (a + b)| &= |(a_n - a) + (b_n - b)| \\
&\leq |a_n - a| + |b_n - b| \\
&< \tfrac{\epsilon}{2} + \tfrac{\epsilon}{2} = \epsilon.
\end{aligned}
$$

$\square$

LS3. *If* $\lim_{n \to \infty} a_n = a$ *and* $\lim_{n \to \infty} b_n = b$, *then*

$$
\lim_{n \to \infty} (a_n b_n) = ab.
$$

Proof. By the hypotheses,

1. For every $\epsilon > 0$, there is N such that for all $n > N$, $|a - a_n| < \epsilon$, and

2. For every $\epsilon > 0$, there is N such that for all $n > N$, $|b - b_n| < \epsilon$.

Let $\epsilon > 0$ be given. Then,

1. there is N_1 such that for all $n > N_1$, $|a - a_n| < \dfrac{\epsilon}{2(|b| + 1)}$, and

2. there is N_2 such that for all $n > N_2$, $|b - b_n| < \dfrac{\epsilon}{2\left(\dfrac{\epsilon}{2(|b| + 1)} + |a| \right)}$.

(Here, we use $|b| + 1$ instead of $|b|$ since b might be 0.) Let N be the maximum of N_1 and N_2. Then for all $n > N$, we have

$$
|a - a_n| < \frac{\epsilon}{2(|b| + 1)} \qquad \text{and} \qquad |b - b_n| < \frac{\epsilon}{2\left(\dfrac{\epsilon}{2(|b| + 1)} + |a| \right)}.
$$

Hence

$$
\begin{aligned}
|a_n b_n - ab| &= |a_n b_n - a_n b + a_n b - ab| \\
&= |a_n(b_n - b)| + |(a_n - a)b| \\
&\leq |a_n||b_n - b| + |a_n - a||b| \\
&= |(a_n - a) + a||b_n - b| + |a_n - a||b| \\
&\leq (|a_n - a| + |a|)|b_n - b| + |a_n - a||b| \\
&< \left(\frac{\epsilon}{2(|b| + 1)} + |a|\right)\frac{\epsilon}{2\left(\dfrac{\epsilon}{2(|b| + 1)} + |a|\right)} + \frac{\epsilon}{2(|b| + 1)}|b| \\
&< \frac{\epsilon}{2} + \frac{\epsilon}{2(|b| + 1)}(|b| + 1) \\
&= \frac{\epsilon}{2} + \frac{\epsilon}{2} \\
&= \epsilon.
\end{aligned}
$$

$\square$

LS4. If $\lim_{n \to \infty} a_n = a \neq 0$, then

$$
\lim_{n \to \infty} \frac{1}{a_n} = \frac{1}{a}.
$$

Proof. Let $\epsilon > 0$ be given. Let ϵ' be the minimum of $\frac{a^2}{2}\epsilon$ and $|a|/2$. Then, by the hypothesis there is an N such that for all $n > N$,

$$
|a_n - a| < \epsilon'.
$$

For these values of $n > N$, we have $|a_n| \geq \frac{|a|}{2}$, for if $|a_n| < \frac{|a|}{2}$, we would have

$$
|a| = |(a - a_n) + a_n| \leq |a - a_n| + |a_n| < \frac{|a|}{2} + \frac{|a|}{2} = |a|,
$$

which is a contradiction. Hence, we have

$$
\begin{aligned}
\left|\frac{1}{a_n} - \frac{1}{a}\right| &= \frac{|a_n - a|}{|a||a_n|} \\
&\leq \frac{2}{a^2}|a_n - a| \\
&< \frac{2}{a^2}\frac{a^2}{2}\epsilon \\
&= \epsilon.
\end{aligned}
$$

$\square$

LS5. If $a_n \leq 0$ for all $n > N$ for some N, and if $\lim_{n \to \infty} a_n = a$, then $a \leq 0$.

Proof. Suppose $a > 0$. By the second hypothesis, for every $\epsilon > 0$ there is an N' such that for $n > N'$, $|a_n - a| < \epsilon$. This will apply if $\epsilon = a$, so we will have that for $n > N'$, $|a_n - a| < a$; the latter implies $0 < a_n < 2a$. Let N^* be the maximum of N and N'. Then for $n > N^*$, we have both $a_n \leq 0$ and $0 < a_n < 2a$, a contradiction. Hence, $a \leq 0$. □

LS8. If a_n is defined for all $n > N$ for some N, and if $\lim_{n \to \infty} a_n$ exists, then there is a g, independent of n, such that

$$|a_n| \leq g \text{ for all } n > N.$$

Proof. Let $a = \lim_{n \to \infty} a_n$. Then there is $N' > N$ such that for all $n > N'$,

$$|a_n - a| < 1.$$

Then

$$
\begin{aligned}
|a_n| &= |a + (a_n - a)| \\
&\leq |a| + |a_n - a| \\
&< |a| + 1.
\end{aligned}
$$

Hence, if g is defined by

$$g = |a| + 1 + M$$

where M is the maximum of $|a_n|$ for $N < n < N'$, then for all $n > N$, $|a_n| < g$. □

LS9. If $a_n \leq b_n \leq c_n$ for $n \geq N$ for some N, and if

$$\lim_{n \to \infty} a_n = \lim_{n \to \infty} c_n = l,$$

then $\lim_{n \to \infty} b_n$ exists and $\lim_{n \to \infty} b_n = l$.

Proof. Let $\epsilon > 0$ be given. Then there are N_1 and N_2 such that

1. for $n > N_1$, $|a_n - l| < \epsilon$, and
2. for $n > N_2$, $|c_n - l| < \epsilon$.

Let N' be the maximum of N, N_1 and N_2. Then for $n > N'$, we have

$$l - \epsilon < a_n \leq b_n \leq c_n < l + \epsilon,$$

from which it follows that $|b_n - l| < \epsilon$. $\square$

LS10. If $|x| < 1$, then

$$\lim_{n \to \infty} x^n = 0.$$

Proof. If $x = 0$, the result is follows by LS1. So suppose $x \neq 0$. Then $0 < |x| < 1$. It follows that

$$\frac{1}{|x|} > 1,$$

so let

$$p = \frac{1}{|x|} - 1.$$

Then $p > 0$ and, in addition, since

$$\frac{1}{|x|} = 1 + p,$$

we have

$$|x| = \frac{1}{1 + p}.$$

Also

$$
\begin{aligned}
(1+p)^n &> (1+p)^n - 1 = \\
(1+p)^n + \textstyle\sum_{i=1}^{n-1}(1+p)^i - \sum_{i=1}^{n-1}(1+p)^i - 1 &= \\
(1+p)^n + \textstyle\sum_{i=1}^{n-1}(1+p)^i - \sum_{i=1}^{n-1}(1+p)^i - (1+p)^0 &= \\
(1+p)^n + \textstyle\sum_{i=1}^{n-1}(1+p)^i - \sum_{i=0}^{n-1}(1+p)^i & \\
(1+p)[(1+p)^{n-1} + \textstyle\sum_{i=0}^{n-2}(1+p)^i] - \sum_{i=0}^{n-1}(1+p)^i &= \\
(1+p)[\textstyle\sum_{i=0}^{n-1}(1+p)^i] - \sum_{i=0}^{n-1}(1+p)^i &= \\
(1+p-1)[\textstyle\sum_{i=0}^{n-1}(1+p)^i] &= \\
p[\textstyle\sum_{i=0}^{n-1}(1+p)^i] &= \\
p[\underbrace{(1+p)^{n-1} + (1+p)^{n-2} + \cdots + 1}_{n}] &\geq
\end{aligned}
$$

pn.

It follows that

$$0 < |x|^n = \left(\frac{1}{1+p}\right)^n = \frac{1}{(1+p)^n} < \frac{1}{pn}.$$

Now let $\epsilon > 0$ be given, and let N be the first integer greater than $\frac{1}{p\epsilon}$. Then

$$\epsilon > \frac{1}{pN},$$

so for $n > N$,

$$\epsilon > \frac{1}{pn} > |x|^n = |x^n|.$$

□

LS11.

$$\lim_{n \to \infty} n = \infty.$$

Proof. Let $M > 0$ be given. Let N be the least integer greater than M. Then for every $n > N$, $n > M$, and the conclusion follows by Definition 9.1.3. □

LS12. If $\lim_{n \to \infty} a_n = \pm\infty$ and $\lim_{n \to \infty} b_n = b$, or if $\lim_{n \to \infty} a_n = a$ and $\lim_{n \to \infty} b_n = \pm\infty$, or if $\lim_{n \to \infty} a_n = \pm\infty$ and $\lim_{n \to \infty} b_n = \pm\infty$, then

$$\lim_{n \to \infty} (a_n + b_n) = \pm\infty.$$

Proof. This is really six results, one for each set of premises.

1. The hypotheses are $\lim_{n \to \infty} a_n = \infty$ and $\lim_{n \to \infty} b_n = b$. By the hypotheses,

   (a) for every $M > 0$, there is N such that for $n > N$, $a_n > M$,
   (b) for every $\epsilon > 0$, there is N such that for $n > N$, $|b_n - b| < \epsilon$.

   Let $M > 0$ be given. Then there are N_1 and N_2 such that

   (a) For $n > N_1$, $a_n > M + |b| + 1$; and
   (b) For $n > N_2$, $|b_n - b| < \epsilon$, where if $b = 0$, $\epsilon = \frac{1}{2}$ and if $b \neq 0$, ϵ is the minimum of $\frac{1}{2}$ and $|b|$.[1]

   Let N be the maximum of N_1 and N_2. Then for $n > N$,

   $$a_n + b_n > M + |b| + 1 + b_n.$$

---

[1] The last condition guarantees that if b is not 0, then b and b_n have the same sign.

This is clearly greater than M if b_n is 0 or positive, so suppose b_n is negative. Then b is either 0 or negative. If b is 0, then $|b_n - b| = |b_n| < \frac{1}{2}$, so, since b_n is negative, $b_n > -\frac{1}{2}$, and

$$M + |b| + 1 + b_n = M + 1 - |b_n| > M + 1 - \frac{1}{2} > M.$$

If b is negative, then $|b| = -b$ and $|b - b_n| < \frac{1}{2}$. Hence,

$$b - \frac{1}{2} < b_n < b + \frac{1}{2},$$

or

$$b_n > -|b| - \frac{1}{2}.$$

Hence,

$$M + |b| + 1 + b_n > M + |b| + 1 - |b| - \frac{1}{2} = M + \frac{1}{2} > M.$$

In either case, $a_n + b_n > M$.

2. The hypotheses are $\lim_{n \to \infty} a_n = -\infty$ and $\lim_{n \to \infty} b_n = b$. By the hypotheses,

(a) for every $M < 0$, there is N such that for $n > N$, $a_n < M$,

(b) for every $\epsilon > 0$, there is N such that for $n > N$, $|b_n - b| < \epsilon$.

Let $M < 0$ be given. Then there are N_1 and N_2 such that

(a) For $n > N_1$, $a_n < M - |b| - 1$; and

(b) For $n > N_2$, $|b_n - b| < \epsilon$, where if $b = 0$, $\epsilon = \frac{1}{2}$ and if $b \neq 0$, ϵ is the minimum of $\frac{1}{2}$ and $|b|$.[2]

Let N be the maximum of N_1 and N_2. Then for $n > N$,

$$a_n + b_n < M - |b| - 1 + b_n.$$

This is clearly less than M if b_n is negative or 0, so suppose b_n is positive. Then b is 0 or positive. If b is 0, then $|b_n - b| = |b_n| < \frac{1}{2}$, so, since b_n is positive, $b_n < \frac{1}{2}$, and

$$M - |b| - 1 + b_n = M - 1 + |b_n| < M - 1 + \frac{1}{2} < M.$$

---

[2]The last condition guarantees that if b is not 0, then b and b_n have the same sign.

If b is positive, then $|b| = b$ and $|b - b_n| < \frac{1}{2}$. Hence,

$$b - \frac{1}{2} < b_n < b + \frac{1}{2},$$

or

$$b_n < b + \frac{1}{2}.$$

Hence,

$$M - |b| - 1 + b_n < M - b - 1 + b + \frac{1}{2} = M - \frac{1}{2} < M.$$

In either case, $a_n + b_n < M$.

3. The hypotheses are $\lim_{n \to \infty} a_n = a$ and $\lim_{n \to \infty} b_n = \infty$. This is like the first case with the a_n and b_n interchanged.

4. The hypotheses are $\lim_{n \to \infty} a_n = a$ and $\lim_{n \to \infty} b_n = -\infty$. This is like the second case with the roles of a_n and b_n interchanged.

5. The hypotheses are $\lim_{n \to \infty} a_n = \infty$ and $\lim_{n \to \infty} b_n = \infty$. From the hypotheses,

   (a) For every $M > 0$, there is $N > 0$ such that if $n > N$, $a_n > M$,
   (b) For every $M > 0$, there is $N > 0$ such that if $n > N$, $b_n > M$.

   Let $M > 0$ be given. Then there are N_1 and N_2 such that

   (a) if $n > N_1$, $a_n > \frac{M}{2}$, and
   (b) if $n > N_2$, $b_n > \frac{M}{2}$.

   Let N be the maximum of N_1 and N_2. Then for $n > N$,

   $$a_n + b_n > \frac{M}{2} + \frac{M}{2} = M.$$

6. The hypotheses are $\lim_{n \to \infty} a_n = -\infty$ and $\lim_{n \to \infty} b_n = -\infty$. From the hypotheses,

   (a) For every $M < 0$, there is $N > 0$ such that if $n > N$, $a_n < M$,
   (b) For every $M < 0$, there is $N > 0$ such that if $n > N$, $b_n < M$.

   Let $M < 0$ be given. Then there are N_1 and N_2 such that

   (a) if $n > N_1$, $a_n < \frac{M}{2}$, and
   (b) if $n > N_2$, $b_n < \frac{M}{2}$.

Let N be the maximum of N_1 and N_2. Then for $n > N$,

$$a_n + b_n < \frac{M}{2} + \frac{M}{2} = M.$$

□

LS13. If $\lim_{n \to \infty} a_n = \pm\infty$ and $\lim_{n \to \infty} b_n = b > 0$, or $\lim_{n \to \infty} a_n = a > 0$ and $\lim_{n \to \infty} b_n = \pm\infty$, then

$$\lim_{n \to \infty} (a_n b_n) = \pm\infty.$$

Proof. This is four results, but the second and fourth follow from the first and third respectively by symmetry.

1. Suppose $\lim_{n \to \infty} a_n = \infty$ and $\lim_{n \to \infty} b_n = b > 0$. Then, by the hypotheses,

   (a) For every $M > 0$, there is an N such that for $n > N$, $a_n > M$, and

   (b) For every $\epsilon > 0$, there is an N such that for $n > N$, $|b_n - b| < \epsilon$.

   Let $M > 0$ be given. Then

   (a) There is N_1 such that for all $n > N_1$, $a_n > \frac{2M}{b}$, and

   (b) There is N_2 such that for all $n > N_2$, $|b_n - b| < \frac{b}{2}$.

   This last assumption means that for $n \geq N_2$, $\frac{b}{2} < b_n < \frac{3b}{2}$. Let N be the maximum of N_1 and N_2. Then for $n > N$, $a_n b_n > \frac{2M}{b} \frac{b}{2} = M$.

2. Suppose $\lim_{n \to \infty} a_n = -\infty$ and $\lim_{n \to \infty} b_n = b > 0$. Then by the hypotheses,

   (a) For every $M < 0$, there is an N such that for $n > N$, $a_n < M$, and

   (b) For every $\epsilon > 0$, there is an N such that for $n > N$, $|b_n - b| < \epsilon$.

   Let $M < 0$ by given. Then

   (a) There is N_1 such that for all $n > N_1$, $a_n < \frac{2M}{b}$, and

   (b) There is N_2 such that for all $n > N_2$, $|b_n - b| < \frac{b}{2}$.

As in the previous case, for $n \geq N$, $\frac{b}{2} < b_n < \frac{3b}{2}$. Let N be the maximum of N_1 and N_2. Then for $n > N$, since $\frac{2M}{b} < 0$, from $b_n > \frac{b}{2}$ we get $\frac{2M}{b} b_n < \frac{2M}{b} \frac{b}{2}$, and hence $a_n b_n < \frac{2M}{b} b_n < \frac{2M}{b} \frac{b}{2} = M$.

$\square$

LS14. If $\lim_{n \to \infty} a_n = \pm\infty$ and $\lim_{n \to \infty} b_n = \pm\infty$, then

$$\lim_{n \to \infty} (a_n b_n) = \infty.$$

Proof. This is two results.

1. The hypotheses are $\lim_{n \to \infty} a_n = \infty$ and $\lim_{n \to \infty} b_n = \infty$. Then by the hypotheses,

   (a) For every $M > 0$, there is N such that for $n > N$, $a_n > M$, and

   (b) For every $M > 0$, there is N such that for $n > N$, $b_n > M$.

   Let $M > 0$ be given. Then

   (a) There is N_1 such that for $n > N_1$, $a_n > \sqrt{M}$, and
   (b) There is N_2 such that for $n > N_2$, $b_n > \sqrt{M}$.

   Let N be the maximum of N_1 and N_2. Then for $n > N$,

   $$a_n b_n > \sqrt{M}\sqrt{M} = M.$$

2. The hypotheses are $\lim_{n \to \infty} a_n = -\infty$ and $\lim_{n \to \infty} b_n = -\infty$. Then by the hypotheses,

   (a) For every $M < 0$, there is N such that for $n > N$, $a_n < M$, and

   (b) For every $M < 0$, there is N such that for $n > N$, $b_n < M$.

   Let $M > 0$ be given. Then

   (a) There is N_1 such that for $n > N_1$, $a_n < -\sqrt{M}$, and
   (b) There is N_2 such that for $n > N_2$, $b_n < -\sqrt{M}$.

   Let N be the maximum of N_1 and N_2. Then for $n > N$,

   $$a_n b_n > (-\sqrt{M})(-\sqrt{M}) = M.$$

$\square$

LS15. If $\lim_{n \to \infty} a_n = a$ and $\lim_{n \to \infty} b_n = \pm\infty$, then

$$\lim_{n \to \infty} \frac{a_n}{b_n} = 0.$$

Proof. This is two results.

1. The second hypothesis is $\lim_{n \to \infty} b_n = \infty$. By the hypotheses,

    (a) For every $\epsilon > 0$, there is N such that for $n > N_1$, $|a_n - a| < \epsilon$, and

    (b) For every $M > 0$, there is N such that for $n > N$, $b_n > M$.

    Let $\epsilon > 0$ be given. Then

    (a) There is N_1 such that for all $n > N_1$, $|a_n - a| < |a|$, and

    (b) There is N_2 such that for all $n > N_2$, $b_n > \frac{2|a|}{\epsilon}$.

    Then for $n > N_2$, $\frac{1}{b_n} < \frac{\epsilon}{2|a|}$, so $\frac{1}{|b_n|} < \frac{\epsilon}{2|a|}$. Let N be the maximum of N_1 and N_2. Then for $n > N$,

$$
\begin{aligned}
\left| \frac{a_n}{b_n} \right| &= \frac{|(a_n - a) + a|}{|b_n|} \\
&\leq \frac{|a_n - a| + |a|}{|b_n|} \\
&= |a_n - a| \frac{1}{|b_n|} + |a| \frac{1}{|b_n|} \\
&< |a| \frac{\epsilon}{2|a|} + |a| \frac{\epsilon}{2|a|} \\
&= \frac{\epsilon}{2} + \frac{\epsilon}{2} \\
&= \epsilon.
\end{aligned}
$$

2. The second hypothesis is $\lim_{n \to \infty} b_n = -\infty$. By the hypotheses,

    (a) For every $\epsilon > 0$, there is N such that for $n > N$, $|a_n - a| < \epsilon$, and

    (b) For every $M > 0$, there is N such that for $n > N$, $b_n < M$.

    Let $\epsilon > 0$ be given. Then

    (a) There is N_1 such that for all $n > N$, $|a_n - a| < |a|$, and

    (b) There is N_2 such that for all $n > N_2$, $b_n < -\frac{2|a|}{\epsilon}$.

Then for $n > N_2$, $\frac{1}{b_n} > -\frac{\epsilon}{2|a|}$, so $\frac{1}{|b_n|} < \frac{\epsilon}{2|a|}$. Let N be the maximum of N_1 and N_2. Then for $n > N$,

$$
\begin{aligned}
\left|\frac{a_n}{b_n}\right| &= \frac{|(a_n - a) + a|}{|b_n|} \\
&\leq \frac{|a_n - a| + |a|}{|b_n|} \\
&= |a_n - a|\frac{1}{|b_n|} + |a|\frac{1}{|b_n|} \\
&< |a|\frac{\epsilon}{2|a|} + |a|\frac{\epsilon}{2|a|} \\
&= \frac{\epsilon}{2} + \frac{\epsilon}{2} \\
&= \epsilon.
\end{aligned}
$$

$\square$

LS17.

$$
\lim_{n \to \infty} a_n = \pm\infty
$$

if and only if

$$
\lim_{n \to \infty} (-a_n) = \mp\infty.
$$

Proof. This is really two results.

1. The equivalence is

$$
\lim_{n \to \infty} a_n = \infty
$$

if and only if

$$
\lim_{n \to \infty} (-a_n) = -\infty.
$$

If we assume $\lim_{n \to \infty} a_n = \infty$, then by the definition for every $M > 0$, there is N such that for all $n > N$, $a_n > M$. Let $M < 0$ be given. Then there is N such that for $n > N$, $a_n > -M$. It follows that $-a_n < M$.

Conversely, suppose $\lim_{n \to \infty}(-a_n) = -\infty$, Let $M > 0$ be given. Then by the hypothesis, there is N such that for $n > N$, $-a_n < -M$. Then $a_n > M$.

2. The equivalence is

$$
\lim_{n \to \infty} a_n = -\infty
$$

if and only if

$$\lim_{n \to \infty} (-a_n) = \infty.$$

If we assume $\lim_{n \to \infty} a_n = -\infty$, let $M > 0$ be given. Then there is N such that for $n > N$, $a_n < -M$. It follows that $-a_n > M$.

Conversely, assume $\lim_{n \to \infty}(-a_n) = \infty$ and let $M < 0$ be given. By the assumption, there is N such that for all $n > N$, $-a_n > -M(> 0)$. Then $a_n < M$.

$\square$

9.1.1 Exercises

Exercise 9.1. Show that if $\lim_{n \to \infty} a_n = l_1$ and $\lim_{n \to \infty} a_n = l_2$ then $l_1 = l_2$.

Exercise 9.2. Prove the rest of the properties of limits of sequences given in Example 8.2.1. I.e., prove LS6, LS7, LS16, LS18 ... LS34.

9.2 Constructing Proofs in the Theory of Limits

Studying the above proofs may not be enough to learn how to construct such proofs. For it is rarely possible to see immediately how such a proof should be put together when only the sequence and the limit are known.

The key to constructing these proofs is to start with scratch work that goes backward. Start by writing the goal of the proof, namely

$$|a_n - l| < \epsilon.$$

Then use algebraic manipulation to find a value of N such that if $n > N$ then the goal is true. This may take some ingenuity. The procedure is illustrated in the following examples.

Example 9.2.1. *Prove that*

$$\lim_{n \to \infty} \frac{1}{n^3} = 0.$$

SCRATCH WORK. *Given $\epsilon > 0$, we want, for large values of n*

$$\left| \frac{1}{n^3} - 0 \right| < \epsilon.$$

Since $n > 0$, this will hold if

$$\frac{1}{n^3} < \epsilon.$$

This is equivalent to

$$n^3 > \frac{1}{\epsilon},$$

which, in turn, is equivalent to

$$n > \frac{1}{\sqrt[3]{\epsilon}}.$$

So we can let N be the smallest positive integer satisfying

$$N > \frac{1}{\sqrt[3]{\epsilon}}.$$

PROOF. *Let $\epsilon < 0$ be given, and let N be the smallest integer satisfying*

$$N > \frac{1}{\sqrt[3]{\epsilon}}.$$

Then if $n > N$, we have

$$n > \frac{1}{\sqrt[3]{\epsilon}}.$$

It follows that

$$n^3 > \frac{1}{\epsilon},$$

and hence

$$\frac{1}{n^3} < \epsilon.$$

But since n is positive, this implies that

$$\left| \frac{1}{n^3} - 0 \right| = \frac{1}{n^3} < \epsilon,$$

as desired.

Example 9.2.2. *Prove that*

$$\lim_{n \to \infty} \frac{5n}{3n - 2} = \frac{5}{3}.$$

SCRATCH WORK. *Given $\epsilon > 0$, we want for large values of n,*

$$\left| \frac{5n}{3n - 2} - \frac{5}{3} \right| < \epsilon.$$

Now we can get by algebraic manipulation

$$\left|\frac{5n}{3n-2} - \frac{5}{3}\right| = \left|\frac{15n - (15n - 10)}{3(3n - 2)}\right|$$

$$= \left|\frac{10}{3(3n - 2)}\right|$$

$$= \frac{|10|}{|3||3n - 2|}$$

$$= \frac{10}{3|3n - 2|}.$$

Hence, our goal is equivalent to

$$\frac{10}{3|3n - 2|} < \epsilon.$$

From this it follows that

$$\frac{10}{3\epsilon} < |3n - 2|.$$

Since we can assume that $n > 2/3$, $3n - 2 > 0$, so $|3n - 2| = 3n - 2$. Hence,

$$\frac{10}{3\epsilon} < 3n - 2.$$

It follows that

$$n > \frac{10}{9\epsilon} + \frac{2}{3}.$$

So we let N be the smallest integer greater than $10/9\epsilon + 2/3$.

PROOF. *Let $\epsilon > 0$ be given. Let N be the smallest integer greater than $10/9\epsilon + 2/3$. Note that since $\epsilon > 0$, we have that $N > 2/3$. Then if $n > N$, we have $n > 2/3$ and so $3n - 2 > 0$ and $3n - 2 = |3n - 2|$. Then*

$$n > \frac{10}{9\epsilon} + \frac{2}{3}$$

$$3n > \frac{10}{3\epsilon} + 2$$

$$3n - 2 > \frac{10}{3\epsilon}$$

$$|3n - 2| > \frac{10}{3\epsilon}$$

$$\epsilon > \frac{10}{3|3n - 2|},$$

Hence,

$$\left| \frac{5n}{3n-2} - \frac{5}{3} \right| = \left| \frac{15n - (15n - 10)}{3(3n-2)} \right|$$

$$= \left| \frac{10}{3(3n-2)} \right|$$

$$= \frac{|10|}{|3||3n-2|}$$

$$= \frac{10}{3|3n-2|}$$

$$< \epsilon,$$

as desired.

Example 9.2.3. *Prove that*

$$\lim_{n \to \infty} \frac{1}{n^{1/3}} = 0.$$

SCRATCH WORK: *Given $\epsilon > 0$, we want, for large values of n,*

$$\left| \frac{1}{n^{1/3}} - 0 \right| < \epsilon.$$

Note that

$$\left| \frac{1}{n^{1/3}} - 0 \right| = \frac{1}{|n|^{1/3}},$$

so that

$$\left| \frac{1}{n^{1/3}} - 0 \right| < \epsilon$$

is equivalent to

$$\frac{1}{|n|^{1/3}} < \epsilon.$$

This, in turn, is equivalent to

$$|n^{1/3}| > \frac{1}{\epsilon},$$

which is equivalent to

$$|n| > \frac{1}{\epsilon^3}$$

which, since n is clearly positive, is equivalent to

$$n > \frac{1}{\epsilon^3}.$$

PROOF: *Let $\epsilon > 0$ be given. Let N be the smallest integer greater than $1/\epsilon^3$. Assume $n > N$. Then $n > 1/\epsilon^3$. Since n is clearly positive, $n = |n|$, so we have*

$$|n| > \frac{1}{\epsilon^3}$$

$$|n|^{1/3} > \frac{1}{\epsilon}$$

$$\epsilon > \frac{1}{|n|^{1/3}}$$

$$= \left| \frac{1}{n^{1/3}} - 0 \right|,$$

as desired.

Example 9.2.4. *Prove that*

$$\lim_{n \to \infty} [\sqrt{4n^2 + n} - 2n] = \frac{1}{4}.$$

SCRATCH WORK: *Given $\epsilon > 0$, we want, for large values of n,*

$$\left| (\sqrt{4n^2 + n} - 2n) - \frac{1}{4} \right| < \epsilon.$$

Now

$$\left| (\sqrt{4n^2 + n} - 2n) - \frac{1}{4} \right| = \left| \frac{4(\sqrt{4n^2 + n} - 2n) - 1}{4} \right|$$

$$= \left| \frac{4\sqrt{4n^2 + n} - (8n + 1)}{4} \right|.$$

And now we might appear to be stuck. But we are not stuck: we can rationalise the numerator by multiplying numerator and denominator by

$(4\sqrt{4n^2 + n} + (8n + 1))$. *This gives us*

$$\left| \frac{(4\sqrt{4n^2 + n} - (8n + 1))(4\sqrt{4n^2 + n} + (8n + 1))}{4(4\sqrt{4n^2 + n} + (8n + 1))} \right|$$

$$= \left| \frac{16(4n^2 + n) - (8n + 1)^2}{4(4\sqrt{4n^2 + n} + (8n + 1))} \right|$$

$$= \left| \frac{(64n^2 + 16n) - (64n^2 + 16n + 1)}{16\sqrt{4n^2 + n} + 32n + 4} \right|$$

$$= \frac{|-1|}{\left| 16\sqrt{4n^2 + n} + 32n + 4 \right|}$$

$$= \frac{1}{16\sqrt{4n^2 + n} + 32n + 4}$$

where the last step is justified by the fact that every term of the denominator is positive. Now we want this to be less than ϵ. We will have this if this is less than something else which is less than ϵ. Furthermore, making the denominator smaller will make the fraction larger: $1/3 < 1/2$. Hence, we have

$$\frac{1}{16\sqrt{4n^2 + n} + 32n + 4} < \frac{1}{16\sqrt{4n^2}}$$

$$= \frac{1}{32n}.$$

So it is sufficient to choose n so that

$$\frac{1}{32n} < \epsilon.$$

This is equivalent to

$$32n > \frac{1}{\epsilon},$$

or

$$n > \frac{1}{32\epsilon}.$$

PROOF: *Let $\epsilon > 0$ be given, and let N be the smallest integer greater than $1/(32\epsilon)$. Then if $n > N$, we have*

$$n > \frac{1}{32\epsilon},$$

so

$$32n > \frac{1}{\epsilon},$$

or

$$16(2n) > \frac{1}{\epsilon},$$

or

$$16\sqrt{4n^2} > \frac{1}{\epsilon}.$$

Now

$$16\sqrt{4n^2 + n} + 32n + 4 > 16\sqrt{4n^2} > \frac{1}{\epsilon}.$$

Hence,

$$\frac{1}{16\sqrt{4n^2 + n} + 32n + 4} < \epsilon.$$

This means that

$$\left| (\sqrt{4n^2 + n} - 2n) - \tfrac{1}{4} \right| =$$
$$\left| \frac{4(\sqrt{4n^2 + n} - 2n) - 1}{4} \right| =$$
$$\left| \frac{4\sqrt{4n^2 + n} - (8n + 1)}{4} \right| =$$
$$\left| \frac{(4\sqrt{4n^2 + n} - (8n + 1))(4\sqrt{4n^2 + n} + (8n + 1))}{4(4\sqrt{4n^2 + n} + (8n + 1))} \right| =$$
$$\left| \frac{16(4n^2 + n) - (8n + 1)^2}{4(4\sqrt{4n^2 + n} + (8n + 1))} \right| =$$
$$\left| \frac{(64n^2 + 16n) - (64n^2 + 16n + 1)}{16\sqrt{4n^2 + n} + 32n + 4} \right| =$$
$$\left| \frac{|-1|}{16\sqrt{4n^2 + n} + 32n + 4} \right| =$$
$$\frac{1}{16\sqrt{4n^2 + n} + 32n + 4} <$$
$$\epsilon,$$

as desired.

Example 9.2.5. *Prove that*

$$\lim_{n \to \infty} \frac{5n^3 + 4n}{n^3 - 5} = 5.$$

SCRATCH WORK. *Given $\epsilon > 0$, we want, for large values of n,*

$$\left| \frac{5n^3 + 4n}{n^3 - 5} - 5 \right| < \epsilon.$$

Now

$$\frac{5n^3 + 4n}{n^3 - 5} - 5 = \frac{4n + 25}{n^3 - 5},$$

So we want

$$\left| \frac{4n + 25}{n^3 - 5} \right| < \epsilon.$$

Since we are concerned with large positive values of n, we can drop the absolute value signs:

$$\frac{4n + 25}{n^3 - 5} < \epsilon.$$

If we follow the previous examples here, we will try to "solve" for n, but here this is difficult. Furthermore, we do not need the smallest value of N such that for $n > N$, we have

$$\left| \frac{5n^3 + 4n}{n^3 - 5} - 5 \right| < \epsilon.$$

Hence, we can estimate. We can do this by constructing a fraction larger than $(4n + 25)/(n^3 - 5)$ and letting that fraction be less than ϵ. To do this, we want to make the numerator bigger and the denominator smaller. Now if $n > 1$, we have $4n + 25 \leq 29n$. Also, if $n \geq 2$, $n^3 - 5 \geq n^3/2$. Hence, we have

$$\frac{4n + 25}{n^3 - 5} \leq \frac{29n}{(n^3/2)} = \frac{58n}{n^3} = \frac{58}{n^2},$$

so we want

$$\frac{58}{n^2} < \epsilon.$$

This is equivalent to

$$n > \sqrt{\frac{58}{\epsilon}}.$$

So we can let N be the smallest integer satisfying both $N \geq 3$ and

$$N > \sqrt{\frac{58}{\epsilon}}.$$

PROOF. *Let $\epsilon > 0$ be given. Let N be the smallest integer satisfying $N \geq 3$ and*

$$N > \sqrt{\frac{58}{\epsilon}}.$$

Assume that $n > N$. Then $n > 3$ and

$$n > \sqrt{\frac{58}{\epsilon}}.$$

Then we have

$$\frac{58}{n^2} < \epsilon.$$

Since $n > 3$, we have $29n \geq 4n + 25$ and $n^3/2 \leq n^3 - 5$. Hence,

$$
\begin{aligned}
\left| \frac{5n^3 + 4n}{n^3 - 5} - 5 \right| &= \left| \frac{4n + 25}{n^3 - 5} \right| \\
&= \frac{4n + 25}{n^3 - 5} \\
&\leq \frac{29n}{n^3/2} \\
&= \frac{58}{n^2} \\
&< \epsilon,
\end{aligned}
$$

as desired.

9.2.1 Exercises

Exercise 9.3. Prove the following:

1.
$$\lim_{n \to \infty} \frac{(-1)^n}{n} = 0.$$

2.
$$\lim_{n \to \infty} \frac{2n - 1}{3n + 2} = \frac{2}{3}.$$

3.
$$\lim_{n \to \infty} \frac{n + 6}{n^2 - 6} = 0.$$

4.
$$\lim_{n \to \infty} \left[\sqrt{n^2 + 1} - n \right] = 0.$$

5.
$$\lim_{n \to \infty} \left[\sqrt{n^2 + n} - n \right] = \frac{1}{2}.$$

Exercise 9.4. Determine the limits of the following sequences and then prove your claims.

1. $a_n = n/(n^2 + 1)$.

2. $b_n = (7n - 19)/(3n + 7)$.

3. $c_n = (4n + 3)/(7n - 5)$.

4. $d_n = (2n + 4)/(5n + 2)$.

5. $s_n = (1/n)\sin n$.

9.3 Limits of Functions: Theory

Let us now do for limits of functions what we have already done in this chapter for limits of sequences.

One way to do this is to use the model of the method of exhaustion and use a double proof by contradiction. Consider the function

$$f(x) = \frac{x^2 - 4}{x - 2}$$

from Section 8.1. If the limit of $f(x)$ as x approaches 2 is not 4, then it must be some l which is either greater than 4 or less than 4.

Suppose $l < 4$. Let $\epsilon = 4 - l$. Then $l = 4 - \epsilon$. Let $c = 2 - \frac{\epsilon}{2}$. Then since $c \neq 2$, f is defined at c, and is equal to

$$g(c) = c + 2 = 2 - \frac{\epsilon}{2} + 2 = 4 - \frac{\epsilon}{2} > l.$$

Since f is clearly an increasing function, this is clearly a contradiction.

Suppose $l > 4$. Then let $\epsilon = l - 4$. Then $l = 4 + \epsilon$. Let $c = 2 + \frac{\epsilon}{2}$. Then, once again, since $c \neq 2$, f is defined at c and equals

$$g(c) = c + 2 = 2 + \frac{\epsilon}{2} + 2 = 4 + \frac{\epsilon}{2} < l,$$

another contradiction.

Since l is neither less than 4 nor greater than 4, it must be 4.

We can eliminate these double proofs by contradiction in much the same way we did for limits of sequences. Saying that $f(x)$ gets close to 4 is similar to saying that for every $\epsilon > 0$, $|f(x) - 4| < \epsilon$. Then saying that x gets close to 2 but is not equal to 2 should be saying that $0 < |x - 2| < \delta$ for some δ. This suggests that corresponding to Definition 9.1.1 is as follows:

Definition 9.3.1. *The function $f(x)$ approaches the* limit l *as x approaches c, in symbols*

$$\lim_{x \to c} f(x) = l,$$

if, for every $\epsilon > 0$, there is a $\delta > 0$ such that if $0 < |x - c| < \delta$, then $|f(x) - l| < \epsilon$.

Then we can prove

$$\lim_{x \to 2} f(x) = g(2) = 4$$

as follows: let $\epsilon > 0$ be given. Let $\delta = \epsilon$, and suppose $0 < |x - 2| < \delta$. Then $|x - 2| < \epsilon$. Also, since if $0 < |x - 2| < \delta$ we have $x \neq 2$,

$$\left| \frac{x^2 - 4}{x - 2} - 4 \right| = |(x + 2) - 4| = |x - 2| < \epsilon,$$

which is what we needed to show.

The corresponding definitions of one-sided limits are as follows:

Definition 9.3.2. *The function $f(x)$ approaches the limit l as x approaches c from the left [from the right], in symbols*

$$\lim_{x \to c^-} f(x) = l \qquad \left[\lim_{x \to c^+} f(x) = l \right],$$

if, for every $\epsilon > 0$, there is a $\delta > 0$ such that if $0 < c - x < \delta$ [$0 < x - c < \delta$], then $|f(x) - l| < \epsilon$.

The corresponding definition for a function having a vertical asymptote and getting arbitrarily large [arbitrarily large in absolute value but negative] is

Definition 9.3.3. *The function $f(x)$ goes to infinity [goes to negative infinity] as x approaches c, in symbols*

$$\lim_{x \to c} f(x) = \infty \qquad \left[\lim_{x \to c} f(x) = -\infty \right],$$

if, for every $M > 0$ [$M < 0$] there is a $\delta > 0$ such that if $0 < |x - c| < \delta$, then $f(x) > M$ [$f(x) < M$].

The one-sided versions of these definitions are as follows:

Definition 9.3.4. *The function $f(x)$ approaches infinity as x approaches c from below [from above], in symbols*

$$\lim_{x \to c^-} f(x) = \infty \qquad \left[\lim_{x \to c^+} f(x) = \infty \right],$$

if, for every $M > 0$ there is a $\delta > 0$ such that if $0 < c - x < \delta$ $[0 < x - c < \delta]$ then $f(x) > M$. The function $f(x)$ approaches negative infinity as x approaches c from below [from above], in symbols

$$\lim_{x \to c^-} f(x) = -\infty \qquad \left[\lim_{x \to c^+} f(x) = -\infty\right],$$

if, for every $M < 0$ there is a $\delta > 0$ such that if $0 < c - x < \delta$ $[0 < x - c < \delta]$ then $f(x) < M$.

The definition of a function has a horizontal asymptote as x gets large (gets very negative) is as follows:

Definition 9.3.5. *A function $f(x)$ has a* horizontal asymptote *as x gets large [gets very negative], in symbols*

$$\lim_{x \to \infty} f(x) = l \qquad \left[\lim_{x \to -\infty} f(x) = l\right],$$

if, for every $\epsilon > 0$ there is an $M > 0$ $[M < 0]$ such that if $x > M$ $[x < M]$, $|f(x) - l| < \epsilon$.

Finally, the definition of a function $f(x)$ increasing or decreasing without limit as x gets very large (gets very negative) is as follows:

Definition 9.3.6. *A function $f(x)$* increases without limit *as x gets very large [gets very negative], in symbols*

$$\lim_{x \to \infty} f(x) = \infty \qquad \left[\lim_{x \to -\infty} f(x) = \infty\right],$$

if, for every $M_1 > 0$ there is an $M_2 > 0$ $[M_2 < 0]$ such that if $x > M_2$ $[x < M_2]$, $f(x) > M_1$. A function $f(x)$ decreases without limit *as x gets very large [gets very negative], in symbols*

$$\lim_{x \to \infty} f(x) = -\infty \qquad \left[\lim_{x \to -\infty} f(x) = -\infty\right],$$

if, for every $M_1 < 0$ there is an $M_2 > 0$ $[M_2 < 0]$ such that if $x > M_2$ $[x < M_2]$, $f(x) < M_1$.

Constructing proofs with these definitions is similar to constructing proofs of limits of sequences. Start with $|f(x) - l| < \epsilon$ or $f(x) > M$ or $f(x) < M$, this last case for negative M, and try to draw conclusions about $|x - c|$, $x - c$, $x > c$, $x > M$, or $x < M$, this last case for negative M.

Example 9.3.7. *Prove that*

$$\lim_{x \to 0} x^2 \sin\left(\frac{1}{x}\right) = 0.$$

(Note that this function is not defined at $x = 0$).

SCRATCH WORK. *We want*

$$\left| x^2 \sin\left(\frac{1}{x}\right) - 0 \right| < \epsilon.$$

Now

$$\left| \sin\left(\frac{1}{x}\right) \right| \leq 1,$$

so

$$\left| x^2 \sin\left(\frac{1}{x}\right) - 0 \right| \leq |x^2|,$$

so the condition we want will be satisfied if we have

$$|x^2| < \epsilon.$$

This is equivalent to

$$|x|^2 < \epsilon,$$

and it will follow from

$$|x| = |x - 0| < \sqrt{\epsilon}.$$

PROOF. *Let*

$$f(x) = x^2 \sin\left(\frac{1}{x}\right).$$

Then since $|\sin y| \leq 1$ for every y, we have

$$|f(x)| \leq x^2.$$

Now let $\epsilon > 0$ be given. Let $\delta = \sqrt{\epsilon}$. Then if $0 < |x - 0| = |x| < \delta$, we have
$|f(x) - 0| = |f(x)| \leq x^2 < \delta^2 = (\sqrt{\epsilon})^2 = \epsilon.$

Example 9.3.8. *Let $f(x) = \sqrt{x}$. Prove that*

$$\lim_{x \to 4} f(x) = 2.$$

SCRATCH WORK. *We want*

$$|\sqrt{x} - 2| < \epsilon.$$

We might appear to be stuck here, but if we treat this as a fraction with denominator 1, we can rationalise the denominator by multiplying both numerator and denominator by $|\sqrt{x} + 2|$. Then we want

$$|\sqrt{x} - 2| = \frac{|\sqrt{x} - 2||\sqrt{x} + 2|}{|\sqrt{x} + 2|}$$

$$= \frac{|x - 4|}{|\sqrt{x} + 2|}$$

$$< \epsilon.$$

This means that we want

$$|x - 4| < |\sqrt{x} + 2|\epsilon,$$

and so we want

$$\delta \le |\sqrt{x} + 2|\epsilon.$$

Now the condition $|x - 4| < \delta$ implies $\sqrt{4 - \delta} + 2 < \sqrt{x} + 2 = |\sqrt{x} + 2|$ whenever $\delta < 4$, so we want

$$\delta \le (\sqrt{4 - \delta} + 2)\epsilon.$$

It is sufficient to take equality here. This implies that

$$\delta - 2\epsilon = \epsilon\sqrt{4 - \delta},$$

which we must solve for δ. Squaring both sides, we get

$$\delta^2 - 4\delta\epsilon + 4\epsilon^2 = 4\epsilon^2 - \delta\epsilon^2.$$

Subtracting $4\epsilon^2$ from both sides, gathering all terms on the left, and simplifying, we get

$$\delta^2 - \delta(4\epsilon - \epsilon^2) = 0,$$

from which it follows that

$$\delta[\delta - (4\epsilon - \epsilon^2)] = 0.$$

Since we need $\delta > 0$, we must have

$$\delta = 4\epsilon - \epsilon^2.$$

PROOF: *Let $\epsilon > 0$ be given. Let*

$$\delta = 4\epsilon - \epsilon^2.$$

Then a simple calculation shows that

$$\sqrt{4 - \delta} + 2 = 4 - \epsilon.$$

Now suppose $0 < |x - 4| < \delta$. *Then* $\sqrt{4 - \delta} + 2 < \sqrt{x} + 2 = |\sqrt{x} + 2|$. *Hence,*

$$
\begin{aligned}
|f(x) - 2| &= |\sqrt{x} - 2| \\
&= \frac{(|\sqrt{x} - 2|)(|\sqrt{x} + 2|)}{|\sqrt{x} + 2|} \\
&= \frac{|x - 4|}{|\sqrt{x} + 2|} \\
&< \frac{\delta}{\sqrt{4 - \delta}} \\
&= \frac{4\epsilon - \epsilon^2}{4 - \epsilon} \\
&= \epsilon.
\end{aligned}
$$

We can now prove some of the properties of limits of functions from Definition 8.1.1.

Theorem 9.3.9. *If* l, m, c, *and* k *are quantities, and if*

$$\lim_{x \to c} f(x) = l, \qquad and \qquad \lim_{x \to c} g(x) = m,$$

and if r *is an integer, then*

LF1.
$$\lim_{x \to c} x = c.$$

LF2.
$$\lim_{x \to c} k = k.$$

LF3.
$$\lim_{x \to c} (f(x) + g(x)) = l + m,$$

LF4.
$$\lim_{x \to c} (f(x)g(x)) = lm,$$

LF5.
$$\lim_{x \to c} \frac{1}{f(x)} = \frac{1}{l} \qquad provided\ that\ l \neq 0,$$

LF6. *If $l \geq 0$ whenever n is even, then*

$$\lim_{x \to c} \sqrt[n]{f(x)} = \sqrt[n]{l}.$$

LF7. [3] *If $g(x) \leq f(x) \leq h(x)$ for all x except possibly for $x = c$ in some open interval which contains c, and if*

$$\lim_{x \to c} g(x) = \lim_{x \to c} h(x) = l,$$

then

$$\lim_{x \to c} f(x) = l.$$

LF8. *If $f(x) \leq g(x)$ for all x except possibly for $x = c$ in some open interval which contains c, then*

$$\lim_{x \to c} f(x) \leq \lim_{x \to c} g(x),$$

provided that both limits exist.

LF9. *If l is a value, then*

$$\lim_{x \to c} f(x) = l \quad \textit{only if} \quad \lim_{x \to c^-} f(x) = \lim_{x \to c^+} f(x) = l.$$

LF10. *If c is a value and if*

$$\lim_{x \to c} f(x) = \infty,$$

then

$$\lim_{x \to c} \frac{1}{f(x)} = 0.$$

LF11. *If c is a value and if*

$$\lim_{x \to c} f(x) = 0,$$

and if, for all x in an open interval containing c, $f(x) \geq 0$, then

$$\lim_{x \to c} \frac{1}{f(x)} = \infty.$$

Proof. LF1. Let $\epsilon > 0$ be given. Let $\delta = \epsilon$. Then, since $f(x) = x$ for all x, we have that if $0 < |x - c| < \epsilon$, then $|f(x) - c| = |x - c| < \epsilon$.

---

[3]This is also referred to as the squeeze theorem.

LF2. Let $\epsilon > 0$ be given. Then, no matter what δ is, since $f(x) = k$ for all x, we have

$$|f(x) - k| = |k - k| = 0 < \epsilon.$$

LF3. Let $\epsilon > 0$ be given. Then there are δ_1 and δ_2 such that

1. if $0 < |x - c| < \delta_1$, then $|f(x) - l| < \frac{\epsilon}{2}$, and
2. if $0 < |x - c| < \delta_2$, then $|g(x) - m| < \frac{\epsilon}{2}$.

Let δ be the smaller of δ_1 and δ_2. Then for $0 < |x - c| < \delta$, we have

$$
\begin{aligned}
|(f(x) + g(x)) - (l + m)| &= |(f(x) - l) + (g(x) - m)| \\
&\le |f(x) - l| + |g(x) - m| \\
&< \frac{\epsilon}{2} + \frac{\epsilon}{2} \\
&= \epsilon.
\end{aligned}
$$

LF4. Let $\epsilon > 0$ be given. Then by hypothesis there are δ_1 and δ_2 such that

1. if $0 < |x - c| < \delta_1$, then $|f(x) - l| < \dfrac{\epsilon}{2\left(\dfrac{\epsilon}{2(|l| + 1)} + |m|\right)}$, and

2. if $0 < |x - c| < \delta_2$, then $|g(x) - m| < \dfrac{\epsilon}{2(|l| + 1)}$.

(Here, $|l| + 1$ is used instead of $|l|$ since l might be 0.) Note that the second of these implies that $|g(x)| < \dfrac{\epsilon}{2(|l| + 1)} + |m|$. Let δ be the minimum of δ_1 and δ_2. Then

$$
\begin{aligned}
&|f(x)g(x) - lm| = \\
&|f(x)g(x) - lg(x) + lg(x) - lm| = \\
&|(f(x) - l)g(x) + l(g(x) - m)| \le \\
&|f(x) - l||g(x)| + |l||g(x) - m| < \\
&\frac{\epsilon}{2\left(\dfrac{\epsilon}{2(|l| + 1)} + |m|\right)}\left(\frac{\epsilon}{2(|l| + 1)} + |m|\right) + |l|\frac{\epsilon}{2(|l| + 1)} < \\
&\frac{\epsilon}{2} + (|l| + 1)\frac{\epsilon}{2(|l| + 1)} = \\
&\frac{\epsilon}{2} + \frac{\epsilon}{2} = \\
&\epsilon.
\end{aligned}
$$

LF5. Let $\epsilon > 0$ be given. Let ϵ' be the minimum of $|l|/2$ and $(l^2\epsilon)/2$. By hypothesis, there is $\delta > 0$ such that if $0 < |x - c| < \delta$, then

$$|f(x) - l| < \frac{l^2\epsilon}{2}$$

and

$$|f(x) - l| < \frac{|l|}{2}.$$

We also have $|f(x)| \geq |l|/2$, since otherwise we would have

$$|l| = |l - f(x) + f(x)| \leq |l - f(x)| + |f(x)| < \frac{|l|}{2} + \frac{|l|}{2} = |l|,$$

a contradiction. It follows that

$$\frac{1}{|f(x)|} \leq \frac{2}{|l|}.$$

Hence, if $0 < |x - c| < \delta$, then

$$\begin{aligned}
\left| \frac{1}{f(x)} - \frac{1}{l} \right| &= \frac{|f(x) - l|}{|l||f(x)|} \\
&= \frac{|f(x) - l|}{|l||f(x)|} \\
&\leq \frac{2}{|l|} \frac{|f(x) - l|}{|l|} \\
&< \frac{2}{l^2} |f(x) - l| \\
&< \frac{2}{l^2} \frac{l^2 \epsilon}{2} \\
&= \epsilon.
\end{aligned}$$

LF6. There are three cases.

Case 1. $l = 0$. This case is left as an exercise.

Case 2. $l > 0$. Let $\epsilon > 0$ be given. Then there is $\delta > 0$ such that if $0 < |x - c| < \delta$, we have

$$|f(x) - l| < \epsilon|(\sqrt[n]{l})^{n-1}|$$

and

$$|f(x) - l| < \frac{|l|}{2}.$$

From the second of these, it follows that $f(x)$ and l have the same sign.

It follows that for these values of x,

$$\left| \sqrt[n]{f(x)} - \sqrt[n]{l} \right| =$$
$$\left| \frac{(\sqrt[n]{f(x)} - \sqrt[n]{l})((\sqrt[n]{f(x)})^{n-1} + (\sqrt[n]{f(x)})^{n-2}\sqrt[n]{l} + \ldots + (\sqrt[n]{l})^{n-1})}{(\sqrt[n]{f(x)})^{n-1} + (\sqrt[n]{f(x)})^{n-2}\sqrt[n]{l} + \ldots + (\sqrt[n]{l})^{n-1}} \right| =$$
$$\frac{|f(x) - l|}{|(\sqrt[n]{f(x)})^{n-1} + (\sqrt[n]{f(x)})^{n-2}\sqrt[n]{l} + \ldots + (\sqrt[n]{l})^{n-1}|} \le$$
$$\frac{|f(x) - l|}{|(\sqrt[n]{l})^{n-1}|} <$$
$$\frac{\epsilon|(\sqrt[n]{l})^{n-1}|}{|\sqrt[n]{l})^{n-1}|} =$$
$$\epsilon.$$

Case 3. $l < 0$. Then n is odd. Let $\epsilon > 0$ be given. Then there is $\delta > 0$ such that if $0 < |x - c| < \delta$, we have

$$|f(x) - l| < \epsilon|(\sqrt[n]{l})^{n-1}|$$

and

$$|f(x) - l| < \frac{|l|}{2}.$$

From the second of these, it follows that $f(x)$ and l have the same sign, both negative, and $|f(x)| = -f(x)$ and $|l| = -l$. It follows that if $0 < |x - c| < \delta$,

$$\left| \sqrt[n]{f(x)} - \sqrt[n]{l} \right| = \left| \sqrt[n]{-|f(x)|} - \sqrt[n]{-|l|} \right|$$
$$= \left| -\sqrt[n]{|f(x)|} + \sqrt[n]{|l|} \right|$$
$$= \left| \sqrt[n]{|f(x)|} - \sqrt[n]{|l|} \right|,$$

and we can prove this less than ϵ by Case 2 above.

LF7. Let $\epsilon > 0$ be given. By hypothesis, we have

1. There is $\delta_1 > 0$ such that if $0 < |x - c| < \delta_1$, then $|g(x) - l| < \epsilon$, and

2. There is $\delta_2 > 0$ such that if $0 < |x - c| < \delta_2$, then $|h(x) - l| < \epsilon$.

Let δ be the smaller of δ_1 and δ_2. Then if $0 < |x - c| < \delta$, Then $|g(x) - l| < \epsilon$ and $|h(x) - l| < \epsilon$. It follows that

$$l - \epsilon < g(x) \le f(x) \le h(x) < l + \epsilon,$$

from which it follows that $|f(x) - l| < \epsilon$.

LF8. Suppose that

$$\lim_{x \to c} f(x) = l \quad \text{and} \quad \lim_{x \to c} g(x) = m,$$

and suppose $l > m$. Let $\epsilon = (l - m)/2 > 0$. By hypothesis,

1. there is δ_1 such that if $0 < |x - c| < \delta_1$, then $|f(x) - l| < \epsilon$, and
2. there is δ_2 such that if $0 < |x - c| < \delta_2$, then $|g(x) - m| < \epsilon$.

Let δ be the smaller of δ_1 and δ_2. Then for $0 < |x - c| < \delta$, we have both $|f(x) - l| < \epsilon$ and $|g(x) - m| < \epsilon$. Then,

$$l - \epsilon = l - \frac{l - m}{2} = \frac{2l - l + m}{2} = \frac{l + m}{2}$$

and

$$m + \epsilon = m + \frac{l - m}{2} = \frac{2m + l - m}{2} = \frac{l + m}{2},$$

so $l - \epsilon = m + \epsilon$. Now the conditions $|f(x) - l| < \epsilon$ and $|g(x) - m| < \epsilon$ imply that $g(x) < m + \epsilon = l - \epsilon < f(x)$, contradicting the hypothesis that $f(x) \le g(x)$.

LF9. Suppose that

$$\lim_{x \to c} f(x) = l.$$

Let $\epsilon > 0$ be given. By hypothesis, there is $\delta > 0$ such that if $0 < |x - c| < \delta$, we have $|f(x) - l| < \epsilon$. It follows that if $0 < c - x < \delta$ and if $0 < x - c < \delta$, we have $|f(x) - l| < \epsilon$, and so

$$\lim_{x \to c^-} f(x) = \lim_{x \to c^+} f(x) = l.$$

LF10. Let $\epsilon > 0$ be given. By hypothesis, there is $\delta > 0$ such that for $0 < |x - c| < \delta$,

$$f(x) > \frac{1}{\epsilon}.$$

It follows that for these values of x, $f(x) > 0$, so $|f(x)| = f(x)$, and

$$\left| \frac{1}{f(x)} - 0 \right| = \frac{1}{f(x)} < \epsilon.$$

LF11. Let $M > 0$ be given. By hypothesis, there is $\delta > 0$ such that if $0 < |x - c| < \delta$, $f(x) \ge 0$ and

$$|f(x) - 0| = f(x) < \frac{1}{M}.$$

Then for these same values of x,

$$\frac{1}{f(x)} > M.$$

□

Limits of functions can be connected to limits of sequences as follows:

Theorem 9.3.10. *The function f has a limit l at $x = c$, i.e.*

$$\lim_{x \to c} f(x) = l,$$

if and only if for every sequence x_n for which no $x_n = c$ and such that

$$\lim_{n \to \infty} x_n = c,$$

we have

$$\lim_{n \to \infty} f(x_n) = l.$$

Proof. ($\Rightarrow$:) Let $\epsilon > 0$ be given. By the first hypothesis, there is $\delta > 0$ such that if $0 < |x - c| < \delta$, then $|f(x) - l| < \epsilon$. Now let $\{x_n\}$ be any sequence of terms different from c such that $\lim_{n \to \infty} x_n = c$. Then there is N such that for $n > N$, $|x_n - c| < \delta$. It follows that for $n > N$, $|f(x_n) - l| < \epsilon$.

($\Leftarrow$:) Suppose

$$\lim_{x \to c} f(x) \neq l.$$

Then, by the negation of the definition, there is $\epsilon > 0$ such that for all $\delta > 0$, there is $x \neq c$ such that

$$0 < |x - c| < \delta \qquad \text{and} \qquad |f(x) - l| \geq \epsilon.$$

In particular, for each positive integer n, there is a point $x_n \neq c$ such that

$$0 < |x_n - c| < \frac{1}{n} \qquad \text{and} \qquad |f(x_n) - l| \geq \epsilon.$$

Then these x_n form a sequence for which

$$\lim_{n \to \infty} x_n = c$$

but it is false that

$$\lim_{n \to \infty} f(x_n) = l.$$

□

Corollary 9.3.11. *Suppose that there are two sequences, $\{x_n\}$ and $\{y_n\}$ such that*

$$\lim_{n \to \infty} x_n = c \qquad and \qquad \lim_{n \to \infty} y_n = c,$$

and suppose there is no value of n with either $x_n = c$ or $y_n = c$. Suppose that

$$\lim_{n \to \infty} f(x_n) = l_1 \qquad and \qquad \lim_{n \to \infty} f(y_n) = l_2.$$

If $l_1 \neq l_2$, then

$$\lim_{x \to c} f(x)$$

does not exist.

Proof. Suppose there is an l such that $\lim_{x \to c} f(x) = l$, then by Theorem 9.3.10, $\lim_{n \to \infty} f(x_n) = l$ and $\lim_{n \to \infty} f(y_n) = l$. But $\lim_{n \to \infty} f(x_n) = l_1$ and $\lim_{n \to \infty} f(y_n) = l_2$. Hence by Exercise 9.1, $l = l_1$ and $l = l_2$. Thus, $l_1 = l_2$. Contradiction. $\qquad \square$

- $\lim_{n \to \infty} a_n = a$ if and only if, for every $\epsilon > 0$, there is a natural number N such that for each $n > N$, $|a_n - a| < \epsilon$.
- $\lim_{n \to \infty} a_n = \infty$ if, for every $M > 0$, there is an N such that for each $n > N$, $a_n > M$.
- $\lim_{n \to \infty} a_n = -\infty$, if, for every $M < 0$, there is an N such that for each $n > N$, $a_n < M$.
- $\lim_{x \to c} f(x) = l$, if, for every $\epsilon > 0$, there is a $\delta > 0$ such that if $0 < |x - c| < \delta$, then $|f(x) - l| < \epsilon$.
- $\lim_{x \to c^-} f(x) = l$, if, for every $\epsilon > 0$, there is a $\delta > 0$ such that if $0 < c - x < \delta$, then $|f(x) - l| < \epsilon$.
- $\lim_{x \to c^+} f(x) = l$, if, for every $\epsilon > 0$, there is a $\delta > 0$ such that if $0 < x - c < \delta$, then $|f(x) - l| < \epsilon$.
- $\lim_{x \to c} f(x) = \infty$, if, for every $M > 0$ there is a $\delta > 0$ such that if $0 < |x - c| < \delta$, then $f(x) > M$.
- $\lim_{x \to c} f(x) = -\infty$, if, for every $M < 0$ there is a $\delta > 0$ such that if $0 < |x - c| < \delta$, then $f(x) < M$.
- $\lim_{x \to c^-} f(x) = \infty$ if, for every $M > 0$ there is a $\delta > 0$ such that if $0 < c - x < \delta$ then $f(x) > M$.
- $\lim_{x \to c^+} f(x) = \infty$, if, for every $M > 0$ there is a $\delta > 0$ such that if $0 < x - c < \delta$ then $f(x) > M$.
- $\lim_{x \to c^-} f(x) = -\infty$, if, for every $M < 0$ there is a $\delta > 0$ such that if $0 < c - x < \delta$ then $f(x) < M$.
- $\lim_{x \to c^+} f(x) = -\infty$, if, for every $M < 0$ there is a $\delta > 0$ such that if $0 < x - c < \delta$ then $f(x) < M$.
- $\lim_{x \to \infty} f(x) = l$, if, for every $\epsilon > 0$ there is an $M > 0$ such that if $x > M$, $|f(x) - l| < \epsilon$.
- $\lim_{x \to -\infty} f(x) = l$, if, for every $\epsilon > 0$ there is an $M < 0$ such that if $x < M$, $|f(x) - l| < \epsilon$.
- $\lim_{x \to \infty} f(x) = \infty$, if, for every $M_1 > 0$ there is an $M_2 > 0$ such that if $x > M_2$, $f(x) > M_1$.
- $\lim_{x \to -\infty} f(x) = \infty$, if, for every $M_1 > 0$ there is an $M_2 < 0$ such that if $x < M_2$, $f(x) > M_1$.
- $\lim_{x \to \infty} f(x) = -\infty$, if, for every $M_1 < 0$ there is an $M_2 > 0$ such that if $x > M_2$, $f(x) < M_1$.
- $\lim_{x \to -\infty} f(x) = -\infty$, if, for every $M_1 < 0$ there is an $M_2 < 0$ such that if $x < M_2$, $f(x) < M_1$.

Figure 9.1: Definition of the Limit of Functions and Sequences

9.3.1   Exercises

Exercise 9.5. Prove the parts of axioms LF1–LF11 which are not proved in Theorem 9.3.9.

Exercise 9.6.   1. Prove directly from the definition that

$$\lim_{x \to 2} x^2 = 4.$$

2. Prove directly from the definition that for every value c,

$$\lim_{x \to c} |x| = |c|.$$

3. Prove directly from the definition that

$$\lim_{x \to 2} (5x - 11) = -1.$$

4. Prove directly from the definition that

$$\lim_{x \to 1} (x^2 + x - 1) = 1.$$

5. Prove directly from the definition that

$$\lim_{x \to 1} (x - 3x^2) = -2.$$

6. Prove directly from the definition that

$$\lim_{x \to 4} (\sqrt{x}) = 2.$$

7. Prove directly from the definition that

$$\lim_{x \to -2} x^3 = -8.$$

8. Prove directly from the definition that

$$\lim_{x \to 1} \frac{4}{3x + 2} = \frac{4}{5}.$$

Exercise 9.7.   1. Prove that the limit of a function is unique; i.e., that a function has at most one limit as $x \to c$.

2. Suppose that f is a function defined on an open interval containing c except possibly at c itself.

(a) Suppose that $\lim_{x \to c} f(x)$ exists. Prove that $\lim_{x \to c} |f(x)|$ exists and $\lim_{x \to c} |f(x)| = |\lim_{x \to c} f(x)|$.

(b) Suppose that $\lim_{x \to c} |f(x)|$ exists. Give an example to show that $\lim_{x \to c} f(x)$ may not exist.

(c) Suppose that $\lim_{x \to c} |f(x)| = 0$. Prove that $\lim_{x \to c} f(x) = 0$.

3. Find
$$\lim_{x \to 2^+} \frac{1}{x^2 - 2x}$$
and prove using the definition that this limit is correct.

4. Find
$$\lim_{x \to \infty} \frac{\sin x}{x}$$
and prove using the definition that this limit is correct.

Exercise 9.8. Let f be an increasingly monotonic function defined on $[a, b]$ (i.e., for all $x, y \in [a, b]$, if $x \leq y$ then $f(x) \leq f(y)$). Let A be the set of points in $[a, b]$ at which f is discontinuous (i.e., let $A = \{c \in [a, b] : \lim_{x \to c^+} f(x) \neq \lim_{x \to c^-} f(x)\}$, where here for uniformity, we agree that $\lim_{x \to b^+} f(x) = f(b)$ and $\lim_{x \to a^-} f(x) = f(a)$). Prove that A is countable. (This exercise is taken from [25].)

Chapter 10

Number Systems

In the last two chapters, we did not say much about the magnitudes that we were using in dealing with limits. This was left ambiguous on purpose: we were looking at both an intuitive approach and some results from Archimedes and Euclid, and then we were looking at the derivative as introduced by Newton and Leibniz.[1] However, we cannot leave this ambiguous indefinitely. In this chapter, we will take this up in detail. First, however, we need to look at what constituted the standards of rigour in mathematics from Greek times to the 19th century.

10.1 Standards of Rigour in Mathematics

Up through the eighteenth century, Euclid's *Elements* was taken as the standard for rigorous proof in mathematics. Considering that it was a book that was written almost 2.5 millennia ago, it is an impressive achievement. However, in the early nineteenth century, some mathematicians began to question whether the deductive structure of the *Elements* was sufficient. In this section we look at some of the reasons why people started to doubt some of the reasoning in the *Elements*.

---

[1]Moreover, the definition of these magnitudes and of real numbers was not clear until after the 1870s. Even the adjective *Real* was not coined before Descartes in the 17th century although back in the 8th/9th century, Abu Kamil ibn Aslam (an Egyptian mathematician known also as al-hasib al-masri, meaning the Egyptian calculator, 850-930) did merge the old concept of number with that of magnitude and was the first to use magnitudes as solutions to quadratic equations.

10.1.1   Possible to Prove in Euclidian Geometry that Every Triangle is Isosceles?

One reason for doubting whether the deductive structure of the Elements was sufficient is in the "proof" below that any triangle is *isosceles*; i.e., that two of its sides are equal. The argument depends on some facts about triangles. One is that the sum of the angles of any triangle is always two right angles. The other two facts are given in the following propositions from Book I:

PROPOSITION 4 OF BOOK I OF THE *Elements*.
If two triangles have two sides equal to two sides respectively, and have the angles contained by the equal straight lines equal, then they also have the base equal to the base, the triangle equals the triangle, and the remaining angles equal the remaining angles respectively, namely those opposite the equal sides.

PROPOSITION 26 OF BOOK I OF THE *Elements*.
If two triangles have two angles equal to two angles respectively, and one side equal to one side, namely, either the side adjoining the equal angles, or that opposite one of the equal angles, then the remaining sides equal the remaining sides and the remaining angle equals the remaining angle.

Example 10.1.1. *Here is a proof that every triangle is isosceles based on Euclid's Elements.*

Proof. Let ABC be any triangle. Let the perpendicular bisector of AC be constructed, and let the angle bisector of $\angle ABC$ be constructed, and let the point at which they meet be G. Construct GE perpendicular to AB, GF perpendicular to BC, and GD perpendicular to AC, and also construct AG and CG.

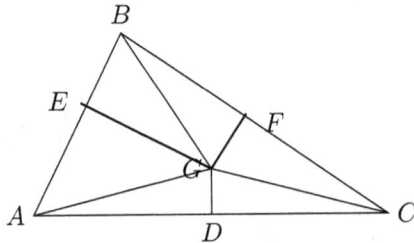

Now, since DG is the perpendicular bisector of AC, $AD = DC$. Also, angles $\angle ADG$ and $\angle GDC$ are both right angles. Then by Proposition 4 of Book I on triangles ADG and GDC, $AG = CG$.

Since BG is the angle bisector of $\angle ABC$, $\angle ABG = \angle GBC$. Also, $\angle BEG = \angle BFG$, since both are right angles by construction, and since BG is common, by Proposition 26 of Book I, it follows that $BE = BF$ and $GE = GF$.

Since $\angle GEA = \angle GFC$ are both right angles, by the Pythagorean triples, $AE^2 = CF^2 = AG^2 - EG^2 = CG^2 - FG^2$. Hence it follows that $AE = CF$.

Now $AB = AE + BE$ and $BC = BF + CF$, it follows that $AB = BC$, and the triangle ABC is isosceles.   □

Since it is obvious that not every triangle is isosceles, there is something wrong here. Can you figure out what it is?

You might think that the fact that G is inside the triangle ABC in the above proof has something to do with the problem or that the above proof is a particular case and that had we made different choices re the point G, then we would not have been able to prove the fallacy. This is not true. Here are other possible cases where again, we will be able to falsely derive that the triangle is isosceles. You will be told why the argument is false for case C1. Can you figure it out for the other cases?

C1. If G was on the line BD (i.e., the extended lines BG and GD are the same), then by applying Proposition 4 of Book I, to the two triangles ABD and BDC, we get that $AB = BC$ and so, ABC would be isosceles.

This is a contradiction, because if triangle ABC is isosceles with sides $AB = BC$, then the perpendicular bisector of the line segment AC and the line that bisects the angle ABC are the very same line, and so these two segments cannot intersect to generate the point G. So the hypotheses of the case is impossible.

C2. Consider the figure below where G is outside the triangle. You can follow the same argument of the proof above to deduce the fallacy that the triangle is isosceles.

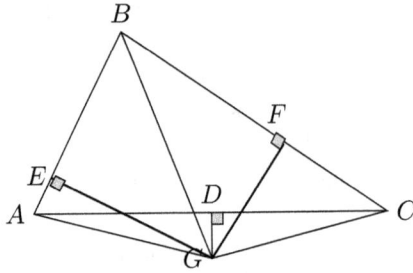

C3. You can also follow the same argument above to deduce the fallacy that the triangle is isosceles, even in the case when the points G and D coincide as can be seen on the diagram below.

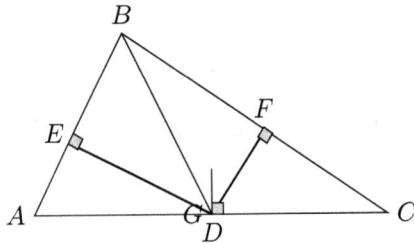

C4. You might think that the problem comes because we always assumed that E and F are inside the triangle. Below is a diagram which shows E and F outside the triangle ABC, but where the fallacy will still be derived by the above argument.

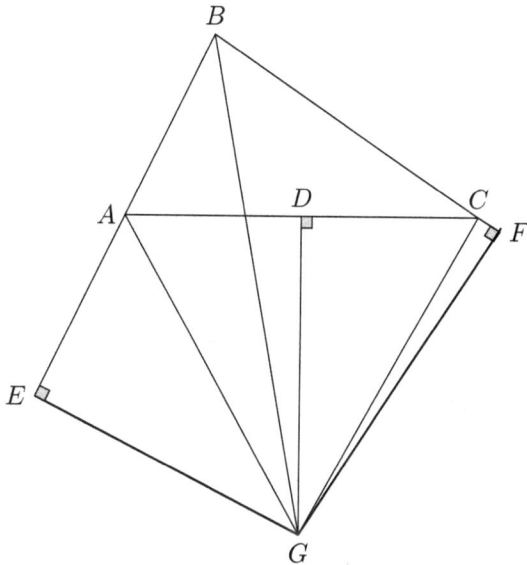

So, the problem is not whether G, E, F occur inside or outside the triangle ABC. The problem is actually "inbetweenness". That is, we always assumed that E and F either both occur inside the triangle or outside it. We did not consider the case where one is outside and one is inside and in that case, say E is the one outside AB, and F is the one inside BC (if E is the one inside and F is the one outside, the proof is similar). Then BA is not the sum of BE and EA, rather a difference is used, whereas a sum is used for $BC = BF + FC$. This can be seen in the following diagram:

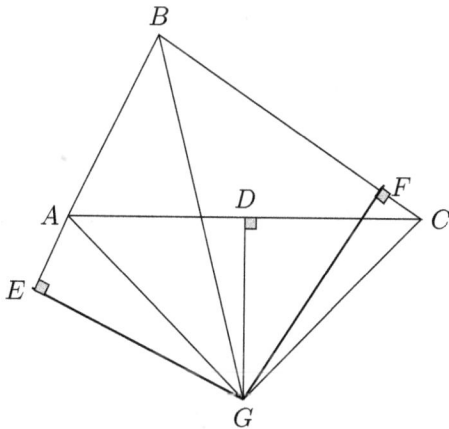

10.1.2   A Problem in the Proof of Proposition 1 of Book I of Euclid's Elements

It turns out that there is a problem in the proof of Proposition 1 of Book I, which is given in Figure 7.15. At first glance, there does not appear to be any doubt that the construction given there constructs the desired equilateral triangle and that the proof proves that it is an equilateral triangle. However, there is a gap in the proof. There is, in fact, no proof that the point C exists.

To see this, let us construct a *model* of geometry in which all of the postulates and axioms are satisfied but Proposition 1 is not.

Example 10.1.2. *To define the model, we need to specify the meaning of some of the key words, such as "point", "line", "plane", etc. In this model, while real numbers will be used for magnitudes, only rational numbers will be allowed as coordinates. The various key words are defined as follows:*

- *The* points *will be triples (p, q, r) of rational numbers. That is:*

- *The* lines *are sets of points of the form (p, q, r) satisfying:*

    - *either $p = a$ and $q = b$, where a and b are rational numbers,*
    - *or $p = a$ and $r = mq + b$, where a, b, and m are rational numbers,*
    - *or $q = m_1 p + b_1$ and $r = m_2 p + b_2$, where $m_1, b_1, m_2,$ and b_2 are rational numbers.*

- *The* planes *are sets of points of the form (p, q, r) satisfying:*

    - *either $p = a$, where a is a rational number,*
    - *or $q = mp + b$, where m and b are rational numbers,*
    - *or $r = m_1 p + m_2 q + b$, where m_1, m_2, and b are rational numbers.*

- *The* distance *between (p_1, q_1, r_1) and (p_2, q_2, r_2) is given by*

$$\sqrt{(p_1 - p_2)^2 + (q_1 - q_2)^2 + (r_1 - r_1)^2}.$$

- *To define the* measurement *of angles, define*

$$(p_1, q_1, r_1) - (p_2, q_2, r_2) = (p_1 - p_2, q_1 - q_2, r_1 - r_2)$$
$$(p_1, q_1, r_1) \cdot (p_2, q_2, r_2) = p_1 p_2 + q_1 q_2 + r_1 r_2$$
$$\| (p, q, r) \| = \sqrt{p^2 + q^2 + r^2}$$

then the measure of $\angle ABC$ is given by

$$\cos^{-1} \frac{(A - B) \cdot (C - B)}{\| A - B \| \cdot \| C - B \|}$$

With these definitions, it is possible to show, using Euclid's definitions, that the postulates all hold. A circle with centre at (p, q, r) with radius ρ can be defined by taking the points of the sphere defined by $(x - p)^2 + (y - q)^2 + (z - r)^2 = \rho^2$ that are also in a plane through (p, q, r). Postulate 5 is more difficult, but it is possible to show that it is satisfied. Thus, we have here a model of Euclid's postulates.

But now let us try to carry out the construction of Proposition 1 when the point A is $(0, 0, 0)$ and B is $(1, 0, 0)$. This means that A is the origin and B is the point on the x axis whose x-coordinate is 1. The x-coordinate of C is clearly $\frac{1}{2}$. This means that we have the situation of Figure 10.1. To calculate the y-coordinate h of C, we use the Pythagorean Theorem:

$$\left(\frac{1}{2}\right)^2 + h^2 = 1^2,$$

and so $h = \sqrt{3}/2$, which is not a rational number. So in our model, the point C does not exist.

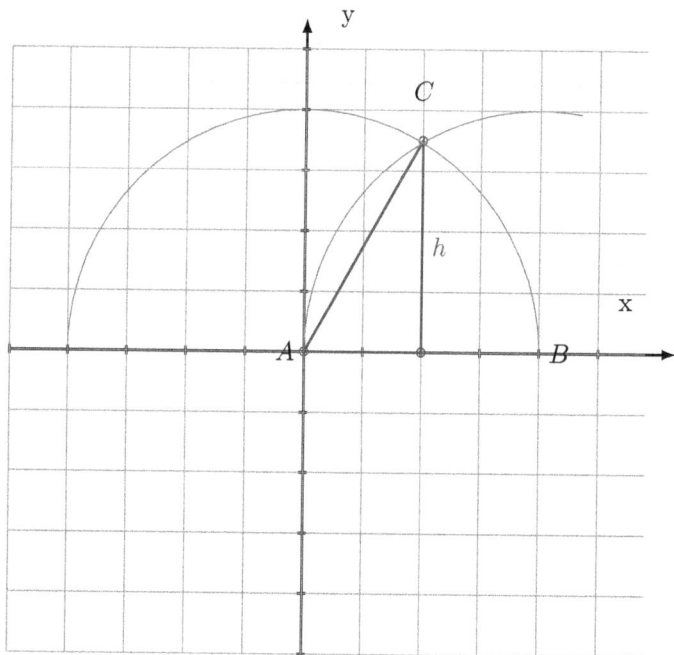

Figure 10.1: Construction of Book I, Proposition 1.

10.1.3   Euclid's Postulate 5 is Less Obvious Than the Other Postulates

Another defect in the *Elements* that began to make itself felt by the early nineteenth century was connected with Postulate 5. The postulate is equivalent to the so-called "Parallel Postulate" which states that given a line and a point not on that line, there is exactly one line through the point parallel to the given line. This postulate had always been felt to be less obvious than the others, and over the millennia since the *Elements* first appeared, various attempts were made to prove it from the other postulates. None of these attempts was successful. One of the most important such attempts was by the Italian mathematician Giovanni Saccheri,[2] who, in his book *Euclides ab omni naevo vindicatus* (Euclid freed of every fleck), (published in Milan in 1733) tried to prove the postulate by assuming it false and deriving a contradiction. His assumption was that given a line and a point not on the line, there is more than one line parallel to the given line. He succeeded in deriving some results that seemed so absurd that he felt he had succeeded in proving his assumption false. But, in fact, he had not. In the early nineteenth century, three mathematicians, namely Carl Friedrich Gauss,[3] János Bolyai,[4] and Nicolai Lobachevsky[5] realised independently that what Saccheri had done was to define a new kind of geometry, which is now called *non-Euclidean geometry*, or, more precisely, *hyperbolic geometry*. Later in the century, Bernhard Riemann[6] developed another form of non-Euclidean geometry, *elliptical geometry*, in which there are no parallel lines. At first these non-Euclidean geometries seemed very strange, but in the second half of the nineteenth century, models of both kinds of non-Euclidean geometry defined within Euclidean geometry showed that if Euclidean geometry is consistent, then so are both kinds of non-Euclidean geometry.

Example 10.1.3. *In the model of Example 10.1.2, consider only the xy-plane. Here, lines are sets of pairs (p, q) of rational numbers satisfying $ap + bq = c$ for rational numbers a, b, and c. Prove that if any two lines are not*

---

[2]Giovanni Girolamo Saccheri, 1667–1733.

[3]Johann Carl Friedrich Gauss, 1777–1855, is considered one of the greatest mathematicians of all time. In 1831, Farkas Bolyai, the father of János Bolyai, showed his son's work on non-Euclidean geometry to Gauss, who replied that he had come to the same conclusion some years earlier but had been afraid to publish it because it seemed so strange.

[4]János Bolyai, 1802–1860. He did his work on non-Euclidean geometry between 1820 and 1824.

[5]Nikolai Ivanovich Lobachevsky, 1792–1856. He published his work on non-Euclidean geometry in 1829.

[6]Georg Friedrich Bernhard Riemann, 1826–1866.

parallel (i.e., do not have the same slope), then they meet in a point of the model; i.e., they meet in a point with rational coefficients.

10.1.4 Further Logical Inaccuracies

Recall Section 7.3 and in particular Figures 7.12, 7.13 and 7.14 which introduced definitions, postulates and common notions from Book I of Euclid's Elements. Recall also from Figures 7.12 the definitions of a point, a straight line and a trilateral (a triangle).

We said in Section 7.3 that the postulates and the common notions are axioms we take for granted and can use even though they are not proven. However, there were a number of other statements that Euclid used as facts in his *Elements* even though they had neither been proved nor been introduced as postulates. As an example, take the statement of Figure 10.2.

> A straight line that intersects one side of a triangle but does not pass through any vertex of the triangle must intersect one and only one of the other sides.

Figure 10.2: A statement that cannot be proven from Euclid's formulation.

This statement has been used by Euclid and many others after him but it cannot be proven in the *Elements* nor was it asserted as a postulate. In 1882, Moritz Pasch[7] gave the following axiom

> Given three non colinear[a] points A, B, C, and a straight line L which does not contain any of these points, if L intersects the straight line AB, then L must intersect one and only one of the straight lines AC or BC.
>
> ---
> [a]I.e., A, B and C do not all lie on the same straight line.

Figure 10.3: Pasch's axiom

This axiom is clearly equivalent to the statement above it. Pasch used his axiom to prove that Euclid's formulation was not complete in the sense

---

[7]Moritz Pasch, 1843–1930, was a German mathematician who argued that Euclidean geometry must be built on precise axiomatic foundations and careful deductive reasoning rather than rely on physical intuition.

that there are statements that should hold but which cannot be proven from Euclid's formulation.

There are other statements that should hold but which cannot be proven from Euclid's formulation. You might be surprised to learn that the following two statements cannot be proven from Euclid's formulation (see Page 143 of [12]):

1. A straight line passing through the centre of a circle must intersect the circle.

2. Given 3 different points on the same line, one of them is between the other two.

The status of statements like the above one, which are used as facts but are neither introduced as axioms nor proven, is usually classified as logical inaccuracies. Such logical inaccuracies in Euclid's Elements have been addressed in the work of Hilbert who wrote 20 postulates [22] adequate to prove all the theorems in the *Elements*.

10.1.5   Exercises

Exercise 10.1. Similarly to the proof given at the start of Example 10.1.1, prove cases $C2 \cdots C4$ of Section 10.1.1.

Exercise 10.2. In the model of Example 10.1.2, consider only the xy-plane. Here, lines are sets of pairs (p, q) of rational numbers satisfying $ap + bq = c$ for rational numbers a, b, and c. Prove that if any two lines are not parallel (i.e., do not have the same slope), then they meet in a point of the model; i.e., they meet in a point with rational coefficients.

10.2   Algebraic Rules for Quantities

Recall that after the discovery of the incommensurability of the side and diagonal of a square, the ancient Greeks completely separated *numbers*, which are discrete, and *magnitudes*, which are continuous. However, by the sixteenth century, Descartes had showed that only one kind of magnitude is needed. Furthermore, by that time, magnitudes were being treated algebraically, following the tradition begun by the Arab mathematicians during the dark ages in Europe. These are the magnitudes that we have been calling *quantities*, following the language of [3].

10.2.1 Algebraic Manipulation of Equalities

Because quantities are treated algebraically, certain operations can be performed on them. The main operations involved are addition and multiplication. The rules these operations obey are as follows:

F1. CLOSURE OF ADDITION. For any two quantities a and b, there is a unique quantity $a + b$, called the *sum* of a and b.

F2. COMMUTATIVE LAW OF ADDITION. For any two quantities a and b,

$$a + b = b + a.$$

F3. ASSOCIATIVE LAW OF ADDITION. For any three quantities a, b, and c,

$$a + (b + c) = (a + b) + c.$$

F4. IDENTITY OF ADDITION. There is a quantity 0, called *zero* such that for any quantity a,

$$a + 0 = 0 + a = a.$$

F5. ADDITIVE INVERSE. For every quantity a, there is a quantity $-a$, called the *negative of a*, such that

$$a + (-a) = (-a) + a = 0.$$

F6. CLOSURE OF MULTIPLICATION. For any two quantities a and b, there is a unique quantity ab called the *product* of a and b.

F7. COMMUTITATIVE LAW OF MULTIPLICATION. For any two quantities a and b,

$$ab = ba.$$

F8. ASSOCIATIVE LAW OF MULTIPLICATION. For any three quantities a, b, and c,

$$a(bc) = (ab)c.$$

F9. IDENTITY OF MULTIPLICATION. There is a quantity 1, called *one*, such that for any quantity a,

$$a1 = 1a = a.$$

F10. MULTIPLICATIVE INVERSE. For every quantity a except 0, there is a quantity $\frac{1}{a}$, called the *reciprocal* of a, such that

$$a\frac{1}{a} = \frac{1}{a}a = 1.$$

F11. DISTRIBUTIVE LAW. For any three quantities a, b, and c,

$$a(b + c) = ab + ac.$$

It can be shown that all the usual rules of algebraic manipulation of equalities follow from these 11 rules.

10.2.2 Ordering Relation for Inequalities and Archimedes Law

For inequalities, we need the properties of the relation "less than", or what is called the *ordering relation*. Its basic properties are as follows:

OF1. For any two quantities a and b, exactly one of $a < b$, $a = b$, and $b < a$ holds.

OF2. For all quantities a, b, and c, if $a < b$ and $b < c$, then $a < c$.

OF3. For quantities a and b, if $0 < a$ and $0 < b$, then $0 < a + b$.

OF4. For quantities a and b, if $0 < a$ and $0 < b$, then $0 < ab$.

OF5. For quantities a and b, $a < b$ if and only if $0 < b + (-a)$.

When $a < b$ we can also write $b > a$. Furthermore, we write $a \leq b$ instead of ($a < b$ or $a = b$).

It can be shown that the usual rules for manipulation of inequalities follow from F1–F11 and OF1–OF5.

There is one other law that our quantities satisfies. To state it, we need to note that any set of quantities satisfying F1–F11 includes the positive integers. For by F9 it includes 1, and by F1 it includes $1 + 1$, $1 + 1 + 1$, etc. Hence, it includes all positive integers.

> AL. ARCHIMEDES LAW. For any two quantities a and b where $b > a > 0$, there is a positive integer n such that $b < an$.

This law is associated with Archimedes despite the fact that it is mentioned in Euclid's *Elements*, which was written before Archimedes did most of his work.

On relevant collections of quantities, we will show that AL is equivalent to the following so-called Archimedean Property:

> AP. ARCHIMEDEAN PROPERTY. For any two quantities a and b where $a > 0$, there is a positive integer n such that $b < an$.

10.2.3 Archimedean Ordered Fields

Sets of objects which, if called "quantities", satisfy F1–F11, OF1–OF5, and AL, are important enough to be given a name.

Definition 10.2.1 (Field, Ordered Field, Archimedean Ordered Field). *A set of objects called quantities which satisfies F1–F11 is called a* field. *A field in which OF1–OF5 are also satisfied is called an* ordered field. *An ordered field which also satisfies AL is called an* Archimedean ordered field.

The following lemma shows that on an ordered field, AL is equivalent to AP. Its proof is left as an exercise (Exercise 10.4).

Lemma 10.2.2. *Let A be an ordered field. Then, A satisfies AL iff A satisfies AP.*

Among the sets of numbers that we are used to, not all are fields.

Example 10.2.3. *1. The set of positive integers $1, 2, \ldots$ is not a field. F1–F3, F6–F9, and F11 are satisfied, as are OF1–OF2. But there is no additive identity and there are no inverses. However, the cancellation law holds for addition and multiplication:*

(a) If $a + c = b + c$, then $a = b$.

(b) If $ac = bc$, then $a = b$.

In a field, we can derive these cancellation laws by adding the negative of c to both sides for 1(a) and multiplying both sides by the reciprocal of c for 1(b). (Since 0 is not a positive integer, we cannot expect OF3–OF5 to hold.)

2. *The set of* natural numbers[8] *0, 1, 2, . . . is not a field. This adds the additive identity, but negatives and reciprocals are missing. In cancellation law 1(b) it must be assumed that $c \neq 0$. OF3–OF5 hold.*

3. *The set of* integers *. . . , −2, −1, 0, 1, 2, . . . is not a field. This system includes negatives, but not reciprocals. The cancellation law for multiplication, 1(b), holds with the assumption $c \neq 0$, as it does for natural numbers. All of OF1–OF5 hold.*

4. *The set of* positive rational numbers, *which consists of the quotients of all positive integers is not a field.[9] This system has reciprocals but not negatives, and cancellation law 1(a) holds. OF1–OF2 hold; OF3–OF5 cannot hold since 0 is not a positive rational number.*

5. *However, the set of* rational numbers, *which consist of quotients of integers,[10] is a field, and, in fact, an Archimedean ordered field.*

We summarise the status so far of our various collections of numbers as follows:

- None of $\mathbb{N}$ or $\mathbb{Z}$ is a field.

- $\mathbb{Q}$ is a field, is an ordered field and is an Archimedean ordered field.

Our quantities must include the set of rational numbers.

10.2.4   Exercises

Exercise 10.3. Show that any Archimedean ordered field contains all the rational numbers.

Exercise 10.4. Prove Lemma 10.2.2.

10.3   Which Field can be the Quantities? Axiom of Completeness

---

[8]Some authors identify the natural numbers with the positive integers.

[9]More precisely, a positive rational number is an equivalence class of formal quotients under the equivalence relation of cross multiplication, see Definitions 3.2.1 and 3.2.12.

[10]Where the denominator is not 0.

10.3.1 Archimedean Ordered Fields are not sufficient for Quantities

We have said in Section 10.2 that quantities should form an Archimedean ordered field. The smallest such field available for the quantities is the field of rational numbers.

Consider the following obvious theorem:

> *Suppose a_n is defined for all $n > N$, and suppose that for all n, $a_n \leq a_{n+1}$. Suppose there is a g, independent of n, such that for $n > N$,*
>
> $$a_n \leq g.$$
>
> *Then $\lim_{n \to \infty} a_n$ exists, and*
>
> $$a_N \leq \lim_{n \to \infty} a_n \leq g.$$

This theorem is intuitively obvious, especially if we look at it geometrically.

Let us consider this theorem further. Let l be the limit of the sequence $\{a_n\}$. We know from the theorem that $a_N \leq l \leq g$. But we know more than that:

1. $l \geq a_n$ for all n. For suppose there is some N such that $l < a_N$. Then for all $n > N$, $l < a_n$. Also, if we let $\epsilon = a_N - l$, then for all $n > N$, $a_n - l \geq \epsilon$. Hence, there is an $\epsilon > 0$ such that for every N, there is an $n > N$ such that $a_n - l \geq \epsilon$. This is the negation of the definition that the limit of a_n is l, and so is a contradiction. Hence, we must have $l \geq a_n$ for all n.

2. There is no quantity $l' < l$ which is greater than or equal to all the a_n. For suppose there is. Let $\epsilon = l - l'$. Then for all n, $|a_n - l| = l - a_n > \epsilon$, another contradiction as above.

Thus, l is the least quantity which is greater than or equal to all the a_n.

We use some vocabulary in connection with these properties:

Definition 10.3.1. *(Bounded above, (least) upper bound, supremum, maximum)* *Let A be a collection of quantities.*

- *If g is a quantity which has the property that for all quantities a of A, $a \leq g$, then we say that g is an* upper bound *of A.*

- *If A has an upper bound, we say that A is* bounded above.

- If g is the smallest quantity such that for all quantities a of A, $a \leq g$, then we say that g is the least upper bound of A. It is also called the supremum of A. The supremum of A is also denoted $\sup A$.

- We say that a quantity c is maximum of A iff ($c \in A$ and $\forall x \in A$, $x \leq c$).

Thus, the theorem quoted at the start of this section says that:

> If $\{a_n\}$ is a nondecreasing sequence which has an upper bound g, then it has a least upper bound, which is the limit l of the sequence.

Example 10.3.2. *Now this theorem is not satisfied if the only quantities allowed are rational numbers. For consider the sequence $\{a_n\}$ defined as follows: $a_1 = 1$ and*

$$a_{n+1} = \frac{2a_n + 2}{a_n + 2}.$$

The following hold:

1. For all $n \geq 1$, we have $a_n \geq 1$. The proof is by induction on $n \geq 1$ noting that $a_{n+1} = 1 + \dfrac{a_n}{a_n + 2}$.

2. $a_n^2 < 2$ for all n, since $a_1^2 = 1^2 = 1 < 2$ and, if $a_n^2 < 2$ we have

$$
\begin{aligned}
a_{n+1}^2 &= \left(\frac{2a_n + 2}{a_n + 2}\right)^2 \\
&= \frac{4(a_n + 1)^2}{(a_n + 2)^2} \\
&= \frac{2(a_n + 2)^2 - 2(2 - a_n^2)}{(a_n + 2)^2} \\
&= 2 - \frac{2(2 - a_n^2)}{(a_n + 2)^2} \\
&< 2
\end{aligned}
$$

Hence the property holds by induction on $n \geq 1$.

3. For all $n \geq 1$, we have $a_{n+1} > a_n$. This is because by 2 above $a_n^2 < 2$ for all n and

$$
\begin{aligned}
a_n + \frac{2 - a_n^2}{a_n + 2} &= \frac{a_n(a_n + 2) + (2 - a_n^2)}{a_n + 2} \\
&= \frac{2a_n + 2}{a_n + 2} \\
&= a_{n+1}.
\end{aligned}
$$

4. By 1 above $a_n \geq 1$ and by 2 above $a_{n+1}^2 = 2 - \dfrac{2(2 - a_n^2)}{(a_n + 2)^2}$ and hence:

$$2 - a_{n+1}^2 = \frac{2}{(a_n + 2)^2}(2 - a_n^2) \leq \frac{2}{9}(2 - a_n^2).$$

This means that the passage from $2 - a_n^2$ to $2 - a_{n+1}^2$ involves, at each step, subtracting more than half of the previous difference, and from this it follows that

$$\lim_{n \to \infty} a_n^2 = 2.$$

But the sequence $\{a_n\}$ has no limit in the rational numbers. Clearly, it has an upper bound, namely 1.5. However, it has no least upper bound. For suppose b is an upper bound for this sequence, which means that $b^2 > 2$, and let

$$c = \frac{b(b^2 + 6)}{3b^2 + 2}.$$

Then

$$
\begin{aligned}
c^2 - 2 &= \frac{b^2(b^2 + 6)^2}{(3b^2 + 2)^2} - 2 \\
&= \frac{b^2(b^2 + 6)^2}{(3b^2 + 2)^2} - \frac{2(3b^2 + 2)^2}{(3b^2 + 2)^2} \\
&= \frac{b^2(b^4 + 12b^2 + 36) - 2(9b^4 + 12b^2 + 4)}{(3b^2 + 2)^2} \\
&= \frac{b^6 + 12b^4 + 36b^2 - 18b^4 - 24b^2 - 8}{(3b^2 + 2)^2} \\
&= \frac{b^6 - 6b^4 + 12b^2 - 8}{(3b^2 + 2)^2} \\
&= \frac{(b^2 - 2)^3}{(3b^2 + 2)^2} \\
&> 0,
\end{aligned}
$$

so $c^2 > 2$ and c is an upper bound of the a_n, and

$$
\begin{aligned}
c - b &= \frac{b(b^2 + 6)}{(3b^2 + 2)} - b \\
&= \frac{b(b^2 + 6)}{(3b^2 + 2)} - \frac{b(3b^2 + 2)}{3b^2 + 2} \\
&= \frac{b^3 + 6b - 3b^3 - 2b}{(3b^2 + 2)} \\
&= \frac{-2b^3 + 4b}{(3b^2 + 2)} \\
&= \frac{2b(2 - b^2)}{(3b^2 + 2)} \\
&< 0,
\end{aligned}
$$

so $c < b$. This shows that for every rational upper bound of the set of rational numbers whose square is less than 2, there is a smaller rational upper bound for that set. This proves that the sequence $\{a_n\}$ has no limit if the only quantities are rational numbers.

Example 10.3.3. *Now this theorem is not satisfied if the only quantities allowed are rational numbers. For consider the sequence $\{a_n\}$ defined as follows: $a_1 = 1$ and*

$$a_{n+1} = \frac{2a_n + 2}{a_n + 2}.$$

Now we have that $a_{n+1} > a_n$ as long as $a_n^2 < 2$, since

$$
\begin{aligned}
a_n + \frac{2 - a_n^2}{a_n + 2} &= \frac{a_n(a_n + 2) + (2 - a_n^2)}{a_n + 2} \\
&= \frac{2a_n + 2}{a_n + 2} \\
&= a_{n+1}.
\end{aligned}
$$

But $a_n^2 < 2$ for all n, since $a_1^2 = 1^2 = 1 < 2$ and, if $a_n^2 < 2$ we have

$$
\begin{aligned}
2 - \frac{2(2 - a_n^2)}{(a_n + 2)^2} &= \frac{2(a_n + 2)^2 - 2(2 - a_n^2)}{(a_n + 2)^2} \\
&= \frac{4(a_n + 1)^2}{(a_n + 2)^2} \\
&= \left(\frac{2a_n + 2}{a_n + 2} \right)^2 \\
&= a_{n+1}^2,
\end{aligned}
$$

which tells us that $a_{n+1}^2 < 2$. Finally, it follows from the last result and the fact that $a_n \geq 1$ that

$$2 - a_{n+1}^2 = \frac{2}{(a_n + 2)^2}(2 - a_n^2) \leq \frac{2}{9}(2 - a_n^2).$$

This means that the passage from $2 - a_n^2$ to $2 - a_{n+1}^2$ involves, at each step, subtracting more than half of the previous difference, and from this it follows that

$$\lim_{n \to \infty} a_n^2 = 2.$$

But this sequence has no limit in the rational numbers. Clearly, it has an upper bound, namely 1.5. However, it has no least upper bound. For suppose b is an upper bound for this sequence, which means that $b^2 > 2$, and let

$$c = \frac{b(b^2 + 6)}{3b^2 + 2}.$$

Then

$$c^2 - 2 = \frac{b^2(b^2+6)^2}{(3b^2+2)^2} - 2$$

$$= \frac{b^2(b^2+6)^2}{(3b^2+2)^2} - \frac{2(3b^2+2)^2}{(3b^2+2)^2}$$

$$= \frac{b^2(b^4+12b^2+36) - 2(9b^4+12b^2+4)}{(3b^2+2)^2}$$

$$= \frac{b^6 + 12b^4 + 36b^4 - 36b^2 - 18b^4 - 24b^2 - 8}{(3b^2+2)^2}$$

$$= \frac{b^6 - 6b^4 + 12b^2 - 8}{(3b^2+2)^2}$$

$$= \frac{(b^2-2)^3}{(3b^2+2)^2}$$

$$> 0,$$

so $c^2 > 2$ and c is an upper bound of the a_n, and

$$c - b = \frac{b(b^2+6)}{3b^2+2} - b$$

$$= \frac{b(b^2+6)}{3b^2+2} - \frac{b(3b^2+2)}{3b^2+2}$$

$$= \frac{b^3 + 6b - 3b^3 - 2b}{3b^2+2}$$

$$= \frac{-2b^3 + 4b}{3b^2+2}$$

$$= \frac{2b(2-b^2)}{3b^2+2}$$

$$< 0,$$

so $c < b$. This shows that for every rational upper bound of the set of rational numbers whose square is less than 2, there is a smaller rational upper bound for that set. *This proves that the sequence $\{a_n\}$ has no limit if the only quantities are rational numbers.*

10.3.2   Ordered Fields and the Axiom of Completeness

Suppose the set of quantities is such that *every* set of quantities that has an upper bound has a least upper bound. Then the theorem quoted in Section 10.3.1 is easily proved: the least upper bound of all the terms a_n

for $n > N$ is the required limit. Thus, our quantities will have to be an Archimedean ordered field that satisfies the following property:[11]

> AC. AXIOM OF COMPLETENESS. Every nonempty set of quantities that has an upper bound has a least upper bound.

Definition 10.3.4 (Real Numbers $\mathbb{R}$). *From now on, we will assume that our quantities form an ordered field that satisfies the Axiom of Completeness AC, and we will refer to them as real numbers and denote their collection by $\mathbb{R}$.*

10.3.3   Exercises

Exercise 10.5. Let S be a nonempty set of real numbers that is bounded above. Prove that if the least upper bound of S belongs to S, then it is a maximum element of S. *Hint*: Your proof should be very short.

Exercise 10.6. Let S be a non empty set of real numbers. Show that S has a maximum iff S has a least upper bound which is an element of S.

Exercise 10.7. Let S be a finite set. Prove that the least upper bound of S belongs to S. Note that it follows from this and the previous exercise that if S is finite, the least upper bound of S is a maximum element of S.

10.4   Simple Consequences of the Axiom of Completeness

10.4.1   Upper versus Lower Bounds

What we have said about upper bounds and least upper bounds can be inverted to deal with lower bounds.

Definition 10.4.1 (Bounded below, Bounded, (greatest) lower bound, infinum, minimum).

- *If g is a real number such that, for all real numbers a of a collection $A \subseteq \mathbb{R}$, $g \le a$, then g is said to be a* lower bound *for A.*

---

[11]The Axiom of Completeness is also known as Dedekind Completeness, after Dedekind who showed in 1872 that the real numbers are exactly those numbers that satisfy F1–F11 and OF1, OF5 and this axiom.

- *If A has a lower bound, we say that A is bounded below.*

- *If A is both bounded below and bounded above, we simply say that A is bounded.*

- *If g is a lower bound for A, and if there is no real number greater than g which is also a lower bound for A, then g is said to be the greatest lower bound for A. It is also called the* infimum *of A, and denoted* inf A.

- *We say that a real number g is minimum of A iff (g ∈ A and ∀a ∈ A, g ≤ a).*

Theorem 10.4.2. *Every nonempty set of real numbers that has a lower bound has a greatest lower bound.*

Proof. Let A be a set of real numbers with a lower bound g. Then for each a in A, $a \geq g$. Let A' be the set of negatives of real numbers in A; i.e., b is in A' if and only if $b = -a$ for some a in A. Since for each a in A we have $a \geq g$, it follows that $-a \leq -g$. Hence, for each b in A', $b \leq -g$, and so $-g$ is an upper bound for A'. By the Axiom of Completeness AC, A' has a least upper bound, say h. Then for each b in A', $b \leq h$, and h is the smallest real number with this property. This means that for each a in A, $-a \leq h$. Hence, $a \geq -h$. Now $-h$ is the greatest lower bound of A; for suppose there were a greater lower bound for A, say h'. Then we would have that for every a in A, $a \geq h'$ and $h' > -h$. Then we would also have $-h' < h$ and, for each a in A, $-a \leq -h'$, which would mean that $-h'$ would be an upper bound for A' less than h, contradicting the fact that h is the least upper bound for A'. $\qquad\square$

Example 10.4.3. *Let* $A = \{\frac{1}{a} : a \in \mathbb{R} \text{ and } a > 0\}$.

- *A has no minimum because for every $\frac{1}{a} \in A$, we have $\frac{1}{a+1} \in A$ and $\frac{1}{a+1} < \frac{1}{a}$.*

- *A has a lower bound 0 because for all $x \in A$, $0 < x$.*

- *By Theorem 10.4.2, A has a greatest lower bound.*

10.4.2 The Axiom of Completeness and the Archimedean Property

Recall that $\mathbb{Q}$ is an Archimedean ordered field which can be seen by the Examples of Section 10.3.1 to not satisfy the Axiom of Completeness. Recall also that by Definition 10.3.4, $\mathbb{R}$ is an ordered field which does satisfy the Axiom of Completeness. Actually $\mathbb{R}$ is not only an ordered field, it is also an Archimedean ordered field. As it turns out, AP (and hence its equivalent AL on an ordered field, see Lemma 10.2.2) can be proved from the Axiom of Completeness.

Theorem 10.4.4 (Completeness implies the Archimedean Property). *Assume a and b are real numbers such that $a > 0$. There is a positive integer n such that $an > b$.*

Proof. Assume the theorem is false. Then there are real numbers a, b where $a > 0$ such that for every positive integer n, $an \leq b$. Let A be the set of all real numbers of the form an for a positive integer n. Then b is an upper bound for A. By the Axiom of Completeness, A has a least upper bound, say c. Then, since $a > 0$, $c < c + a$, from which it follows that $c - a < c$. Since c is the least upper bound of A, it follows that $c - a$ cannot be an upper bound for A. This means that there is a positive integer N such that $c - a < aN$. But then $c < a + aN = a(N + 1)$. Since $a(N + 1)$ is in A, this implies that c is not an upper bound of A, which is a contradiction. Hence, the theorem must be true. $\qquad\square$

10.4.3 The Density of the Rationals

The following theorem shows that we can approximate real numbers by rational numbers.

Theorem 10.4.5 (Density of rationals). *If a and b are any two real numbers with $a < b$, then there is a rational number r such that $a < r < b$.*

Proof. We need to show that there are integers m and n, where $n > 0$, such that $a < \frac{m}{n} < b$, and for this we need $an < m < bn$. Now since $a < b$, it follows that $b - a > 0$. By theorem 10.4.4, there is a positive integer n such that $(b - a)n > 1$. This means that $bn - an > 1$, and so it seems reasonable that there must be an integer m between an and bn, from which it follows that $an < m < bn$. Hence, there is a rational number $r = \frac{m}{n}$ such that $a < r < b$.

But this is not really a proof. For a real proof, note that:

- We said above that by theorem 10.4.4, there is a positive integer n such that $bn - an > 1$.

- Since n is positive and $a < b$ then $an < bn$.

- Let $c = \max\{|an|, |bn|\}$. By Theorem 10.4.4, there is a positive integer k such that $c < k$. Hence

$$-k < an < bn < k.$$

Then the set of all integers j such that $-k < j \le k$, and $an < j$ is finite and nonempty. Let m be the least such j; i.e., m is the least j such that $-k < j \le k$, and $an < j$. It follows that $an < m$ but $m - 1 < an$. Also, we have

$$m = (m - 1) + 1 \le an + 1 < an + (bn - an) = bn,$$

and so it follows that $an < m < bn$. And from this, it follows that $a < \frac{m}{n} < b$. $\qquad\square$

10.4.4 Exercises

Exercise 10.8. Show that if a is a real number, then there exists an integer $m > 0$ such that $a < m$.

Exercise 10.9. Let S be a nonempty bounded set of real numbers.

1. Prove that the greatest lower bound of S is less than or equal to the least upper bound of S. *Hint:* This is almost obvious; your proof should be short.

2. What can you say about S if its greatest lower bound equals its least upper bound?

Exercise 10.10. If we look back at Example 10.4.3, we recall that the collection A of that example had a lower bound 0 and a greatest lower bound. However, we did not give such a greatest lower bound. Show that 0 is a greatest lower bound of A.

Exercise 10.11. Let S and T be nonempty bounded sets of real numbers.

1. Prove that if S is a subset of T, then the greatest lower bound of T is less than or equal to the greatest lower bound of S, which is less than or equal to the least upper bound of S, which is less than or equal to the least upper bound of T. In other words, prove that if $S \subseteq T$, then $\inf T \le \inf S \le \sup S \le \sup T$.

2. Prove that $\sup(S \cup T) = \max\{\sup S, \sup T\}$. *Note:* In this part, do *not* assume that $S \subseteq T$.

Exercise 10.12. Let S be a nonempty set of real number that is bounded above, and let β be the least upper bound of S. Prove that for every $\epsilon > 0$, there exists an element x of S such that $x > \beta - \epsilon$.

Exercise 10.13. Let x be a real number. Prove that for each $\epsilon > 0$, there exists a rational number q such that $|x - q| < \epsilon$.

Exercise 10.14. Prove that if $a < b$, then there is an irrational number x such that $a < x < b$. *Hint*: First show that if $q \neq 0$ is rational, then $q + \sqrt{2}$ and $q\sqrt{2}$ are irrational.

Exercise 10.15. Let S and T be nonempty subsets of the real numbers with the property that for any $s \in S$ and any $t \in T$, $s \leq t$.

1. Show that S is bounded above and T is bounded below.

2. Prove that $\sup S \leq \inf T$.

3. Give an example of such sets S and T where $S \cap T$ is nonempty.

4. Give an example of such sets S and T where $\sup S = \inf T$ and $S \cap T$ is the empty set.

Exercise 10.16. Prove that if $a > 0$, then there is an integer n such that $1/n < a < n$.

Exercise 10.17. Let A and B be nonempty bounded sets of real numbers, and let S be the set of all sums $a + b$ where $a \in A$ and $b \in B$.

1. Prove that $\sup S = \sup A + \sup B$.

2. Prove that $\inf S = \inf A + \inf B$.

Exercise 10.18. For real numbers a and b, prove that if $a \leq b + 1/n$ for all positive integers n, then $a \leq b$.

Exercise 10.19. For each real number a, let the set S_a be the set of all rational numbers less than a. Prove that for each real number a, $\sup S_a = a$.

10.5   Zeno's Paradoxes

We are now in a position to give a resolution of each of the four paradoxes of Zeno given in Section § 2.2.

1. *Dichotomy.* "There is no motion, because what moves must arrive at the middle of its course before it reaches the end." In other words, to leave the room, you first have to get halfway to the door, then you have to get halfway from that point to the door, etc. No matter how close you are to the door, you have to go half the remaining distance before proceeding.

   Our modern solution is to consider the sequence of times, where time is regarded as being given by a real number. If the object is moving at constant speed, the times involved in going to the halfway point and then doing that again, *ad infinitum* form an increasing convergent sequence, and so the terms of the sequence never reach the limit. but since the time represented by the limit clearly comes eventually, it is simply not true that what moves never reaches the end.

2. *Achilles.* "The slower in a race will never be overtaken by the quicker; because the pursuer must first reach the starting point of the pursued, so that the slower must always be some distance ahead."

   This paradox is also resolved these days by considering a sequence of times, which is also a convergent sequence.

3. *Arrow.* "The flying arrow is at rest," because a thing is at rest when occupying its own space at a given time, as the arrow does at every instant of its alleged flight.

   In modern terms, this does not contradict in any way the possibility of motion. Today we regard a position in space as given by a triple (x, y, z) of real numbers. If we measure the position of the arrow by its tip, that position changes with time, and in the case of motion, that motion is expressed by a continuous function of each of the coordinates with respect to time. Each instant is represented by one real number of the time coordinate. Since the functions expressing the motion are continuous, there are no sudden jumps, and so these functions represent motion as we expect it to. The arrow is at rest at each instant, but it is in motion because at different instants, it is in different positions.

4. *Stadium.* This argument concerns 2 rows of bodies equal in number which pass one another on a race course as they move in opposite directions. This is illustrated in the passage from

$$
\begin{array}{ccccccc}
& & & A & A & A & A & A & A \\
B_6 & B_5 & B_4 & B_3 & B_2 & B_1 & & & \\
& & & & & & C_1 & C_2 & C_3 & C_4 & C_5 & C_6
\end{array}
$$

to

$$
\begin{array}{cccccc}
A & A & A & A & A & A \\
B_6 & B_5 & B_4 & B_3 & B_2 & B_1 \\
C_1 & C_2 & C_3 & C_4 & C_5 & C_6
\end{array}
$$

B_1 will have passed alongside 3 A's and 6 C's during the same time and without changing. The same would be true of C_1. Zeno's idea was to show that a time is equal to its double.

In modern terms, we have no problem with the relative velocity of the Bs to the Cs being twice the relative motion of each to the As. After all, the calculus is about motion and rates of change, and so we have learned to recognise the possibility of different rates of change, or different velocities.

Chapter 11

Continuity and Derivatives

> If we could find characters or signs appropriate for expressing all
> our thoughts as definitely and as exactly as arithmetic expresses
> numbers or geometric analysis expresses lines, we could in all
> subjects in so far as they are amenable to reasoning accomplish
> what is done in Arithmetic and Geometry.   Leibniz [28]

In this chapter, we introduce the continuity and derivatives concepts and
show how continuity is preserved under different function operations and
give a number of properties of derivatives such as the Quotient rule and the
Chain rule.

11.1 Magnitudes

As we mentioned in Chapter 7, the ancient Greeks separated numbers, which
are discrete, from continuous magnitudes. They did not use fractions to
approximate continuous magnitudes, and they had different kinds of magnitudes for lengths, areas, volumes, angles, etc. They never multiplied two
lengths to get another length.

By the 17th century, this had changed. The idea of using fractions to
approximate continuous magnitudes developed first in the Arab world during
the middle ages, and came to Europe in the 16th and 17th centuries. This
idea of using fractions to approximate continuous magnitudes would have
later on been assumed by both Newton and Leibniz.

Also, before Newton and Leibniz introduced the calculus (which they did
in the second half of the 17th century), Descartes had published a ruler-and-
compass construction for multiplying two lengths to get a length. Recall

that in Chapter 7 we stated that for the Greeks, two length magnitudes could be multiplied, but the result was not another length magnitude, but an area magnitude. Similarly for the Greeks, an area magnitude multiplied by a length magnitude was a volume magnitude. Hence, Descartes' ruler-and-compass construction for *multiplying* two lengths to get a length was innovative and it allowed Algebra to be a science concerned with numbers rather than geometric magnitudes. Here is an example of how Descartes' ruler-and-compass construction would allow to multiply and divide lengths to obtain lengths.

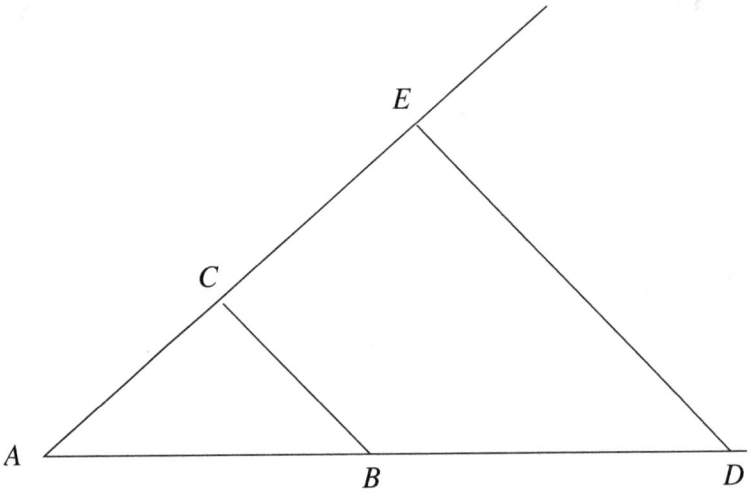

Figure 11.1: Descartes' Construction

Example 11.1.1. *Let the two length magnitudes be a and b. On the line AB, construct the point B so that the length AB is a. On a line AC through A and at an angle to AB, let the length of AC be a unit, and construct E on the same line so that the length of AE is b (see Figure 11.1). Join C and B with line segment BC, and construct a line through E parallel to BC; let this line intersect the extension of AB at D. Then triangles ABC and ADE are similar; i.e., $\angle ACB = \angle AED$ and $\angle ABC = \angle ADE$, and*

$$AC \text{ is to } AE \text{ as } AB \text{ is to } AD \text{ as } BC \text{ is to } DE.$$

$$\text{Hence, } AE \text{ is to } AC \text{ as } AD \text{ is to } AB,$$

$$\text{or } AD \text{ is to } a \text{ as } b \text{ is to the unit. Hence, } AD = ab.$$

Using the modern interpretation,

$$AC : AE = AB : AD = BC : DE$$

$$AE : AC = AD : AB$$

$$b : 1 = AD : a \text{ and hence, } AD = ab.$$

Essentially the same construction can be used to divide any two magnitudes. This time let AD be a, and, as before, let AC be the unit and AE be b. Then AB will be the length $\frac{a}{b}$. Using the modern interpretation,

$$AE : AC = AD : AB$$

$$b : 1 = a : AB \text{ and hence, } AB = \frac{a}{b}.$$

For this reason, the different kinds of magnitudes the ancient Greeks had kept separate could now be considered together. Furthermore, they were now being treated algebraically. And, thanks to Descartes, the ideas behind analytic geometry were now commonplace.

To emphasise that magnitudes were no longer considered as a separate kind, let us refer to them as *quantities*.

11.1.1   Exercises

Exercise 11.1. Read Appendix A.1 and based on the text, think of a couple of ways to double the area of a given square.

Exercise 11.2. In Euclid's *Elements* you learn that given a square S, you can build a square T with a specific ratio to S. For example, assume S has side a, to build a square T whose area is double that of S, you need to find the side x of the new square which is one mean proportional between a and $2a$. In modern notation, this amounts to finding the side x of the new square T such that

$$a : x = x : 2a.$$

1. Explain why this formula will produce a new square T whose area is double that of S.

2. Which formula should we use if we want the new square T to have 3 times the area of that of S.

3. Repeat the exercise but where the area of T is three quarters of that of S.

Exercise 11.3. Based on the above Exercise 11.2, assume cube S has side a, and build a cube T whose volume is double that of S. Hint, here, you need to find the side x of T and for this you will need two mean proportionals between a and $2a$. Use a temporary variable y such that $a : x = x : y = ...$

Repeat the exercise to find a cube T whose volume is three quarters that of S.

11.2 Continuity

Any students who have taken calculus will be familiar with the definition of what it means for a function to be continuous. Basically, the change in the values of the output of a continuous function is small when the change in the values of the input is also small. A function is continuous at a point means the graph of the function will not have any hole at the points at which it is continuous.

Definition 11.2.1 (Continuous function). *A function $f(x)$ is continuous at a quantity c if c is in the domain of f and*

$$\lim_{x \to c} f(x) = f(c).$$

In view of the definition of the limit of a function, this means that for every $\epsilon > 0$, there is a $\delta > 0$ such that for all x satisfying $|x-c| < \delta$, $|f(x)-f(c)| < \epsilon$.

If an interval I is in the domain of f, then f is said to be continuous on *I if f is continuous at c for every c in I. If c happens to be an endpoint of the interval I, then only the appropriate one-sided limit is considered.*

Remark 11.2.2. *It is worth recalling Definition 4.1.16 where we stated that a* closed *interval is a set of the form $[a, b] = \{x : a \le x \le b\}$. An* open *interval is one of the form $(a, b) = \{x : a < x < b\}$. A half open* interval *is one of one of the two forms $[a, b) = \{x : a \le x < b\}$ and $(a, b] = \{x : a < x \le b\}$.*

The following theorem should be familiar. Its proof is left as an exercise (see Exercise 11.8).

Theorem 11.2.3. *Let I be an interval which contains c and let f be a function whose domain includes I. The following statements are equivalent:*

1. *The function f is continuous at c.*

2. *The function f has a limit at c and $\lim_{x \to c} f(x) = f(c)$. (If c is an endpoint of I, then an appropriate one-sided limit is used here.)*

3. *The sequence $\{f(x_n)\}$ converges to $f(c)$ for each sequence $\{x_n\}$ in I that converges to c.*

If k is a constant and f is a function, then define $(kf)(x) = kf(x)$. Furthermore, define the sum, difference, product and division of functions as follows:

$$(f + g)(x) = f(x) + g(x) \qquad (f - g)(x) = f(x) - g(x)$$
$$(fg)(x) = f(x)g(x) \qquad (\tfrac{f}{g})(x) = \frac{f(x)}{g(x)}.$$

The next theorem is concerned with the continuity of the sum, product, difference and division of continuous functions. Its proof is left as an exercise (see Exercise 11.9).

Theorem 11.2.4. *Let f and g be functions defined on a domain that includes an interval which contains c, and let k be a constant. If f and g are continuous at c, then the functions $f + g$, $f - g$, kf, and fg are continuous at c. If moreover $g(c) \neq 0$, then $\frac{f}{g}$ is continuous at c.*

Corollary 11.2.5. *Let f and g be continuous functions defined on an interval I and let k be a constant. Then the functions $f + g$, $f - g$, kf and fg are continuous on I. If moreover $g(x) \neq 0$ for all x in I, then $\frac{f}{g}$ is continuous on I.*

Recall Definition 4.2.21 where we defined the composition function $f \circ g$ of two functions f and g. We repeat this definition here.

Definition 11.2.6. *Let I be an interval, let c be an element of I, let g be a function whose domain includes I, and let f be defined on an interval J that includes the image $g(I) = \{g(x) : x \in I\}$. The function $f \circ g$, which is defined by $(f \circ g)(x) = f(g(x))$, is called the* composition *of f and g.*

The following theorem states that the composition of two continuous functions is continuous. See Exercise 11.10.

Theorem 11.2.7. *Let I be an interval, let c be an element of I, let g be a function whose domain includes I, and let f be defined on an interval J that includes the image $g(I) = \{g(x) : x \in I\}$. If g is continuous at c and f is continuous at $g(c)$, then $f \circ g$, which is defined by $(f \circ g)(x) = f(g(x))$, is continuous at c. Hence, if g is continuous on I and f is continuous on J, then $f \circ g$ is continuous on I.*

Recall the definition of a polynomial function 4.1.11. Note also that a rational function $f(x)$ is of the form $\frac{p(x)}{q(x)}$ where $p(x)$ and $q(x)$ are polynomials. The following theorem states that polynomial functions and rational functions are continuous. See Exercise 11.11.

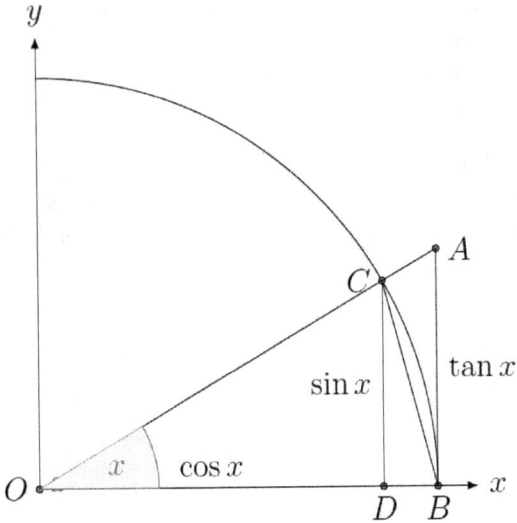

Figure 11.2: Diagram for $(\sin x)/x$

Theorem 11.2.8.   *1. A polynomial is continuous at every quantity.*

2. A rational function is continuous at every quantity for which it is defined.

Now we move to the trigonomic functions:

Theorem 11.2.9. *The six trigonometric functions are continuous at every quantity for which they are defined.*

This theorem will follow from the following lemmas:

Lemma 11.2.10. $|\sin x| \leq |x|$ *for all* x.

Proof. It is clear from Figure 11.2 that if $0 < x < \pi/2$, $\sin x < x$. As x increases beyond $\pi/2$, $x > 1$, while $\sin x \leq 1$, so for all positive values of x, $\sin x < x$. We also have $\sin 0 = 0$. As for negative numbers, $\sin(-x) = -(\sin x)$, and if x is negative, $|x| = -x$ and $|\sin x| = -\sin x$. This shows that for all values of x, $|\sin x| \leq |x|$. $\qquad\square$

Lemma 11.2.11. *For all values of a and b,* $|\sin a - \sin b| \leq |a - b|$.

Proof. Let $x = (a+b)/2$ and $y = (a-b)/2$. It is easy to verify that $x+y = a$ and $x - y = b$. Now for any x and y (and hence for these two values of x and y),

$$
\begin{aligned}
\sin(x + y) - \sin(x - y) &= (\sin x \cos y + \cos x \sin y) - (\sin x \cos y - \cos x \sin y) \\
&= \sin x \cos y + \cos x \sin y - \sin x \cos y + \cos x \sin y \\
&= 2 \sin y \cos x.
\end{aligned}
$$

It follows that

$$
\begin{aligned}
|\sin a - \sin b| &= \left| 2 \sin \frac{(a - b)}{2} \cos \frac{(a + b)}{2} \right| \\
&= 2 \left| \sin \frac{a - b}{2} \right| \left| \cos \frac{a + b}{2} \right| \\
&\leq 2 \frac{|a - b|}{2} \cdot 1 \\
&= |a - b|.
\end{aligned}
$$

$\qquad\square$

Lemma 11.2.12. *The function* $\sin$ *is continuous for all values.*

Proof. We need to prove that for every value c,

$$\lim_{x \to c} \sin x = \sin c.$$

To use the definition, let $\epsilon > 0$ be given and let $\delta = \epsilon$. For values of x for which $|x - c| < \delta = \epsilon$, we have, by Lemma 11.2.11,

$$|\sin x - \sin c| \leq |x - c| < \epsilon.$$

$\square$

Lemma 11.2.13. *The functions* cos, tan, cot, sec, *and* csc *are all continuous for all values for which they are defined.*

Proof. We have that $\cos x = \sin(\pi/2 - x)$, and is thus the composition of two functions that are continuous everywhere. Since $\tan x = (\sin x)/(\cos x)$ and $\cot x = (\cos x)/(\sin x)$, each is the quotient of two continuous functions and is thus continuous at all values for which the denominator is not 0; i.e., at every value at which both are defined. The same is true of $\sec x = 1/\cos x$ and $\csc x = 1/\sin x$. $\square$

11.2.1   Exercises

Exercise 11.4. Use the definition of continuity to prove that the given function is continuous everywhere.

1. $f(x) = 6x - 11$.

2. $g(x) = |x|$.

3. $h(x) = x^2$.

Exercise 11.5. Let a and b be constants with $a \neq 0$. Use the definition of continuity to prove that the function f defined by $f(x) = ax + b$ is continuous everywhere.

Exercise 11.6. Use the definition of continuity to prove that the function f defined by $f(x) = \sqrt{x}$ is continuous for all nonnegative values of x.

Exercise 11.7. Let f be a function defined on a closed interval $[a, b]$ that is continuous at a value c in $[a, b]$, and suppose that $f(c) > 0$. Prove that there exist a positive value m and an interval $[u, v]$ which is included in $[a, b]$ such that c is in $[u, v]$ and $f(x) \geq m$ for all x in $[u, v]$.

Exercise 11.8. Prove Theorem 11.2.3.

Exercise 11.9. Prove Theorem 11.2.4.

Exercise 11.10. Prove Theorem 11.2.7.

Exercise 11.11. Prove Theorem 11.2.8.

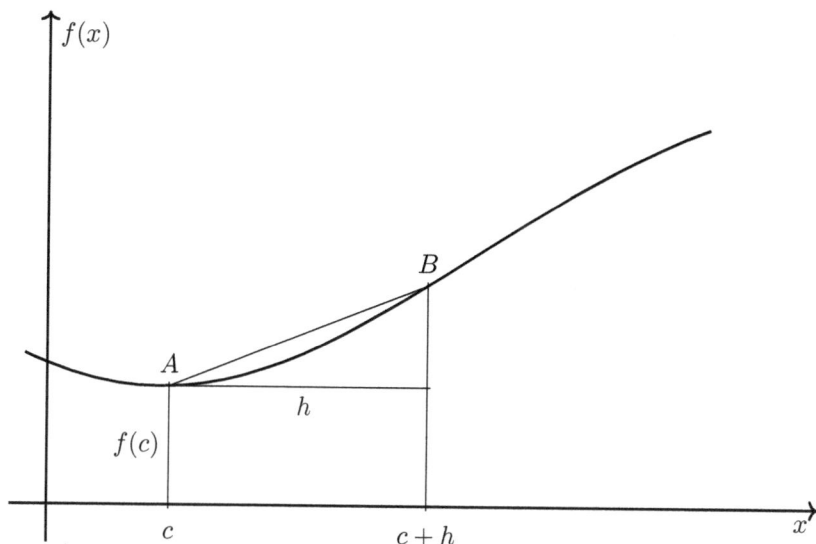

Figure 11.3: The derivative and the slope of the function

11.3 Derivatives

We can now start doing calculus. We start with the definition of the *derivative*. The derivative measures the steepness/slope/rate of change of the graph of a function. The derivative is also the slope of the tangent line to the graph of the function at a given point. For example, in the graph of Figure 11.3, the smaller h gets, the more AB approximates a tangent to the graph of the function.

Definition 11.3.1. *Let $f(x)$ be a function defined on an interval which includes c. The* derivative *of f at $x = c$ is defined by*

$$f'(c) = \lim_{h \to 0} \frac{f(c+h) - f(c)}{h}$$

if the limit exists. (If c is an endpoint of the interval, the appropriate one-sided limit is used here.) Other notations for $f'(c)$ are $\frac{dy}{dx}\big|_{x=c}$ if $y = f(x)$, or, if the value of c is clear from the context, $\frac{dy}{dx}$, and $Df\big|_{x=c}$, or, if the value of c is clear from the context, Df. The derivative of f, $f'(x)$, is the *function defined by*

$$f'(x) = \lim_{h \to 0} \frac{f(x+h) - f(x)}{h}$$

for each value of x for which the limit is defined. Other notations are $\frac{df}{dx}$ or Df. (If x is an endpoint of an interval on which f is defined, the appropriate one-sided limit is used.) If f has a derivative at c, then f is said to be differentiable at c. If f' exists for a set of values x, then f is said to be differentiable for those values of x.

Example 11.3.2.   • If $f(x) = x^2$ then

$$f'(c) = \lim_{h \to 0} \frac{f(c+h) - f(c)}{h} = \lim_{h \to 0} \frac{(c+h)^2 - c^2}{h} =$$

$$\lim_{h \to 0} \frac{h^2 + 2ch}{h} = h + 2c = 2c.$$

• If $f(x) = |x|$ then

$$f'(0) = \lim_{h \to 0} \frac{f(0+h) - f(0)}{h} = \lim_{h \to 0} \frac{|0+h| - f|0|}{h} = \lim_{h \to 0} \frac{|h|}{h}.$$

Hence, $\lim_{h \to 0+} \frac{f(0+h) - f(0)}{h} = \lim_{h \to 0+} \frac{|h|}{h} = 1$

and $\lim_{h \to 0-} \frac{f(0+h) - f(0)}{h} = \lim_{h \to 0-} \frac{|h|}{h} = -1.$

Therefore, $\lim_{h \to 0} \frac{f(0+h) - f(0)}{h}$ does not exist and $|x|$ has no derivative at 0.

We can now prove many of the familiar properties of the derivative.

Theorem 11.3.3. *A function f is differentiable at a value c, iff*

$$f'(c) = \lim_{x \to c} \frac{f(x) - f(c)}{x - c}.$$

Proof. Let $h = x - c$. Note that $h \to 0$ iff $x \to c$. Hence $\lim_{x \to c} \frac{f(x) - f(c)}{x - c} = \lim_{h \to 0} \frac{f(c+h) - f(c)}{h}.$ □

Since derivatives are defined in terms of limits, Theorem 9.3.10 implies the following theorem:

Theorem 11.3.4. *The function f is differentiable at c with $f'(c) = l$ if and only if for each sequence $\{x_n\}$ for which no $x_n = c$ such that $\lim_{n \to \infty} x_n = c$, we have*

$$\lim_{n \to \infty} \frac{f(x_n) - f(c)}{x_n - c} = l.$$

The following theorem states that any differentiable function is also continuous.

Theorem 11.3.5. *If a function is differentiable at a value c, then it is continuous at c.*

Proof. By hypothesis,

$$f'(c) = \lim_{x \to c} \frac{f(x) - f(c)}{x - c}$$

exists. Then for all x in the domain of f such that $x \neq c$, we have

$$f(x) = (x - c)\frac{f(x) - f(c)}{x - c} + f(c).$$

Hence, since

$$\lim_{x \to c}(x - c) = \lim_{x \to c} x - \lim_{x \to c} c = c - c = 0,$$

we have

$$
\begin{aligned}
\lim_{x \to c} f(x) &= \lim_{x \to c}\left((x - c)\frac{f(x) - f(c)}{x - c} + f(c)\right) \\
&= \left(\lim_{x \to c}(x - c)\right)\left(\lim_{x \to c}\frac{f(x) - f(c)}{x - c}\right) + \lim_{x \to c} f(c) \\
&= 0f'(c) + f(c) \\
&= f(c),
\end{aligned}
$$

which proves that f is continuous at c. $\qquad\square$

The opposite direction of this theorem does not hold however. Take for example the absolute value function $|x|$. This function is continuous at 0 but is not differentiable at 0 (see Exercise 11.15).

Theorem 11.3.6. *If $f(x) = k$ for all x, then $f'(x) = 0$.*

Proof. We have

$$
\begin{aligned}
f'(x) &= \lim_{h \to 0} \frac{f(x + h) - f(x)}{h} \\
&= \lim_{h \to 0} \frac{k - k}{h} \\
&= \lim_{h \to 0} \frac{0}{h} \\
&= 0.
\end{aligned}
$$

$\qquad\square$

Theorem 11.3.7. *If $f(x) = x$ for all x, then $f'(x) = 1$.*

Proof. We have

$$
\begin{aligned}
f'(x) &= \lim_{h \to 0} \frac{f(x+h) - f(x)}{h} \\
&= \lim_{h \to 0} \frac{(x+h) - x}{h} \\
&= \lim_{h \to 0} \frac{h}{h} \\
&= 1.
\end{aligned}
$$

$\square$

The next theorem is concerned with the derivative of the sum of two differentiable functions.

Theorem 11.3.8. *If $f(x) = g_1(x) + g_2(x)$, then $f'(x) = g_1'(x) + g_2'(x)$, provided that both derivatives exist.*

Proof. We have

$$
\begin{aligned}
f'(x) &= \\
\lim_{h \to 0} \frac{f(x+h) - f(x)}{h} &= \\
\lim_{h \to 0} \frac{(g_1(x+h) + g_2(x+h)) - (g_1(x) + g_2(x))}{h} &= \\
\lim_{h \to 0} \frac{(g_1(x+h) - g_1(x)) + (g_2(x+h) - g_2(x))}{h} &= \\
\lim_{h \to 0} \left(\frac{g_1(x+h) - g_1(x)}{h} + \frac{g_2(x+h) - g_2(x)}{h} \right) &= \\
\lim_{h \to 0} \frac{g_1(x+h) - g_1(x)}{h} + \lim_{h \to 0} \frac{g_2(x+h) - g_2(x)}{h} &= \\
g_1'(x) + g_2'(x).
\end{aligned}
$$

$\square$

The next theorem is concerned with the derivative of the difference of two differentiable functions.

Theorem 11.3.9. *If $f(x) = g_1(x) - g_2(x)$, then $f'(x) = g_1'(x) - g_2'(x)$, provided that both derivatives exist.*

Proof. We have

$$f'(x) =$$
$$\lim_{h \to 0} \frac{f(x+h) - f(x)}{h} =$$
$$\lim_{h \to 0} \frac{(g_1(x+h) - g_2(x+h)) - (g_1(x) - g_2(x))}{h} =$$
$$\lim_{h \to 0} \frac{(g_1(x+h) - g_1(x)) - (g_2(x+h) - g_2(x))}{h} =$$
$$\lim_{h \to 0} \left(\frac{g_1(x+h) - g_1(x)}{h} - \frac{g_2(x+h) - g_2(x)}{h} \right) =$$
$$\lim_{h \to 0} \frac{g_1(x+h) - g_1(x)}{h} - \lim_{h \to 0} \frac{g_2(x+h) - g_2(x)}{h} =$$
$$g_1'(x) - g_2'(x).$$

□

The next theorem is concerned with the derivative of the product of two differentiable functions.

Theorem 11.3.10. *If $f(x) = g_1(x)g_2(x)$, then $f'(x) = g_1(x)g_2'(x) + g_1'(x)g_2(x)$ provided that both derivatives $g_1'(x)$ and $g_2'(x)$ exist.*

Proof. We have

$$f'(x) =$$
$$\lim_{h \to 0} \frac{g_1(x+h)g_2(x+h) - g_1(x)g_2(x)}{h} =$$
$$\lim_{h \to 0} \frac{g_1(x+h)g_2(x+h) - g_1(x+h)g_2(x) + g_1(x+h)g_2(x) - g_1(x)g_2(x)}{h} =$$
$$\lim_{h \to 0} \left(g_1(x+h) \frac{g_2(x+h) - g_2(x)}{h} + \frac{g_1(x+h) - g_1(x)}{h} g_2(x) \right) =$$
$$\lim_{h \to 0} g_1(x+h) \frac{g_2(x+h) - g_2(x)}{h} + \lim_{h \to 0} \frac{g_1(x+h) - g_1(x)}{h} g_2(x) =$$
$$\left(\lim_{h \to 0} g_1(x+h) \right) \left(\lim_{h \to 0} \frac{g_2(x+h) - g_2(x)}{h} \right)$$
$$+ \left(\lim_{h \to 0} \frac{g_1(x+h) - g_1(x)}{h} \right) \left(\lim_{h \to 0} g_2(x) \right) =$$
$$g_1(x)g_2'(x) + g_1'(x)g_2(x).$$

□

Corollary 11.3.11. *If $f(x) = kg(x)$, then $f'(x) = kg'(x)$ provided that $g'(x)$ exists.*

Corollary 11.3.12. *If n is a positive integer and $f(x) = x^n$, then $f'(x) = nx^{n-1}$.*

Proof. By induction on n.

BASIS: $n = 1$. Then $x^n = x$ and $f'(x) = 1 = 1x^0$ by Theorem 11.3.7.

INDUCTION STEP: Assume that if $g(x) = x^k$ then $g'(x) = kx^{k-1}$, and let $f(x) = x^{k+1}$. Then $f(x) = xg(x)$, and by Theorem 11.3.10, we have

$$
\begin{aligned}
f'(x) &= xg'(x) + 1g(x) \\
&= x(kx^{k-1}) + x^k \\
&= kx^k + x^k \\
&= (k+1)x^k.
\end{aligned}
$$

$\square$

By these two corollaries and Theorem 11.3.8, we have the following corollary:

Corollary 11.3.13. *If*

$$f(x) = a_n x^n + a_{n-1} x^{n-1} + \ldots + a_1 x + a_0,$$

then

$$f'(x) = na_n x^{n-1} + (n-1)a_{n-1} x^{n-2} + \ldots + a_1.$$

The next theorem is concerned with the derivative of the quotient of two differentiable functions.

Theorem 11.3.14 (Quotient Rule). *If $f(x) = g_1(x)/g_2(x)$, then*

$$f'(x) = \frac{g_2(x)g_1'(x) - g_1(x)g_2'(x)}{g_2^2(x)}$$

provided that the two derivatives $g_1'(x)$ and $g_2'(x)$ exist and $g_2(x) \neq 0$.

Proof. Since $g_2(x)$ is differentiable, it is continuous by Theorem 11.3.5. It follows that there is an open interval about 0 such that if h is in that interval, then $g_2(x+h) \neq 0$; for example, there is a $\delta > 0$ such that if $|h| < \delta$, then $|g_2(x+h) - g_2(x)| < |g_2(x)|/2$. It follows that we can calculate the limits involved in taking the derivative with h satisfying $|h| < \delta$ in the following way:

$$f'(x) =$$

$$\lim_{h \to 0} \frac{\dfrac{g_1(x+h)}{g_2(x+h)} - \dfrac{g_1(x)}{g_2(x)}}{h} =$$

$$\lim_{h \to 0} \frac{\dfrac{g_2(x)g_1(x+h) - g_2(x+h)g_1(x)}{g_2(x+h)g_2(x)}}{h} =$$

$$\lim_{h \to 0} \frac{g_2(x)g_1(x+h) - g_2(x)g_1(x) + g_2(x)g_1(x) - g_2(x+h)g_1(x)}{g_2(x+h)g_2(x)h} =$$

$$\lim_{h \to 0} \left(\left(g_2(x) \frac{g_1(x+h) - g_1(x)}{h} - \frac{g_2(x+h) - g_2(x)}{h} g_1(x) \right) \frac{1}{g_2(x+h)g_2(x)} \right) =$$

$$\lim_{h \to 0} \left(g_2(x) \frac{g_1(x+h) - g_1(x)}{h} - \frac{g_2(x+h) - g_2(x)}{h} g_1(x) \right) \frac{1}{\lim\limits_{h \to 0} g_2(x+h)g_2(x)} =$$

$$\lim_{h \to 0} \left(g_2(x) \frac{g_1(x+h) - g_1(x)}{h} - \frac{g_2(x+h) - g_2(x)}{h} g_1(x) \right) \frac{1}{(g_2(x))^2} =$$

$$\left(g_2(x) \lim_{h \to 0} \frac{g_1(x+h) - g_1(x)}{h} - g_1(x) \lim_{h \to 0} \frac{g_2(x+h) - g_2(x)}{h} \right) \frac{1}{(g_2(x))^2} =$$

$$(g_2(x)g_1'(x) - g_1(x)g_2'(x)) \frac{1}{(g(x))^2}.$$

□

The next theorem is concerned with the derivative of the composition of two differentiable functions.

Theorem 11.3.15 (Chain Rule). *Let I be an interval, let g_1 be a function defined on I, and let g_2 be a function defined on an interval J that contains the values $g_1(x)$ for all x in I. If g_1 is differentiable on I and g_2 is differentiable on J, then the composite function $f = g_2 \circ g_1$ is differentiable on I and*

$$f'(x) = g_2'(g_1(x)) \cdot g_1'(x).$$

Discussion It might appear that we can prove this easily using Theorem 11.3.3 as follows: for any c in I,

$$\begin{aligned}
f'(c) &= \lim_{x \to c} \frac{g_2(g_1(x)) - g_2(g_1(c))}{x - c} \\
&= \lim_{x \to c} \frac{g_2(g_1(x)) - g_2(g_1(c))}{g_1(x) - g_1(c)} \cdot \frac{g_1(x) - g_1(c)}{x - c} \\
&= g_2'(g_1(c)) \cdot g_1'(x).
\end{aligned}$$

But this argument will fail if there are values of x arbitrarily close to c such that $g_1(x) = g_1(c)$. The following proof, although less intuitive, avoids this difficulty.

Proof. Fix c in I and define a function F on J by

$$F(x) = \begin{cases} \dfrac{g_2(x) - g_2(g_1(c))}{x - g_1(c)}, & \text{if } x \neq g_1(c); \\ g_2'(g_1(c)), & \text{if } x = g_1(c), \end{cases}$$

Since g_2 is differentiable at $g_1(c)$, the function F is continuous at $g_1(c)$. Since the equality

$$F(x)(x - g_1(c)) = g_2(x) - g_2(g_1(c))$$

is valid for all x in J,[1] it follows that

$$F(g_1(x))(g_1(x) - g_1(c)) = g_2(g_1(x)) - g_2(g_1(c))$$

for all x in I. Since $F \circ g_1$ is continuous at c and g_1 is differentiable at c, we have

$$\begin{aligned} f'(c) &= \lim_{x \to c} \frac{g_2(g_1(x)) - g_2(g_1(c))}{x - c} \\ &= \lim_{x \to c} F(g_1(x)) \cdot \frac{g_1(x) - g_1(c)}{x - c} \\ &= F(g_1(c)) g_1'(c). \end{aligned}$$

This finishes the proof since $F(g_1(c)) = g_2'(g_1(c))$. □

Recall Chapter 4 and our definitions there of a function being one-to-one and of the inverse of a function (see Definitions 4.2.20 and 4.2.21 and the various related lemmas there. We summarise here what we need for this chapter.

Definition 11.3.16. *A function f is said to be* one-to-one *if and only if whenever $f(x) = f(y)$ it follows that $x = y$.*

Definition 11.3.17. *Let f be a function which is one-to-one. Then g is an* inverse function *of f provided that*

$$g(x) = y \qquad \text{if and only if} \qquad f(y) = x.$$

This definition makes sense for the following reason: if f is one-to-one, then there is only one value of x for which $f(x) = y$ for any given y. For suppose there are x_1 and x_2 with $f(x_1) = f(x_2) = y$. Then, since f is one-to-one, $x_1 = x_2$.

Note also that a one-to-one function has only one inverse.

---

[1] If $x = g_1(c)$, then both sides are 0; otherwise, this equation follows by the definition of F.

Definition 11.3.18. *If f is a one-to-one function, its inverse is denoted by* f_{inv}.

This means that

$$f_{\text{inv}}(x) = y \quad \text{if and only if} \quad f(y) = x.$$

In Chapter 4, the inverse of f was denoted f^{-1}.

The next theorem is concerned with the derivative of the inverse of a differentiable function. It is different from the earlier theorems in that it imposes extra conditions that were not imposed when we showed that differentiation was closed under addition, difference, product, quotient and composition of functions. Here, for function inverse, we are imposing the injectivity and continuity of f as well as the continuity of f_{inv} at the concerned point. In Chapter 12, when we learn the Intermediate Value Property which will help us to show in Corollary 12.3.10 that continuity of f_{inv} can be derived from the injectivity and continuity of f. As a preparatory practice, see Exercises 11.21 and 11.22.

Theorem 11.3.19. *Let I and J be intervals and let f be a continuous, one-to-one function defined on I with values in J. If f is differentiable at c, if* f_{inv} *is continuous at f(c), and if* $f'(c) \neq 0$, *then* f_{inv} *is differentiable at f(c) and* $f'_{\text{inv}}(f(c)) = 1/f'(c)$.

Proof. We will use Theorem 11.3.4 to prove that f_{inv} is differentiable at $f(c)$. Let $\{y_n\}$ be any sequence defined on J for which no $y_n = f(c)$ whose limit is $f(c)$. We will prove that the limit of the sequence

$$\left\{ \frac{f_{\text{inv}}(y_n) - f_{\text{inv}}(f(c))}{y_n - f(c)} \right\}$$

is $1/f'(c)$. Let $x_n = f_{\text{inv}}(y_n)$ for each n.

- Clearly $\{x_n\}$ is a sequence defined on I with no $x_n = c$ since otherwise, if for some n, $x_n = c$ then $f_{\text{inv}}(y_n) = c$ and since f is one-to-one, $y_n = f(c)$, a contradiction.

- Since f_{inv} is continuous at $f(c)$, and $\{y_n\}$ converges to $f(c)$, by Theorem 11.2.3, $\{f_{\text{inv}}(y_n)\}$ converges to $f_{\text{inv}}(f(c)) = c$. Hence, $\{x_n\}$ converges to c.

Since f is differentiable at c,

$$\lim_{n \to \infty} \frac{f(x_n) - f(c)}{x_n - c} = f'(c).$$

Since $f'(c) \neq 0$, by LS4

$$\lim_{n \to \infty} \frac{x_n - c}{f(x_n) - f(c)} = \frac{1}{f'(c)}.$$

Now,

$$\lim_{n \to \infty} \frac{f_{\text{inv}}(y_n) - f_{\text{inv}}(f(c))}{y_n - f(c)} = \lim_{n \to \infty} \frac{x_n - c}{f(x_n) - f(c)} = \frac{1}{f'(c)}.$$

By Theorem 11.3.4, f_{inv} is differentiable at $f(c)$ and $f'_{\text{inv}}(f(c)) = 1/f'(c)$. □

Corollary 11.3.20. *Let I and J be intervals and let f be a differentiable, one-to-one function defined on I with values in J. For all x in J for which $f'(f_{\text{inv}}(x)) \neq 0$, if f_{inv} is continuous on x, then f_{inv} is differentiable at x and $f'_{\text{inv}}(x) = 1/f'(f_{\text{inv}}(x))$.*

Note that f' and f'_{inv} are evaluated at different values.

Example 11.3.21. *Let $f(x) = x^3$. Then f is increasing and is thus one-to-one. By Corollary 11.3.12 $f'(x) = 3x^2$. Then Corollary 11.3.20 states that f_{inv}, the cube root function, is differentiable at all $x \neq 0$ and*

$$f'_{\text{inv}}(x) = \frac{1}{3(\sqrt[3]{x})^2} = \frac{1}{3}x^{-(2/3)}.$$

As stated above, we will in Chapter 12 prove the so-called important Intermediate Value Theorem 12.3.2 which states that the image of an interval by a continuous function is also an interval and hence a continuous function on an interval I will have no holes on I. We will use this theorem to prove in Corollary 12.3.10 that if f is continuous one-to-one on interval I then f_{inv} is continuous on $f(I)$. Since we have not done these yet, we include the hypotheses of the continuity of f_{inv} in our statement of Theorem 11.3.19. This theorem will be revised in Corollary 12.3.11 to remove the condition that f_{inv} is continuous.

- f is continuous at c iff $\lim_{x \to c} f(x) = f(c)$ iff $\{f(x_n)\}$ converges to $f(c)$ for each $\{x_n\}$ that converges to c iff $\forall \epsilon > 0$, $\exists \delta > 0$ such that if $|x - c| < \delta$ then $|f(x) - f(c)| < \epsilon$.

- If an interval I is in the domain of f, then f is be *continuous* on I if $\forall c \in I$, f is continuous at c. If c is an endpoint of the interval I, then only the appropriate one-sided limit is considered.

- If f and g are continuous, then $f + g$, $f - g$, kf, fg, $\frac{f}{g}$ and $f \circ g$ are continuous (conditions need to hold for $\frac{f}{g}$ and $f \circ g$).

- Polynomial functions, rational functions and the trigonometric functions are continuous.

- The derivative of f at c, $f'(c) = \lim_{h \to 0} \frac{f(c+h) - f(c)}{h}$.

- If $f'(c)$ exists we say that f is differentiable at c.

- f is differentiable at c, iff
$f'(c) = \lim_{x \to c} \frac{f(x) - f(c)}{x - c} = \lim_{h \to 0} \frac{f(c+h) - f(c)}{h}$ iff
$\lim_{n \to \infty} \frac{f(x_n) - f(c)}{x_n - c} = l$ for each $\{x_n\}$ for which $\lim_{n \to \infty} x_n = c$ and no $x_n = c$.

- If f is differentiable at c, then f is continuous at c.

- If $g_1'(x)$ and $g_2'(x)$ exist then:

  - $(g_1 \heartsuit g_2)'(x) = g_1'(x) \heartsuit g_2'(x)$ when $\heartsuit$ is either $+$ or $-$.
  - $(g_1 g_2)'(x) = g_1(x) g_2'(x) + g_1'(x) g_2(x)$.
  - $(\frac{g_1}{g_2})'(x) = \frac{g_2(x) g_1'(x) - g_1(x) g_2'(x)}{g_2^2(x)}$ assuming $g_2(x) \neq 0$.
  - $(g_2 \circ g_1)'(x) = g_2'(g_1(x)) \cdot g_1'(x)$.

- Let I and J be intervals and let f be a differentiable, one-to-one function defined on I with values in J. For all x in J for which $f'(f_{\text{inv}}(x)) \neq 0$, if f_{inv} is continuous on x, then f_{inv} is differentiable at x and $f_{\text{inv}}'(x) = 1/f'(f_{\text{inv}}(x))$. The condition "$f_{\text{inv}}$ is continuous on x" is not necessary (see in Corollary 12.3.11).

Figure 11.4: Main Concepts of Continuity and Derivatives from Chapter 11

11.3.1   Exercises

Exercise 11.12. Use the definition of the derivative to find $f'(1)$.

1. $f(x) = x^4 - 2x^2$.

2. $f(x) = \sqrt{x}$.

3. $f(x) = x/\sqrt{x^2 + 1}$.

Exercise 11.13. For each function $f(x)$ below, find $f'(x)$.

1. $f(x) = x^4 - 2x^2$.

2. $f(x) = \sqrt{x}$.

3. $f(x) = \sqrt{x^2 + 1}$.

4. $f(x) = x/\sqrt{x^2 + 1}$.

Exercise 11.14. Use the definition of the derivative to find $f'(x)$.

1. $f(x) = x^3 - 5x^2$.

2. $f(x) = \dfrac{1}{x^2}$.

3. $f(x) = \sqrt{x}$.

Exercise 11.15. Show that the absolute value function is continuous at 0 but is not differentiable at 0.

Exercise 11.16. Prove that the function f defined on the closed interval $[0, 1]$ by $f(x) = |x|$ is differentiable on the interval $[0, 1]$. (Read the definition carefully.)

Exercise 11.17. Prove that the function defined by

$$f(x) = \begin{cases} x^2 & \text{if } x \le 1; \\ 2x - 1 & \text{if } x > 1; \end{cases}$$

is differentiable at 1. Is the function continuous at 1?

Exercise 11.18. Suppose that f is defined on an open interval I and that f is differentiable at a point c in I. Prove that

$$\lim_{n \to \infty} \left(n \left(f\left(c + \frac{1}{n}\right) - f(c) \right) \right) = f'(c).$$

Exercise 11.19. Prove that if $f(x) = x^r$ where r is a nonzero rational number (quotient of two integers), then $f'(x) = rx^{r-1}$.

Exercise 11.20. Prove that the function defined by

$$f(x) = \begin{cases} x^r \cos(1/x) & \text{if } x \neq 0; \\ 0 & \text{if } x = 0; \end{cases}$$

is differentiable at 0 if $r = 2$ and not differentiable at 0 if $r = 1$.

Exercise 11.21. We say that a continuous function f on an interval I enjoys IVT on I if the following holds:

If a and b belong to I where $a < b$ and if $\min(f(a), f(b)) < M < \max(f(a), f(b))$ then there is a c such that $a < c < b$ and $f(c) = M$.

Let f be a one-to-one continuous function on interval I which enjoys IVT on I, and let J be the set of $f(x)$ such that $x \in I$ (i.e., $J = f(I)$). Prove the following:

1. f is either strictly increasing or strictly decreasing on I.

2. J is an interval and f_{inv} is continuous on J.

3. If f is differentiable at c and $f'(c) \neq 0$ then f_{inv} is differentiable at $f(c)$ and $f'_{\text{inv}}(f(c)) = \frac{1}{f'(c)}$.

Exercise 11.22. Let f be a one-to-one differentiable function on interval I and assume that f enjoys IVT of Exercise 11.21 on I. Let $J = f(I)$. Show that

$$f'_{\text{inv}}(x) = \frac{1}{f'(f_{\text{inv}}(x))}$$

for all values of x in J for which $f'(f_{\text{inv}}(x)) \neq 0$.

Exercise 11.23.   1. Prove that if $f(x) = \sin x$, then $f'(x) = \cos x$.

2. Prove that if $f(x) = \cos x$, then $f'(x) = -\sin x$.

3. Prove that if $f(x) = \tan x$, then $f'(x) = \sec^2 x$.

4. Prove that if $f(x) = \cot x$, then $f'(x) = -\csc^2 x$.

5. Prove that if $f(x) = \sec x$, then $f'(x) = \sec x \tan x$.

6. Prove that if $f(x) = \csc x$, then $f'(x) = -\csc x \cot x$.

7. Prove that if $f(x) = arcsinx$, then $f'(x) = 1/\sqrt{1 - x^2}$.

8. Prove that if $f(x) = arctanx$, then $f'(x) = 1/(1 + x^2)$.

9. Prove that if $f(x) = arcsecx$, then $f'(x) = 1/(|x|\sqrt{x^2 - 1})$.

Chapter 12

More on Sequences and Continuity

In this chapter we will look at some results about sequences, limits, and continuity that depend on the Axiom of Completeness of the real numbers given in Section 10.3.2. As we shall see, the Axiom of Completeness has significant consequences for the notion of continuity.

12.1 Bounded Sequences

Let us now turn to the intuitively obvious property of limits of sequences quoted at the beginning of 10.3 which was shown in Example 10.3.2 not to hold for the rationals, and which led us to the Axiom of Completeness for the real numbers. We repeat a reformulation of this property here:

> If $\{a_n\}$ is a nondecreasing sequence which has an upper bound g, then it has a least upper bound, which is the limit l of the sequence.

We can now prove this property as a theorem on the real numbers (see Theorem 12.1.2).

First, let us extend our vocabulary. We already know that a sequence of the kind described in the theorem we want to prove can be called *bounded*. We have spoken of sequences that are *bounded above* and those that are *bounded below*. Recall that when a sequence is bounded from above and from below, we simply say it is bounded.

We can also describe the fact that the terms never decrease as we go out the sequence.

Definition 12.1.1 (nondecreasing/nonincreasing/monotone sequence). *A sequence $\{a_n\}$ of real numbers is called a nondecreasing sequence if $a_n \leq a_{n+1}$ for all n. It is called a nonincreasing sequence if $a_n \geq a_{n+1}$ for all n. A sequence that is nondecreasing or nonincreasing is called a monotone sequence.*

Note that if $\{a_n\}$ is a nondecreasing sequence, then $a_m \leq a_n$ whenever $m < n$. Now we can formulate and prove the theorem.

Theorem 12.1.2 (Bounded monotone sequences). *All bounded monotone sequences have limits.*

Proof. We will prove this for nondecreasing sequences, since the proof for nonin- creasing sequences is similar. So suppose $\{a_n\}$ is a nondecreasing sequence. Let A be the set of all real numbers a_n in the sequence, and since A is bounded and not empty, by the Axiom of Completeness (Section 10.3.2), it has a least upper bound, say l. Let $\varepsilon > 0$ be given. Then $l - \varepsilon$ cannot be an upper bound for A, so there is a positive integer N such that $a_N > l - \varepsilon$. Since a_n is nondecreasing, $a_N \leq a_n$ for all $n > N$. Of course, for all n, $a_n \leq l$, and so if $n > N$, $l - \varepsilon < a_n < l < l + \varepsilon$. This latter implies that $|a_n - l| < \varepsilon$. This shows that $\lim_{n \to \infty} a_n = l$. $\square$

Sequences that are not bounded are called *unbounded*.

Theorem 12.1.3 (Unbounded monotone sequences).   *1. If $\{a_n\}$ is an unbounded nondecreasing sequence, then $\lim_{n \to \infty} a_n = \infty$.*

   2. If $\{a_n\}$ is an unbounded nonincreasing sequence, then $\lim_{n \to \infty} a_n = -\infty$.

Proof.   1. Let $\{a_n\}$ be an unbounded nondecreasing sequence, and let $M > 0$ be given. Since the set of all the terms a_n of the sequence is bounded below by a_1, it must be unbounded above. Hence, for some N, we have $a_N > M$. Clearly $n > N$ implies $a_n \geq a_N > M$, and so $\lim_{n \to \infty} a_n = \infty$.

   2. Let $\{a_n\}$ be an unbounded noninccreasing sequence, and let $M < 0$ be given. Since the set of all the terms a_n of the sequence is bounded above by a_1, it must be unbounded below. Hence, for some N, we have $a_N < M$. Clearly $n > N$ implies $a_n \leq a_N < M$, and so $\lim_{n \to \infty} a_n = -\infty$.

   $\square$

Corollary 12.1.4. *If $\{a_n\}$ is a monotone sequence, then the sequence either converges to a real number, diverges to ∞, or diverges to $-\infty$.*

Proof.   • If the sequence is bounded then by Theorem 12.1.2 it converges to a real number.

• If the sequence is unbounded nondecreasing then by Theorem 12.1.3.1, $\lim_{n\to\infty} a_n = \infty$.

• If the sequence is unbounded nonincreasing then by Theorem 12.1.3.1, $\lim_{n\to\infty} a_n = -\infty$.

$\square$

So far, in order to prove using the definition that a sequence has a limit, it is necessary to have the limit. But this is not necessary: if the terms get close enough together as n increases, then the sequence converges. This result is due to Cauchy,[1] who is generally credited with introducing these definitions of limits into analysis.

Definition 12.1.5 (Cauchy Sequences). *A sequence $\{a_n\}$ is called a Cauchy sequence if for each $\varepsilon > 0$, there is a number N such that for any $m, n > N$, $|a_n - a_m| < \varepsilon$.*

Lemma 12.1.6 (Sequences that Converge are Cauchy). *Any sequence that converges to a limit is a Cauchy sequence.*

Proof. Let $\{a_n\}$ be a sequence that converges to the limit a; i.e., suppose $\lim_{n\to\infty} a_n = a$. Let $\varepsilon > 0$ be given. Then, by definition, there is a number N such that for any $n > N$, $|a_n - a| < \frac{\varepsilon}{2}$. Let $n, m > N$. Then

$$|a_n - a_m| = |a_n - a + a - a_m| \leq |a_n - a| + |a_m - a| \leq \frac{\varepsilon}{2} + \frac{\varepsilon}{2} = \varepsilon.$$

$\square$

Lemma 12.1.7 (Cauchy Sequences are Bounded). *Every Cauchy sequence is bounded.*

Proof. Let $\{a_n\}$ be a Cauchy sequence. Applying Definition 12.1.5 with $\varepsilon = 1$, we obtain an integer N such that for $m, n > N$, we have $|a_n - a_m| < 1$. In particular, $|a_n - a_{N+1}| < 1$ for all $n > N$, and it follows that $|a_n| < |a_{N+1}| + 1$ for $n > N$. If M is the maximum of $|a_1|, |a_2|, \cdots, |a_{N-1}|, |a_N| + 1$, then it follows that $|a_n| \leq M$ for all n. $\square$

---

[1]Augustin Louis Cauchy, 1789–1857, French mathematician.

Lemma 12.1.8 (Cauchy Sequences converge). *Every Cauchy sequence of real numbers converges to a limit.*

Proof. Let $\{a_n\}$ be a Cauchy sequence. By Lemma 12.1.7, the sequence $\{a_n\}$ is bounded. For each positive integer n, let b_n be the greatest lower bound of the terms $a_n, a_{n+1}, a_{n+2}, \cdots$; each b_n exists by the Axiom of Completeness (Page 301). Then the sequence $\{b_n\}$ is nondecreasing and bounded. It is increasing because, as n increases, the set of terms of which b_n is the greatest lower bound is getting smaller, and this means that increasing n cannot introduce a new value that is smaller than any in the previous sets. It is bounded because the a_n are bounded, and so b_n must be bounded above by any upper bound of the sequence $\{a_n\}$. ($\{b_n\}$ is bounded from below by b_1.)

Hence, by Theorem 12.1.2, the sequence $\{b_n\}$ has a limit, say b. We will prove that $\{a_n\}$ converges to b. Let $\varepsilon > 0$ be given. By Definition 12.1.5, there is a positive integer N such that for all $m, n > N$, $|a_n - a_m| < \frac{\varepsilon}{2}$. It follows that the parts of both sequences $\{a_n\}$ and $\{b_n\}$ for $n > N$ are contained in the interval from $a_N - \frac{\varepsilon}{2}$ to $a_N + \frac{\varepsilon}{2}$. In other words,

$$a_N - \frac{\varepsilon}{2} < a_n < a_N + \frac{\varepsilon}{2}$$

and

$$a_N - \frac{\varepsilon}{2} < b_n < a_N + \frac{\varepsilon}{2}.$$

By the second of these and LS31 on Page 224,[2]

$$a_N - \frac{\varepsilon}{2} < b < a_N + \frac{\varepsilon}{2}.$$

It follows that for $n > N$,
$$|a_n - b| < \varepsilon,$$

and so b is the limit of $\{a_n\}$. $\square$

Theorem 12.1.9 (A Sequence is Cauchy iff it Converges). *A sequence of real numbers converges if and only if it is a Cauchy sequence.*

Proof. Use Lemmas 12.1.6 and 12.1.8. $\square$

Definition 12.1.10 (Nested Intervals). *A nested sequence of intervals is a sequence $\{I_n\}$ of intervals with the property that for each n, I_{n+1} is contained in I_n.*

---

[2]Use this twice, and in each case let one of the sequences be a constant.

Theorem 12.1.11 (Nested Intervals Theorem). *If $\{[a_n, b_n]\}$ is a nested sequence of closed and bounded intervals, then there exists a real number z which belongs to all the intervals in the sequence. Furthermore, if $\lim_{n\to\infty}(b_n - a_n) = 0$, then the real number z is unique.*

Proof. Since for every n, $[a_{n+1}, b_{n+1}]$ is contained in $[a_n, b_n]$, it follows that

$$a_n \leq a_{n+1} \text{ and } b_{n+1} \leq b_n.$$

Hence, the sequences $\{a_n\}$ and $\{b_n\}$ are monotone. Furthermore, the sequence $\{a_n\}$ is bounded (from above) by b_1 and the sequence $\{b_n\}$ is bounded (from below) by a_1. It follows by Theorem 12.1.2 that the sequences $\{a_n\}$ and $\{b_n\}$ have limits, say a and b respectively. Since for each n, $a_n < b_n$, it follows by LS31 on Page 224, that $a \leq b$, and then any z satisfying $a \leq z \leq b$ is in all the intervals. If $\lim_{n\to\infty}(b_n - a_n) = 0$, then then we have

$$0 = \lim_{n\to\infty}(b_n - a_n) = \lim_{n\to\infty} b_n - \lim_{n\to\infty} a_n = b - a,$$

so $a = b$ and $z = b = a$ is unique. $\qquad\square$

Recall the example on Page 220 of the sequence $1, -1, 1, -1, \cdots, (-1)^{n+1}$. As we saw earlier, this sequence does not converge to a limit. However, if we take only the odd terms, each term is $(-1)^{((2m+1)+1)} = 1$, and the sequence consisting of these odd terms is a constant sequence which does converge, to 1. Similarly, if we take the even terms, each term is $(-1)^{((2m)+1)} = -1$, and again we have a constant sequence which converges, in this case to -1. These two constant sequences are both subsequences of our original sequence.

Definition 12.1.12 (Subsequence). *Let $\{a_n\}$ be a sequence of real numbers, and let $\{p_n\}$ be a strictly increasing sequence of positive integers. Then the sequence $\{a_{p_n}\}$ is called a subsequence of $\{a_n\}$.*

Note that for all natural numbers n, we have $p_n \geq n$ since otherwise, we would have $p_1 < p_2 < \cdots < p_n < n$ and this means there are n distinct naturals before n which is absurd.

Theorem 12.1.13. *Let $\{a_n\}$ be a sequence of real numbers.*

1. *If $\lim_{n\to\infty} a_n = a$, then every subsequence of $\{a_n\}$ also converges to a.*

2. *If $\{a_n\}$ has two subsequences that converge to different limits, then the sequence $\{a_n\}$ does not converge.*

Proof. 1. Suppose that $\{a_n\}$ converges to a, and let $\{a_{p_n}\}$ be any subsequence of $\{a_n\}$. Let $\varepsilon > 0$ be given. Then there is a positive integer N such that for all $n > N$, $|a_n - a| < \varepsilon$. Since $p_n \geq n$ for all n, it follows that for all $n > N$, $|a_{p_n} - a| < \varepsilon$. Hence, $\lim_{n\to\infty} a_{p_n} = a$.

2. Let $\{a_{p_n}\}$ and $\{a_{q_n}\}$ be the two subsequence of $\{a_n\}$ which converge to b and c respectively. Suppose that $\{a_n\}$ converges to a. By 1. above, each of $\{a_{p_n}\}$ and $\{a_{q_n}\}$ converge to a. By Exercise 9.1, $a = b$ and $a = c$. Hence $b = c$. A contradiction.

□

The goal here is to prove that every bounded sequence is like the sequence $1, -1, 1, -1, \cdots, (-1)^{n+1}$ in that it has a subsequence that converges. We need to prove another theorem first:

Theorem 12.1.14. *Every sequence of real numbers has a monotone subsequence.*

Proof. Let $\{a_n\}$ be a sequence of real numbers. Let S be the set of all integers n such that a_n is a lower bound for the set of terms which follow it in the sequence; i.e., a_n is a lower bound for the set $\{a_{n+1}, a_{n+2}, a_{n+3}, \cdots\} = \{a_k : k > n\}$.

- If S is infinite, then S can be expressed as a strictly increasing sequence $\{p_n\}$ of natural numbers, and the sequence $\{a_{p_n}\}$ is a nondecreasing subsequence of $\{a_n\}$.

- If S is finite, then there exists an integer N larger than every integer in S. Let p_1 be any integer greater than N; if S is empty, then $p_1 = 1$. Since p_1 is not in S, a_{p_1} is not a lower bound for the set of terms of the sequence following a_{p_1}; i.e., a_{p_1} is not a lower bound for the set of terms $\{a_{p_1+1}, a_{p_1+2}, a_{p_1+3}, \cdots\}$, so there exists an integer $p_2 > p_1$ such that $a_{p_2} < a_{p_1}$. Similarly, there exists an integer $p_3 > p_2$ such that $a_{p_3} < a_{p_2}$. Continuing this process yields a decreasing subsequence $\{a_{p_n}\}$ of $\{a_n\}$.

□

Now we are ready to prove the Bolzano[3]-Weierstrass[4] Theorem:

Theorem 12.1.15 (Bolzano-Weierstrass Theorem). *Every bounded sequence has a subsequence that converges to a limit.*

Proof. Now let $\{a_n\}$ be any bounded sequence of real numbers. By Theorem 12.1.14, this sequence has a monotone subsequence, and since the original sequence is bounded, so is the monotone subsequence. Hence, this monotone subsequence is a bounded monotone sequence, and by Theorem 12.1.2, it has a limit.

□

---

[3]Bernhard Bolzano, 1781–1848, Czech philosopher and mathematician.
[4]Karl Weierstrass, 1815–1897, German mathematician.

12.1.1 Exercises

Exercise 12.1. 1. Let $\{a_n\}$ be a sequence such that $|a_{n+1} - a_n| < 2^{-n}$ for all positive integers n. Prove that $\{a_n\}$ is a Cauchy sequence and hence a sequence that converges to a limit. *Hint:* Recall the sequence that we looked at in Section 8.2.1 in connection with Zeno's paradox Dichotomy:

$$\frac{1}{2}, \frac{1}{2} + \frac{1}{4}, \cdots, \frac{1}{2} + \frac{1}{4} + \cdots + \frac{1}{2^n}, \cdots$$

Recall that the nth term is equal to

$$1 - \frac{1}{2^n},$$

and that the limit of the sequence is 1. It follows that the limit of the sequence starting with the nth term, which is obtained from the entire sequence by deleting the first $n - 1$ terms, is

$$\frac{1}{2^{n-1}}.$$

2. Is the result in the previous part true if we only assume that $|a_{n+1} - a_n| < \frac{1}{n}$ for all positive integers n?

Exercise 12.2. Let $\{a_n\}$ be a sequence of real numbers and let r be a real number that satisfies $0 < r < 1$. Suppose that

$$|a_{n+1} - a_n| < r^n$$

for $n > 1$. Prove that $\{a_n\}$ is a Cauchy sequence and hence a sequence that converges to a limit.

Exercise 12.3. Let $\{a_n\}$ be a sequence of real numbers and let r be a real number that satisfies $0 < r < 1$. Suppose that

$$|a_{n+1} - a_n| \leq r|a_n - a_{n-1}|$$

for $n > 1$. Prove that $\{a_n\}$ is a Cauchy sequence and hence a sequence that converges to a limit.

Exercise 12.4. Let a_0 and a_1 be distinct real numbers. For each $n > 1$, let

$$a_n = \frac{(a_{n-1} + a_{n-2})}{2}.$$

Use Exercise 12.3 to prove that the sequence $\{a_n\}$ is a Cauchy sequence. Express the limit of the sequence in terms of a_0 and a_1.

Exercise 12.5. Let S be a bounded nonempty set of real numbers and suppose that $\sup S \notin S$. Prove that there is a nondecreasing sequence $\{s_n\}$ of elements of S such that $\lim_{n \to \infty} s_n = \sup S$.

Exercise 12.6. Let $s_1 = 1$ and $s_{n+1} = (s_n + 1)/3$ for $n \geq 1$.

1. Find s_2, s_3, and s_4.

2. Use mathematical induction to show that $s_n > 1/2$ for all n.

3. Show that $\{s_n\}$ is a nonincreasing sequence.

4. Show that $\{s_n\}$ converges and find $\lim_{n \to \infty} s_n$.

Exercise 12.7. Let $t_1 = 1$ and $t_{n+1} = [1 - 1/(n+1)^2]t_n$ for $n \geq 1$.

1. Show that $\{t_n\}$ converges.

2. Use induction to show that $t_n = (n+1)/(2n)$ for all $n > 1$.

3. Find the limit of the sequence and prove that it is the limit.

Exercise 12.8. Prove that all of the following are equivalent.

1. The Completeness Axiom (see Page 301).

2. Every bounded monotone sequence of real numbers converges (Theorem 12.1.2).

3. Every Cauchy sequence of real numbers converges (Lemma 12.1.8).

4. The Nested Intervals Theorem 12.1.11.

5. The Bolzano-Weierstrass Theorem 12.1.15.

12.2   Tails of Sequences

We can do better than the proof of Theorem 12.1.15 in describing the converging subsequence(s) of a sequence. For this, we need to consider some facts about sequences more closely.

One crucial fact about limits of sequences is that if a sequence has a limit, that limit remains unchanged if we change a finite number of terms at the beginning of the sequence. It does not matter how big a finite number of terms is changed in this way. Whether ten terms or a billion terms are changed, there is no difference in the limit. The limit depends only on the terms from a certain point in the sequence on.

Furthermore, the terms in a sequence from a certain point on have already occurred in some of our proofs, for example in the proof of Theorem 12.1.14. It seems worth it to give these sets a name:

Definition 12.2.1 (Tail of a Sequence). *Let $\{a_n\}$ be a sequence. Then the tail of $\{a_n\}$ determined by N is defined to be*

$$T_{a_n,N} = \{a_{N+1}, a_{N+2}, \cdots\} = \{a_n : n > N\}.$$

If the sequence is clear from the context, we will denote this T_N.

It is clear that a tail of a sequence is always nonempty and that if a sequence has a limit, then each of its tails has the same limit:

Theorem 12.2.2. *For any N, $T_{a_n,N}$ is a subsequence of $\{a_n\}$ which is not empty. Furthermore, if*

$$\lim_{n \to \infty} a_n = a,$$

then

$$\lim_{n \to \infty} T_{a_n,N} = a.$$

∎

It should also be clear that if $N_1 > N_2$, then $T_{N_1} \subseteq T_{N_2}$; i.e., the tails do not get bigger as N increases. Note that this means that we have

$$T_1 \supseteq T_2 \supseteq T_3 \supseteq \cdots$$

If a sequence is bounded, each of its tails is nonempty and bounded, and so by the axiom of completeness of Page 301, has a least upper bound and by Theorem 10.4.2 has a greatest lower bound. Let u_N be the greatest lower bound of T_N and let v_N be the least upper bound. For the reasons given in the proof of Theorem 12.1.14, we have $u_1 \leq u_2 \leq u_3 \leq \cdots$ and $v_1 \geq v_2 \geq v_3 \geq \cdots$. Furthermore, for every n, $u_n \leq v_m$ for every m and $v_n \geq u_m$ for every m:

$$u_1 \leq u_2 \leq \cdots v_3 \leq v_2 \leq v_1.$$

Hence, the sequences u_n and v_n are bounded monotone sequences, and it follows by Theorem 12.1.2 that they have limits. Let u be the limit of u_n and v be the limit of v_n. Then we have

$$u_1 \leq u2 \leq \cdots u \leq v \cdots v3 \leq v2 \leq v1.$$

Also, if $\lim_{n \to \infty} a_n$ exists, and we call it a, then $u \leq a \leq v$, so

$$u_1 \leq u_2 \leq \cdots u \leq a \leq v \leq v_3 \leq v_2 \leq v1.$$

These numbers u and v are useful whether or not the sequence a_n has a limit, and so we give them names: u is called $\liminf a_n$ and v is called $\limsup a_n$.

Definition 12.2.3. *Let $\{a_n\}$ be a sequence of real numbers. We define*

1. $\lim \sup a_n = \lim_{N \to \infty} \sup T_{a_n, N}$,

2. $\lim \inf a_n = \lim_{N \to \infty} \inf T_{a_n, N}$,

By the above discussion we have:

Theorem 12.2.4. *Let $\{a_n\}$ be a sequence and let $T_1, T_2, \cdots$ be its tails. Then*

$$T_1 \supseteq T_2 \supseteq T_3 \supseteq \cdots$$

Furthermore, if $\{a_n\}$ is bounded then for each natural number N, T_N has a least upper bound $\sup T_N$ and a greatest lower bound $\inf T_N$ such that $\{\inf T_N\}$ and $\{\sup T_N\}$ both converge and we have

$$\inf T_1 \leq \inf T_2 \leq \cdots \lim_{N \to \infty} \inf T_N \leq \lim_{N \to \infty} \sup T_N \cdots \leq \sup T_2 \leq \sup T_1.$$

Hence

$$\inf T_1 \leq \inf T_2 \leq \cdots \lim \inf a_n \leq \lim \sup a_n \cdots \leq \sup T_2 \leq \sup T_1.$$

Example 12.2.5. *Consider the sequence $\{a_n\}$ defined by $a_n = (-1)^{n+1} = 1$, $-1, 1, -1, \cdots$. If N is odd, T_N is the same as the entire sequence, while if N is even, $T_N = \{-1, 1, -1, 1, \cdots\}$. Clearly, for each N,*

$$\inf T_N = -1 \text{ and } \sup T_N = 1.$$

Hence, $\lim \inf a_n = -1$ and $\lim \sup a_n = 1$.

Example 12.2.6. *Consider the sequence $\{a_n\}$ defined by*

$$\begin{aligned} a^{3n} &= 5 + \tfrac{1}{10^n} \text{ if } n \neq 0, \\ a^{3n+1} &= 3, \\ a^{3n+2} &= 2 - \tfrac{1}{10^n}. \end{aligned}$$

The first few terms are

$$3, 1.9, 5.1, 3, 1.99, 5.01, 3, 1.999, 5.001, 3, \cdots$$

The tails are

$$
\begin{aligned}
T_1 &= \{3, 1.9, 5.1, 3, 1.99, 5.01, 3, 1.999, 5.001, 3, \cdots\}, \\
T_2 &= \{1.9, 5.1, 3, 1.99, 5.01, 3, 1.999, 5.001, 3, \cdots\}, \\
T_3 &= \{5.1, 3, 1.99, 5.01, 3, 1.999, 5.001, 3, 1.999, 5.0001, 3, \cdots\}, \\
T_4 &= \{3, 1.99, 5.01, 3, 1.999, 5.001, 3, 1.999, 5.0001, 3, \cdots\}, \\
T_5 &= \{1.99, 5.01, 3, 1.999, 5.001, 3, 1.999, 5.0001, 3, \cdots\}, \\
T_6 &= \{5.01, 3, 1.999, 5.001, 3, 1.999, 5.0001, 3, \cdots\}, \\
T_7 &= \{3, 1.999, 5.001, 3, 1.999, 5.0001, 3, \cdots\}, \\
T_8 &= \{1.999, 5.001, 3, 1.999, 5.0001, 3, \cdots\}, \\
T_9 &= \{5.001, 3, 1.999, 5.0001, 3, \cdots\}, \\
T_9 &= \{3, 1.999, 5.0001, 3, \cdots\}.
\end{aligned}
$$

It should be clear that for every N,

$$
\inf T_N = 1.9 \underbrace{9 \cdots 9}_{k} \ \text{and} \ \sup T_N = 5.\underbrace{0 \cdots 0}_{l} 1 \ \text{where} \ 0 \le k, l < N.
$$

Hence, $\lim \inf a_n = 2$ *and* $\lim \sup a_n = 5$.

It is not always the case that $\lim \sup a_n = \sup T_N$, but it is true that $\lim \sup a_n \le \sup T_N$. Some values of a_n may be much larger than $\lim \sup a_n$. What is true is that $\lim \sup a_n$ is the largest value that infinitely many of the a_n can get close to. Similar remarks apply to $\lim \inf a_n$.

Theorem 12.2.7. *Let* $\{a_n\}$ *be a sequence of real numbers.*

1. *If* $\{a_n\}$ *approaches* a *as a limit;[5] i.e., if* $\lim_{n \to \infty} a_n = a$ *then*

$$
\lim \inf a_n = a = \lim \sup a_n.
$$

2. *If* $\{a_n\}$ *is bounded and* $\lim \inf a_n = \lim \sup a_n$, *then* $\{a_n\}$ *converges and*

$$
\lim_{n \to \infty} a_n = \lim \inf a_n = \lim \sup a_n.
$$

Proof. Throughout the proof, u_N will denote $\inf T_N$, v_N will denote $\sup T_N$, u will denote $\lim_{N \to \infty} u_N = \lim \inf a_n$, and v will denote $\lim_{N \to \infty} v_N = \lim \sup a_n$.

1. Suppose that $\lim_{n \to \infty} a_n = a$, a real number. Let $\varepsilon > 0$ be given. Then there is a natural number N such that $n > N$ implies $|a_n - a| < \varepsilon$. This means that $n > N$ implies $a_n < a + \varepsilon$, and so

$$
v_N = \sup T_N \le a + \varepsilon.
$$

---

[5]Hence, by Theorem 12.1.9, $\{a_n\}$ is a Cauchy sequence, and by Lemma 12.1.7, $\{a_n\}$ is bounded.

Also, $m > N$ implies $v_m \le a + \varepsilon$, and it follows that

$$\limsup a_n = \lim_{n \to \infty} v_m \le a + \varepsilon.$$

Since $\limsup a_n \le a + \varepsilon$ for all $\varepsilon > 0$, no matter how small, it follows that

$$\limsup a_n \le a = \lim_{n \to \infty} a_n.$$

A similar argument shows that $a \le \liminf a_n$. Since $\liminf a_n \le \limsup a_n$, it follows that the three values are all equal:

$$\liminf a_n = \lim a_n = \limsup a_n.$$

2. Suppose that $\liminf a_n = \limsup a_n = a$, where a is a real number. We need to prove that $\lim_{n \to \infty} a_n = a$. Let $\varepsilon > 0$ be given. Since $a = \lim_{N \to \infty} v_N$, there exists a natural number N_0 such that

$$|a - \sup T_{N_0}| < \varepsilon.$$

Thus, $\sup T_{N_0} < a + \varepsilon$, and so

$$\text{For all } n > N_0, \quad a_n < a + \varepsilon.$$

Similarly, since $a = \lim_{N \to \infty} u_N$, there exists N_1 such that $|a - \inf T_{N_1}| < \varepsilon$, and hence $\inf T_{N_1} > a - \varepsilon$. It follows that

$$\text{For all } n > N_1, \quad a_n > a - \varepsilon.$$

Hence, if N is the larger of N_0 and N_1,

$$\text{For all } n > N, \quad a - \varepsilon < a_n < a + \varepsilon.$$

or equivalently,

$$\text{For all } n > N, \quad |a_n - a| < \varepsilon.$$

But this proves that

$$\lim_{n \to \infty} a_n = a,$$

as desired.

$\square$

It is now possible to prove a stronger version of Theorem 12.1.2:

Theorem 12.2.8. *Let $\{a_n\}$ be any bounded sequence. Then $\{a_n\}$ has a monotone subsequence whose limit is* $\lim \sup a_n$ *and it has a monotone subsequence whose limit is* $\lim \inf a_n$.

Remark 12.2.9. *If we allow ∞ to be the least upper bound of an unbounded increasing sequence and $-\infty$ to be the greatest lower bound of an unbounded decreasing sequence, and thus allow ∞ to be a* $\lim \sup$ *and $-\infty$ to be a* $\lim \inf$, *then the condition that the sequence be bounded can be dropped from the theorem.*

Proof. For all positive integers N, let $v_N = \sup T_N$ and let $v = \lim_{N \to \infty} v_n = \lim \sup a_n$. We will prove that $\{a_n\}$ has a monotone subsequence whose limit is v; the case for a monotone subsequence whose limit is $\lim \inf a_n$ is similar. There are two cases.

C1. Suppose that for all positive integers N,

$$v_N \in T_N.$$

This means that $v_1 \in T_1$. Select n_1 so that $a_{n_1} = v_1$ and $n_1 > 1$. Then use the fact that $v_{n_1} \in T_{n_1}$ to select $n_2 > n_1$ so that $a_{n_2} = v_{n_1}$. Continuing, if $n_1, n_2, \cdots, n_k$ have been selected so that

$$n_1 < n_2 < \cdots < n_k$$

and for $j = 2, 3, \cdots, k$,

$$a_{n_j} = v_{n_{j-1}};$$

use the fact that $v_{n_k} \in T_{n_k}$ to select $n_{k+1} > n_k$ so that $a_{n_{k+1}} = v_{n_k}$. We will obtain a nonincreasing sequence $\{v_{n_i}\}$ for positive integers i, nonincreasing because each $v_N = \sup T_N$ and these sets get no bigger as N increases, and it follows that $\{a_{n_i}\}$ is a nonincreasing sequence. Furthermore, $\{v_{n_i}\}$ is a subsequence of $\{v_i\}$ and by Theorem 12.1.13 we have

$$\lim_{i \to \infty} a_{n_i} = \lim_{N \to \infty} v_N = v.$$

C2. Suppose that there exists N_0 such that

(12.1) $$v_{N_0} \notin T_{N_0}.$$

Then clearly $a_n < v_{N_0}$ for $n > N_0$ and $v_N \leq v_{N_0}$ for $N \geq N_0$. In fact, for $N \geq N_0$, $v_N = v_{N_0}$. Otherwise, $v_{N_1} < v_{N_0}$ for some $N_1 > N_0$. Then $\sup T_{N_1} < v_{N_0}$. Now $\{a_n : N_0 < n \leq N_1\} = \{a_{N_0+1}, a_{N_0+2}, \cdots, a_{N_1}\}$ is a finite set, so

$$\sup\{a_n : N_0 < n \leq N_1\} = \max\{a_n : N_0 < n \leq N_1\}.$$

Since $a_n < v_{N_0}$ for $N_0 < n \le N_1$,

$$\max\{a_n : N_0 < n \le N_1\} < v_{N_0}.$$

We have chosen N_1 so that $v_{N_1} < v_{N_0}$. Hence,

$$\max\{a_n : N_0 < n \le N_1\} \cup \{v_{N_1}\} < v_{N_0}.$$

But since $v_{N_1} = \sup T_{N_1} = \sup\{a_{N_1+1}, a_{N_1+2}, \cdots\}$,

$$
\begin{aligned}
v_{N_0} &= \sup T_{N_0} \\
&= \sup\{a_{N_0+1}, a_{N_0+2}, \cdots\} \\
&= \sup\{a_{N_0+1}, a_{N_0+2}, \cdots, a_{N_1}\} \cup \{a_{N_1+1}, a_{N_1+2}, \cdots\} \\
&= \sup\{a_n : N_0 < n \le N_1\} \cup T_{N_1} \\
&= \max\{a_n : N_0 < n \le N_1\} \cup v_{N_1},
\end{aligned}
$$

and this implies $v_{N_0} < v_{N_0}$, which is a contradiction. So for $N \ge N_0$, $v_N = v_{N_0}$. Then $v = \lim_{N \to \infty} v_N = v_{N_0}$, and (12.1) shows that

$$(12.2) \qquad\qquad v = \sup T_{N_0} \notin T_{N_0}.$$

Now we select an increasing sequence $\{b_n\}$ that increases to v. In fact, $b_n = v - \frac{1}{n}$ will do. By (12.2), there is $n_1 > N_0$ so that $a_{n_1} > b_1$. Then $a_{n_1} < v$. Suppose $n_1, n_2, \cdots, n_k$ have been selected so that

$$(12.3) \qquad\qquad n_1 < n_2 < \cdots < n_k$$

and for $j = 1, 2, \cdots, k-1$,

$$(12.4) \qquad\qquad \max\{a_{n_j}, b_{j+1}\} < a_{n_{j+1}} < v.$$

In (12.4), we require that $a_{n_{j+1}} > a_{n_j}$ to assure that the subsequence is nondecreasing and we require $a_{n_{j+1}} > b_{j+1}$ to assure that it has limit v. Since $v = v_{n_k}$, we have

$$\sup\{a_n : n > n_k\} = v$$

and so there exists $n_{k+1} > n_k$ such that $a_{n_{k+1}} > \max\{a_{n_k}, b_{k+1}\}$. Of course $a_{n_{k+1}} < v$ by (12.2), and so (12.3) and (12.4) hold for $k+1$ in place of k and the procedure continues by induction. Since $a_{n_{k+1}} > a_{n_k}$ for all k, $\{a_{n_k}\}$ is a monotone nondecreasing subsequence of $\{a_n\}$. Since $b_k < a_{n_k} < v$ for all k and $\lim_{k \to \infty} b_k = v$, we also have $\lim_{k \to \infty} a_{n_k} = v$.

$\square$

Example 12.2.10. *In Example 12.2.5, the subsequence that converges to 1 is clearly the odd terms and the subsequence that converges to -1 is clearly the even terms.*

Example 12.2.11. *In Example 12.2.6, the subsequence that converges to 2 is clearly $1.9, 1.99, 1.999, 1.9999, \cdots$, or $\{2 - 1/(10^n)\}$, and the subsequence that converges to 5 is clearly $5.1, 5.01, 5.001, 5.0001, \cdots$, or $\{5 + 1/(10^n)\}$.*

Example 12.2.12. *So far we have used* lim *infs and* lim *sups in proofs, but we have not really explored their properties. Although the definition of them is complicated, some of them are easy to calculate.*

Theorem 12.2.13. *Let $\{a_n\}$ be a bounded sequence of real numbers.*

1. *If for all $n > N$, $m \le a_n \le M$, then* $\liminf a_n \ge m$ *and* $\limsup a_n \le M$.

2. *If $\beta > \limsup a_n$, then there is a positive integer N such that for all $n > N$, $a_n < \beta$.*

3. *If $\alpha < \liminf a_n$, then there is a positive integer N such that for all $n > N$, $a_n > \alpha$.*

4. *For every real number c,* $\liminf (c + a_n) = c + \liminf a_n$ *and* $\limsup (c + a_n) = c + \limsup a_n$.

5. *If $c > 0$, then* $\liminf (c\, a_n) = c \liminf a_n$ *and* $\limsup (c\, a_n) = c \limsup a_n$.

6. *If $c < 0$, then* $\liminf (c\, a_n) = c \limsup a_n$ *and* $\limsup (c\, a_n) = c \liminf a_n$.

7. *If $\{a_{p_n}\}$ is any subsequence of $\{a_n\}$, then* $\liminf a_n \le \liminf a_{p_n} \le \limsup a_{p_n} \le \limsup a_n$.

8. *For each $\varepsilon > 0$, both of the sets of integers $\{n : a_n < \liminf a_n + \varepsilon\}$ and $\{n : a_n > \limsup a_n - \varepsilon\}$ are infinite.*

9. $\liminf a_n = -\limsup (-a_n)$ *and* $\limsup a_n = -\liminf (-a_n)$.

Proof.   1. Since for all $n > N$, $m \le a_n$, then $m \le \inf T_N \le \inf T_{N+1} \le \cdots \le \liminf a_n$. Hence $m \le \liminf a_n$.
Similarly, since for all $n > N$, $a_n \le M$, then $\limsup a_n \le \cdots \le \sup T_{N+1} \le \sup T_N$. Hence $\limsup a_n \le M$.

2. Suppose that $\beta > \limsup a_n$. Since the sequence $\{\sup T_N\}$ is decreasing and converges to $\limsup a_n$, there exists a positive integer N such that $\sup T_N < \beta$. It follows that $a_n < \beta$ for all $n > N$.

3. Suppose that $\alpha < \liminf a_n$. Since the sequence $\{\inf T_N\}$ is increasing and converges to $\liminf a_n$, there exists a positive integer N such that $\inf T_N > \alpha$. It follows that $a_n > \alpha$ for all $n > N$.

4. Let $T_{CN} = \{c + a_{N+1}, c + a_{N+2}, \cdots\}$. Then, $\inf T_{CN} = c + \inf T_N$ and $\sup T_{CN} = c + \sup T_N$. Now,

   - $\liminf (c + a_n) =^{Def.\ 12.2.3} \lim_{N \to \infty} \inf T_{CN} = \lim_{N \to \infty} \inf (c + \inf T_N) =^{LS1 + LS2} c + \lim_{N \to \infty} \inf T_N =^{Def.\ 12.2.3} c + \liminf a_n$.

   - $\limsup (c + a_n) =^{Def.\ 12.2.3} \lim_{N \to \infty} \sup T_{CN} = \lim_{N \to \infty} \sup (c + \inf T_N) =^{LS1 + LS2} c + \lim_{N \to \infty} \sup T_N =^{Def.\ 12.2.3} c + \limsup a_n$.

5. Let $T_{CN} = \{c\, a_{N+1}, c\, a_{N+2}, \cdots\}$. Then, $\inf T_{CN} = c \inf T_N$ and $\sup T_{CN} = c \sup T_N$. Now,

   - $\liminf (c a_n) =^{Def.\ 12.2.3} \lim_{N \to \infty} \inf T_{CN} = \lim_{N \to \infty} \inf (c \inf T_N) =^{LS1 + LS3} c \lim_{N \to \infty} \inf T_N =^{Def.\ 12.2.3} c \liminf a_n$.

   - $\limsup (c a_n) =^{Def.\ 12.2.3} \lim_{N \to \infty} \sup T_{CN} = \lim_{N \to \infty} \sup (c \inf T_N) =^{LS1 + LS3} c \lim_{N \to \infty} \sup T_N =^{Def.\ 12.2.3} c \limsup a_n$.

6. Let $T_{CN} = \{c\, a_{N+1}, c\, a_{N+2}, \cdots\}$. For all $n > N$ we have $\inf T_N \leq a_n \leq \sup T_N$ and hence $c \sup T_N \leq c\, a_n \leq c \inf T_N$. Hence

   (12.5) $c \sup T_N \leq \inf T_{cN}$ and $\sup T_{cN} \leq c \inf T_N$.

   Furthermore, for all $n > N$ we have $\inf T_{cN} \leq c a_n \leq \sup T_{cN}$ and hence $\dfrac{\sup T_{cN}}{c} \leq a_n \leq \dfrac{\inf T_{cN}}{c}$. Hence $\dfrac{\sup T_{cN}}{c} \leq \inf T_N$ and $\sup T_N \leq \dfrac{\inf T_{cN}}{c}$. Hence

   (12.6) $\inf T_{cN} \leq c \sup T_N$ and $c \inf T_N \leq \sup T_{cN}$.

   By (12.5) and (12.6), $\inf T_{cN} = c \sup T_N$ and $c \inf T_N = \sup T_{cN}$. Now,

   - $\liminf (c a_n) =^{Def.\ 12.2.3} \lim_{N \to \infty} \inf T_{cN} = \lim_{N \to \infty} c \sup T_N =^{LS1 + LS3} c \lim_{N \to \infty} \sup T_N =^{Def.\ 12.2.3} c \limsup a_n$.

   - $\limsup (c\, a_n) =^{Def.\ 12.2.3} \lim_{N \to \infty} \sup T_{cN} = \lim_{N \to \infty} c \inf T_N =^{LS1 + LS3} c \lim_{N \to \infty} \inf T_N =^{Def.\ 12.2.3} c \liminf a_n$.

7. Recall that $T_{a_n,N} = \{a_{N+1}, a_{N+2}, \cdots\} = \{a_n : n > N\}$ and $T_{a_{p_n},N} = \{a_{p_{N+1}}, a_{p_{N+2}}, \cdots\} = \{a_{p_n} : n > N\}$. Recall also that $p_n \geq n$ for all n. Hence $T_{a_n,N} \supseteq T_{a_{p_n},N}$ for all N. Therefore, $\inf T_{a_n,N} \leq \inf T_{a_{p_n},N}$ and $\sup T_{a_n,N} \geq \sup T_{a_{p_n},N}$. Hence $\lim_{N\to\infty} \inf T_{a_n,N} \leq \lim_{N\to\infty} \inf T_{a_{p_n},N}$ and $\lim_{N\to\infty} \sup T_{a_n,N} \geq \lim_{N\to\infty} \sup T_{a_{p_n},N}$. Hence $\liminf a_n \leq \liminf a_{p_n} \leq \limsup a_{p_n} \leq \limsup a_n$.

8. Suppose that for some $\varepsilon > 0$, the set $\{n : a_n < \liminf a_n + \varepsilon\}$ is finite. Then there is a positive integer N such that for all $n > N$, $a_n \geq \liminf a_n + \varepsilon$. By part 1 of the theorem, $\liminf a_n \geq \liminf a_n + \varepsilon$, a contradiction. The proof for the other set is similar.

9. By 6. above, $\liminf(-a_n) = -\limsup a_n$ and $\limsup(-a_n) = -\liminf a_n$. Hence $-\liminf(-a_n) = \limsup a_n$ and $-\limsup(-a_n) = \liminf a_n$.

$\square$

Definition 12.2.14. *Let $\{a_n\}$ be a bounded sequence of real numbers. A number a is a sequential limit if there is a subsequence of $\{a_n\}$ which converges to a.*

Example 12.2.15. *In Example 12.2.5, $S = \{1, -1\}$ is the set of sequential limits. See Example 12.2.10.*

Example 12.2.16. *In Example 12.2.6, in addition to the convergent subsequences mentioned in Example 12.2.11, there is the subsequence $3, 3, \cdots$, which converges to 3. So $S = \{2, 3, 5\}$ is the set of sequential limits.*

Theorem 12.2.17. *Let $\{a_n\}$ be a bounded sequence of real numbers and let S be the set of sequential limits of $\{a_n\}$. Then the set S contains its greatest lower bound and its least upper bound and*

$$\liminf a_n = \inf S \text{ and } \limsup a_n = \sup S.$$

Proof. Since the sequence $\{a_n\}$ is bounded, then every one of its subsequences is bounded and hence for each subsequence that has a limit, that limit is bounded. Hence the set S is bounded. Furthermore, by Theorem 12.2.8, $S \neq \emptyset$. Hence the numbers $\inf S$ and $\sup S$ exist by the Completeness Axiom and Theorem 10.4.2.

Let s be an element of S and let $\{a_{q_n}\}$ be a subsequence of $\{a_n\}$ which converges to s, i.e., $s = \lim_{n\to\infty} a_{q_n}$. By Theorem 12.2.7 and part 7 of Theorem 12.2.13,

$$\liminf a_n \leq \liminf a_{q_n} = s = \limsup a_{q_n} \leq \limsup a_n.$$

This shows that $\liminf a_n \leq \inf S$ and $\sup S \leq \limsup a_n$. By Theorem 12.2.8, the sequence $\{a_n\}$ has a (monotone) subsequence that converges to $\limsup a_n$ and another (monotone) subsequence that converges to $\liminf a_n$. Hence $\limsup a_n = \sup S$ and $\liminf a_n = \inf S$. Finally, since $\{a_n\}$ has a (monotone) subsequence that converges to $\limsup a_n$, then $\limsup a_n \in S$. Similarly, since $\{a_n\}$ has a (monotone) subsequence that converges to $\liminf a_n$, then $\liminf a_n \in S$. Hence $\inf S \in S$ and $\sup S \in S$. □

Since in this chapter we are concerned with the consequences of the Axiom of Completeness of the real numbers, the next section considers some of these consequences for the notion of continuity.

12.2.1   Exercises

Exercise 12.9. Let $\{a_n\}$ be a sequence and N be a positive integer. We say that a_N is a *floor term* if a_N is a lower bound of $T_{a_n,N}$. For each of the sequences $\{a_n\}$ given below, give all the tails T_N, for each T_N give $\inf T_N$ and $\sup T_N$, all the floor terms, all the sequential limits, and also give $\liminf a_n$ and $\limsup a_n$.

1. $\{a_n\}$ where $a_n = (-1)^{n+1}$.

2. $\{a_n\}$ where $a_n = (-1)^n$.

3. $\{a_n\}$ where $a_n = \frac{1}{n}$.

12.3   More on Continuity

We will now see proofs of two theorems which, while obvious geometrically, were once considered impossible to prove without reference to geometric diagrams. These theorems are the Intermediate Value Theorem 12.3.2 and the Continuous one-to-one is strictly monotonic Theorem 12.3.7. The Intermediate Value Theorem 12.3.2 can actually be proven from the Bolzano's theorem 12.3.1, which Bolzano proved in 1817. Note also that Bolzano's theorem is actually a special case of the Intermediate Value Theorem 12.3.2.

Theorem 12.3.1 (Bolzano Theorem). *Suppose that f is a real-valued function whose domain includes the interval $[a, b]$, and suppose that f is continuous on $[a, b]$. If $f(a)$ and $f(b)$ have opposite signs (i.e., one is strictly negative and the other is strictly positive) then there is a value $c \in (a, b)$ such that $f(c) = 0$.*

Proof. Since $f(a)$ and $f(b)$ have different signs, we do the proof first for $f(a) < 0 < f(b)$. The proof for $f(b) < 0 < f(a)$ follows. Let $S = \{x \in [a, b] : f(x) \leq 0\}$. Since S is nonempty (it contains a) and bounded above (b is an upper bound), S has a least upper bound $c = \sup S$ by the Completeness Axiom. Note that for $x \in [a, b]$:

1. If $f(x) \leq 0$ then $x \leq c \leq b$.

2. Hence if $x > c$ then $f(x) > 0$.

3. Furthermore, since c is the least upper bound such that $f(x) \leq 0$ then there is no $y < c$ such that $f(x) > 0$ for $x > y$.

4. Finally, since f is continuous, $\lim_{x \to c} f(x) = f(c)$.

We show now that the value c satisfies $f(c) = 0$.

- If $f(c) > 0$, then let $\varepsilon = \frac{f(c)}{2} > 0$. Since f is continuous, there is $\delta > 0$ such that $|x - c| < \delta$ implies $|f(x) - f(c)| < \frac{f(c)}{2}$. Hence $c - \delta < x < c + \delta$ implies $\frac{f(c)}{2} < f(x) < \frac{3f(c)}{2}$. Hence for $x \in (c - \delta, c]$ we have $f(x) > 0$. So we found $c - \delta < c$ such that if $x > c - \delta$ then $f(x) > 0$. Contradiction (see item 3 above).

- If $f(c) < 0$, then let $\varepsilon = -\frac{f(c)}{2} > 0$. Since f is continuous, there is $\delta > 0$ such that $|x - c| < \delta$ implies $|f(x) - f(c)| < \frac{-f(c)}{2}$. Hence $c - \delta < x < c + \delta$ implies $\frac{3f(c)}{2} < f(x) < \frac{f(c)}{2}$. Hence for $x \in (c, c + \delta)$ we have $f(x) < 0$. Contradiction (see item 2 above).

Hence $f(c) = 0$. Also, since $f(a)$ and $f(b)$ are both nonzero, $c \in (a, b)$. As for the proof when $f(b) < 0 < f(a)$, let $g(x) = -f(x)$. Then g is defined on the same interval as f and g is continuous on $[a, b]$ and $g(a) < 0 < g(b)$. By what we proved above, there is a $c \in [a, b]$ such that $g(c) = 0$. Hence $f(c) = 0$. $\qquad\square$

Theorem 12.3.2 (Intermediate Value Theorem). *Suppose that f is a real-valued function whose domain includes the interval $[a, b]$, and suppose that f is continuous on $[a, b]$. If v is any number between $f(a)$ and $f(b)$, then there is a value $c \in (a, b)$ such that $f(c) = v$.*

Proof. Note that v is strictly between $f(a)$ and $f(b)$. Let $g(x) = f(x) - v$. Then g is defined and continuous everywhere f is defined and continuous. Then g is a real-valued function whose domain includes the interval $[a, b]$,

g is continuous on $[a, b]$ and 0 is (strictly) between $g(a)$ and $g(b)$. Hence $g(a)$ and $g(b)$ have opposite signs and by Theorem 12.3.1 there is a value $c \in (a, b)$ such that $g(c) = 0$. That is, $f(c) = v$. Since $f(a) \neq v$ and $f(b) \neq v$, $c \in (a, b)$. $\qquad\square$

Corollary 12.3.3. *If f is continuous on an interval I, then the set $J = f(I) = \{f(x) : x \in I\}$ is also an interval. Furthermore, if J is bounded then for any y such that $\inf J < y < \sup J$ we have $y \in J$.*

Proof. We assume that f is a non constant function. I.e., there are $a, b \in I$ such that $f(a) \neq f(b)$.

By Intermediate Value Theorem 12.3.2, if $f(x_0)$ and $f(x_1)$ are in J, and $f(x_0) < y < f(x_1)$ then there is $x \in (\inf(x_0, x_1), \sup(x_0, x_1))$ such that $f(x) = y$. Hence $y \in J$. So we have:

(12.7) $\qquad\qquad y_0, y_1 \in J$ and $y_0 < y < y_1$ imply $y \in J$.

Such a set J must be an interval.

Assume J is bounded. Since J is non empty, we have by Completeness Axiom and Theorem 10.4.2 that $\inf J$ and $\sup J$ exist. And so J is an interval with endpoints $\inf J$ and $\sup J$; $\inf J$ and $\sup J$ may or may not belong to J.

We show that (12.7) implies the following (12.8):

(12.8) $\qquad\qquad \inf J < y < \sup J$ implies $y \in J$.

For any y such that $\inf J < y < \sup J$, it must be the case that there are y_0 and y_1 in J such that $\inf J < y_0 < y < y_1 < \sup J$. Thus, $y \in J$ by (12.7). $\qquad\square$

Example 12.3.4 (Fixed Point of a Continuous Function). *We can now prove the following: if f is a continuous function defined on $[0, 1]$ such that if $x \in [0, 1]$ then $f(x) \in [0, 1]$, then f has a fixed point, i.e., a point $x_0 \in [0, 1]$ such that $f(x_0) = x_0$, so that x_0 is left fixed by f. This says that the graph of f, which is contained in the square formed by the points $(0, 0)$, $(1, 0)$, $(1, 1)$, and $(0, 1)$, crosses the line $y = x$, and is obvious from the geometry. But it can be proved without reference to any diagram: if $f(0) = 0$, then 0 is a fixed point; if not, and if $f(1) = 1$, then 1 is a fixed point, and if both $f(0) \neq 0$ and $f(1) \neq 1$, then $f(0) > 0$ and $f(1) < 1$, and let $g(x) = f(x) - x$. Then g is also a continuous function defined on $[0, 1]$. Also, $g(0) = f(0) - 0 = f(0) > 0$ and $g(1) = f(1) - 1 < 1 - 1 = 0$, so by the Intermediate Value theorem, there is a point $x_0 \in (0, 1)$ such that $g(x_0) = 0$. But then $0 = g(x_0) - f(x_0) - x_0$, so $f(x_0) = x_0$.*

Example 12.3.5 (Existence of mth root of a positive number). *We can also now prove that if $y > 0$ and m is a positive integer, then y has an mth root. For the function $f(x) = x^m$ is continuous. There is $b > 0$ such that $y < b^m$; if $y \leq 1$ let $b = 2$ and if $y > 1$ let $b = y + 1$. Then $f(0) < y < f(b)$, and so the Intermediate Value Theorem implies that $f(x) = y$ for some $x \in (0, b)$. So $y = x^m$ and x is an mth root of y.*

Just like we defined the increasing/decreasing sequences (Definition 12.1.1), we will now define similar concepts for functions.

Definition 12.3.6 (Strictly Increasing/Decreasing/Monotone Functions). *A function f is said to be strictly increasing on an interval I if, for any x_1 and x_2 in I, if $x_1 < x_2$ then $f(x_1) < f(x_2)$. It is strictly decreasing on I if, for any x_1 and x_2 in I, if $x_1 < x_2$ then $f(x_1) > f(x_2)$. It is nondecreasing if, for any x_1 and x_2 in I, if $x_1 < x_2$ then $f(x_1) \leq f(x_2)$. It is nonincreasing if, for any x_1 and x_2 in I, if $x_1 < x_2$ then $f(x_1) \geq f(x_2)$. A function f is said to be strictly monotone on an interval I if it is strictly increasing or strictly decreasing on I. We call f monotone if it is nonincreasing or nondecreasing.*

Note that the function f in Example 12.3.5 is strictly increasing on $[0, \infty)$: if $x_1 < x_2$, then $x_1^m < x_1^m$, or $f(x_1) < f(x_2)$. It follows that f is one-to-one, as defined by Definition 11.3.17, hence each $y \geq 0$ has exactly one mth root, and the notations $\sqrt[m]{y}$ and $y^{1/m}$ are unambiguous. Also, f has an inverse in the sense of Definition 11.3.18. In fact, $f_{\mathrm{inv}}(y) = y^{1/m}$.

A continuous one-to-one function on an interval, is strictly increasing or strictly decreasing on that interval:

Theorem 12.3.7 (Continuous one-to-one is strictly monotonic). *Let f be a continuous, one-to-one function on an interval I. Then f is strictly increasing or strictly decreasing on I.*

Proof. First, we show that

(12.9)   if $a < b < c$ in I, then $f(b)$ is between $f(a)$ and $f(c)$.

If not, then $f(b)$ is either greater than the maximum of $f(a)$ and $f(c)$ or less than their minimum. Suppose that it is greater than the maximum; the other case is similar. Select y so that $f(b) > y > \max\{f(a), f(c)\}$. By the Intermediate Value Theorem applied to $[a, b]$ and $[b, c]$, there are $x_1 \in (a, b)$ and $x_2 \in (b, c)$ such that $f(x_1) = f(x_2) = y$, contradicting the one-to-one property of f.

Now select $a_0 < b_0$ in I and suppose that $f(a_0) < f(b_0)$. We will show that f is strictly increasing on I. (If $f(a_0) > f(b_0)$, a similar argument would show that f is strictly decreasing on I.) By (12.9) we have

$$
\begin{array}{lll}
f(x) < f(a_0) & \text{for } x < a_0 & [\text{ since } x < a_0 < b_0], \\
f(a_0) < f(x) < f(b_0) & \text{for } a_0 < x < b_0, & \\
f(b_0) < f(x) & \text{for } x > b_0 & [\text{ since } a_0 < b_0 < x].
\end{array}
$$

In particular,

$$(12.10) \qquad\qquad f(x) < f(a_0) \text{ for all } x < a_0$$

and

$$(12.11) \qquad\qquad f(a_0) < f(x) \text{ for all } x > a_0.$$

Consider any $x_1 < x_2$ in I. If $x_1 = a_0$ use (12.11). If $x_2 = a_0$ use (12.10). If $x_1 < a_0 < x_2$, then $f(x_1) < f(x_2)$ by (12.10) and (12.11). If $x_1 < x_2 < a_0$, then $f(x_1) < f(a_0)$ by (12.10) and so, by (12.9), we have $f(x_1) < f(x_2)$. Finally (and similarly), if $a_0 < x_1 < x_2$, then $f(a_0) < f(x_2)$ and so $f(x_1) < f(x_2)$. $\qquad\square$

Corollary 12.3.8. *If a one-to-one function f is continuous on an interval I, and if $f(I) = J$, then f_{inv} is either strictly increasing (if f is strictly increasing) or strictly decreasing (if f is strictly decreasing) on J.*

Proof. Note that if f is one-to-one on an interval I and is continuous on I, then $f(I)$ is an interval J by Corollary 12.3.3, and by Definition 11.3.18, f_{inv} is defined on J and $f_{inv}(J) = I$. By Theorem 12.3.7, f is strictly increasing or strictly decreasing. We will prove that if f is strictly increasing on I, then f_{inv} is strictly increasing on J. The proof that if f is strictly decreasing on I then f_{inv} is strictly decreasing on J is similar.

Suppose $y_1 < y_2$ in J. Since f_{inv} is one-to-one, we must either have $f_{inv}(y_1) < f_{inv}(y_2)$ or $f_{inv}(y_1) > f_{inv}(y_2)$. Suppose, for a proof by contradiction, that there are $y_1 < y_2$ in J such that $f_{inv}(y_1) > f_{inv}(y_2)$. Then $x_1 > x_2$, where $y_1 = f(x_1)$ and $y_2 = f(x_2)$ and we have $f(x_1) < f(x_2)$, a contradiction since f is strictly increasing. $\qquad\square$

We can now prove that the inverse of a continuous function is continuous.

Theorem 12.3.9. *Let g be a strictly increasing or strictly decreasing function on an interval J such that $g(J)$ is an interval I. Then g is continuous on J.*

Proof. By hypothesis, g is either strictly increasing or strictly decreasing on J. We will prove the theorem on the assumption that g is strictly increasing;

the case for a strictly decreasing g is similar. Consider x_0 in J. We assume that x_0 is not an endpoint of J; small changes are needed in the proof otherwise. Then $g(x_0)$ is not an endpoint of I and so there exists $\varepsilon_0 > 0$ such that $(g(x_0) - \varepsilon_0, g(x_0) + \varepsilon_0) \subseteq I$.

Let $\varepsilon > 0$ be given. Let $\varepsilon' = min\{\varepsilon, \varepsilon_0\}$. Hence $(g(x_0) - \varepsilon', g(x_0) + \varepsilon') \subseteq (g(x_0) - \varepsilon_0, g(x_0) + \varepsilon_0) \subseteq I$ and there exist $x_1, x_2 \in J$ such that $g(x_1) = g(x_0) - \varepsilon'$ and $g(x_2) = g(x_0) + \varepsilon'$. Clearly, $x_1 < x_0 < x_2$ since g is strictly increasing and $g(x_1) < g(x_0) < g(x_2)$. Also, if $x_1 < x < x_2$, then $g(x_1) < g(x) < g(x_2)$, and hence $g(x_0) - \varepsilon' < g(x) < g(x_0) + \varepsilon'$. Now if we set $\delta = min\{x_2 - x_0, x_0 - x_1\}$, then $|x - x_0| < \delta$ implies $x_1 < x < x_2$ and hence $|g(x) - g(x_0)| < \varepsilon' \leq \varepsilon$. $\qquad \square$

Corollary 12.3.10. *If f is one-to-one and continuous on an interval I and if $f(I) = J$, then f_{inv} is continuous on J.*

Proof. By Corollary 12.3.3, J is interval. By Corollary 12.3.8, f_{inv} is strictly increasing or strictly decreasing on the interval J. But $f_{inv}(J) = I$ is an interval. By Theorem 12.3.9, f_{inv} is continuous on J. $\qquad \square$

This allows us to strengthen Theorem 11.3.19 and Corollary 11.3.20 by removing the hypothesis that f_{inv} is continuous:

Corollary 12.3.11. *Let I and J be intervals and let f be a continuous, one-to-one function defined on I with values in J. If f is differentiable at c and if $f'(c) \neq 0$, then f_{inv} is differentiable at $f(c)$ and $f'_{inv}(f(c)) = 1/f'(c)$. Consequently, if f is differentiable on I, then*

$$f'_{inv}(x) = \frac{1}{f'(f_{inv}(x))}$$

for all values of x in J for which $f'(f_{inv}(x)) \neq 0$.

We can now prove another theorem which is obvious from diagrams by a proof that does not depend on diagrams:

Theorem 12.3.12 (Extreme Value Theorem). *If f is a function whose domain includes $[a, b]$, and if f is continuous on $[a, b]$, then there exist values c and d in $[a, b]$ such that $f(c) \leq f(x) \leq f(d)$ for all x in $[a, b]$.*

Remark 12.3.13. *This theorem has three hypotheses: the function must be continuous, the interval must be closed, and the interval must be bounded. If any one of these hypotheses is not satisfied, then the conclusion of the theorem may be false. The following three examples illustrate this point. In each two of the three hypotheses are satisfied.*

1. The function f defined by $f(x) = 1/x$ for $x \neq 0$ and $f(0) = 0$ does not have a maximum value on $[0, 1]$. ($\lim_{x\to 0} f(x) = \lim_{x\to 0} \frac{1}{x} = \infty$ and f is not continuous at 0.)

2. The function $g(x) = x^2$ does not have a maximum value on the interval $[-1, 2)$. (The interval is not closed.) Note that for all $x \in [-1, 2)$, $g(0) = 0 \leq g(x) \leq 4 = \lim_{x\to 2^-} g(x)$ but there is no $x \in [-1, 2)$ such that $g(x) = 4$.

3. The function $h(x) = x^3$ does not have a maximum value on the interval $[0, \infty)$. (The interval is not bounded.)

Proof. We first prove that the continuous function f is bounded on $[a, b]$ by proving the contrapositive: if f is unbounded on $[a, b]$, then f is not continuous on $[a, b]$. Suppose that f is unbounded on $[a, b]$. Then for each positive integer n, there exists x_n in $[a, b]$ such that $|f(x_n)| > n$. By the Bolzano-Weierstrass Theorem, the sequence $\{x_n\}$ contains a subsequence $\{x_{p_n}\}$ that converges to a value z in $[a, b]$. Since $|f(x_{p_n})| > p_n$ for all n, the sequence $\{f(x_{p_n})\}$ is unbounded and cannot converge. By Theorem 11.2.3, the function f is not continuous at z.

Since f is bounded on $[a, b]$, the set $f([a, b]) = \{f(x) : x \in [a, b]\}$ is bounded. Let α be the greatest lower bound of $f([a, b])$ and let β be its least upper bound, and note that $\alpha \leq f(x) \leq \beta$ for all x in $[a, b]$. We will prove that there exists a value d in $[a, b]$ such that $f(d) = \beta$; the proof that there exists a value c in $[a, b]$ such that $f(c) = \alpha$ is similar. Since β is the least upper bound of $f([a, b])$, for each positive integer n, there exists d_n in $[a, b]$ such that

$$\beta - \frac{1}{n} < f(d_n) \leq \beta.$$

By the Bolzano-Weierstrass Theorem, the sequence $\{d_n\}$ contains a subsequence $\{d_{q_n}\}$ that converges to a value d in $[a, b]$. Since f is continuous at d,

$$f(d) = \lim_{n\to\infty} f(d_{q_n}) = \beta.$$

$\square$

12.3.1   Exercises

Exercise 12.10. Prove the Intermediate Value Theorem 12.3.2 directly, without using the Bolzano Thoeorem 12.3.1.

Exercise 12.11. Prove that if f is a continuous function on a closed interval I then f is bounded on I (i.e., $f(I) = \{f(x) : x \in I\}$ is bounded) and f

attains its bounds (i.e., if g and l are the greatest lower versus least upper bounds of $f(I)$ then there are x and y in I such that $f(x) = g$ and $f(y) = l$). Give an example of a continuous function h on a bounded interval J such that $h(J)$ is not bounded.

Give an example of a continuous function h' on a bounded interval J' such that $h'(J')$ is bounded but does not attain one of its bounds.

Exercise 12.12. 1. Prove that there is a real number $x > 2\pi$ such that $\tan x = x$.

2. Prove that there is some x in $(0, \pi/2)$ such that $\cos x = x$.

3. Suppose that f is a function whose domain is a closed interval $[a, b]$ and whose range is a subset of $[a, b]$. Suppose also that f is continuous on $[a, b]$. Prove that there exists at least one value $x \in [a, b]$ such that $f(x) = x$. (A point with this property is known as a fixed point of f.)

4. Prove that there is a value of x in $(0,1)$ such that $x2^x x = 1$.

5. Suppose that f is a continuous real-valued function of real numbers, and suppose that there are real numbers a and b such that $f(a)f(b) < 0$. Prove that there is a number x between a and b such that $f(x) = 0$.

12.4 Converging Sequences of Functions

We have talked about the convergence of sequences and the limits of functions. One thing we have not discussed is the possibility of convergent sequences of functions.

Suppose that f_n is a function for each natural number n. It is natural to ask if this sequence of functions approaches another function f as a limit. It is easy to see that it is, and the natural definition is as follows:

Definition 12.4.1. *Suppose $\{f_n\}$ is a sequence of functions all of which have the same domain, and suppose f is a function with the same domain. Then the sequence $\{f_n\}$ converges pointwise to f on the common domain if for each x in that domain, $\{f_n(x)\}$ converges to $f(x)$. In other words, for each x in the common domain,*

$$\lim_{n \to \infty} f_n(x) = f(x).$$

It is natural in this connection to ask if the limit function f inherits properties from the functions $\{f_n\}$. In particular, it is natural to ask whether it is true that if each f_n is continuous, then f is continuous. Cauchy, who is credited with introducing our modern definitions of limits and continuity, asserted in his [8] that it is; i.e., that every convergent sequence of continuous functions always has a continuous limit function. His proof went something like this: assuming each f_n is continuous at c, we have

$$|f(x) - f(c)| \leq |f(x) - f_n(x)| + |f_n(x) - f_n(c)| + |f_n(c) - f(c)|.$$

Now $|f(x) - f_n(x)|$ can be made small for $n > N_1$ since the sequence $\{f_n(x)\}$ converges to $f(x)$, $|f_n(c) - f(c)|$ can be made small for $n > N_2$ for the same reason, and $|f_n(x) - f_n(c)|$ can be made small for suitable δ since each $\{f_n\}$ is continuous at c.

But the theorem is, in fact, not true. Consider the following example:

Example 12.4.2. *Let $f_n(x) = x^n$ for each n, and consider the interval $[0, 1]$. Then each of the f_n is defined on $[0, 1]$ and is continuous at every point of that interval. On this interval, $[0, 1]$, the sequence $\{f_n\}$ approaches as a limit the following function f:*

$$f(x) = \begin{cases} 1 & \text{if } x = 1 \\ 0 & \text{otherwise} \end{cases}$$

Clearly, f is not continuous at $x = 1$.

What is wrong here?

Let us try to apply Cauchy's proof to this example. Let $\varepsilon > 0$ be given, and let $x \in [0, 1]$. Then there is N_1 such that if $n > N_1$, then

$$|f(x) - f_n(x)| = \begin{cases} |0 - x^n| = |x^n| < \varepsilon/3 & \text{if } x < 1 \\ |1 - 1^n| = |1 - 1| = 0 < \varepsilon/3 & x = 1 \end{cases}$$

Also, there is $\delta > 0$ such that if $|x - c| < \delta$, then $|f_n(x) - f_n(c)| < \varepsilon/3$. And finally, there is N_2 such that if $n > N_2$, then

$$|f(c) - f_n(c)| = \begin{cases} |0 - c^n| = |c^n| < \varepsilon/3 & \text{if } c < 1 \\ |1 - 1^n| = |1 - 1| = 0 < \varepsilon/3 & c = 1 \end{cases}$$

This does seem to imply that $|f(x) - f(c)| < \varepsilon$.

However, things are not really this simple. The number N_1 depends on the value of x, and that of N_2 depends on the value of c. If $x = 1/2$, then $x^2 = 1/4$, $x^3 = 1/8$, and $x^n = 1/2^n$. But if $x = .99$, then $x^2 = 0.9801$, $x^3 = 0.970299$, and it should be clear that to make $|f(x) - f_n(x)| < \varepsilon/3$ for $n > N_1$ requires a much larger value of N_1 for $x = .99$ than for $x = 1/2$. Furthermore, the value of N_1 that is needed may grow arbitrarily as x

approaches 1. A similar consideration applies to N_2 and the value of c. In order to make Cauchy's argument work here, there must be a value of N_1 with the property that if $n > N_1$, then for every value of $x \in [0, 1]$, $|f(x) - f_n(x)| < \varepsilon/3$. This is not pointwise convergence, but uniform convergence.

Definition 12.4.3. *Let $\{f_n\}$ be a sequence of functions whose domains include an interval I, and let f be another function whose domain includes I. Then the sequence $\{f_n\}$ converges uniformly to f on I if for each $\varepsilon > 0$ there is a positive integer N such that for all $x \in I$ and all $n > N$, $|f(x) - f_n(x)| < \varepsilon$. A sequence of functions $\{f_n\}$ whose domains include an interval I is said to converge uniformly on I if there is a function f whose domain includes I and $\{f_n\}$ converges uniformly to f on I.*

If the convergence of $\{f_n\}$ to f is uniform on an interval, then Cauchy's argument works.

Theorem 12.4.4. *Suppose that f_n for each n and f are functions whose domains include an interval I, suppose that for each n, f_n is continuous at c for some $c \in I$, and suppose that $\{f_n\}$ converges to f uniformly on I. Then f is continuous at c.*

Proof. Let $\varepsilon > 0$ be given. Since $\{f_n\}$ converges uniformly to f on I, there is a positive integer p such that $|f_p(x) - f(x)| < \varepsilon/3$ for all $x \in I$, including $x = c$. Since f_p is continuous at c, there is a $\delta > 0$ such that for all $x \in I$ satisfying $|x - c| < \delta$, $|f_p(x) - f_p(c)| < \varepsilon/3$. Hence, for these values of x,

$$
\begin{aligned}
|f(x) - f(c)| &\leq |f(x) - f_p(x)| + |f_p(x) - f_p(c)| + |f_p(c) - f(c)| \\
&< \tfrac{\varepsilon}{3} + \tfrac{\varepsilon}{3} + \tfrac{\varepsilon}{3} \\
&= \varepsilon.
\end{aligned}
$$

Hence, f is continuous at c. □

Corollary 12.4.5. *Suppose that f_n for each n and f are functions whose domains include an interval I, suppose that for each n, f_n is continuous on I, and suppose that $\{f_n\}$ converges to f uniformly on I. Then f is continuous on I.*

In Example 12.4.2, the sequence $\{f_n\}$ of functions does not converge uniformly to f on $[0, 1]$. This can be seen directly by negating the definition. For a sequence of functions $\{f_n\}$ that converges pointwise to f on an interval, $\{f_n\}$ does not converge uniformly to f if there is an $\varepsilon > 0$ such that for every positive integer N there is a number $x \in I$ and an integer $n > N$ such that $|f_n(x) - f(x)| \geq c$. For the sequence at hand, let $c = 0.5$, and let N be any positive integer. Then the number $\sqrt[N]{0.5} \in (0, 1)$ approaches 1 as N

increases, and $|f_N(x)| = 0.5$. This is sufficient to show that the convergence of $\{f_n\}$ to 0 is not uniform on $[0, 1]$.

Remark 12.4.6. *Does this mean that Cauchy made a mistake in his theorem? Since the late 19th century, most mathematicians have thought so. Not only did Cauchy publish a theorem to which there are counterexamples (some of which were well known before 1821), but he repeated the theorem in [9, 7]. However, newer historical research suggests that this was not really a mistake. Remember that the real number system as we know it had not yet been defined in Cauchy's day. Lakatos, in his [27] explains this by pointing out that Cauchy was following Leibniz in believing in infinitely small quantities, which means that Cauchy believed that Archimedes Law AL is false. This enabled Cauchy to claim that the convergence of $\{f_n\}$ to f had to be pointwise for the infinitesimal quantities as well as the real ones, and he was thus able to argue that this kind of pointwise convergence was equivalent to uniform convergence. The idea of infinitely small quantities was discredited by the work on the arithmetisation of analysis in the late 19th century. However, it was revived by Abraham Robinson, who proved in 1960 using some highly technical results in mathematical logic, that there is a theory of an extension of the real number system with infinitely small numbers that is consistent if the standard theory of real numbers is. Also, Cauchy's argument would fail in this theory. This theory, which is presented in full in [36], is beyond the scope of this course.*

Now let us look at some other properties of uniform convergence.

Theorem 12.4.7. *Suppose that the sequence of functions $\{f_n\}$ converges pointwise to f on an interval I and let*

$$M_n = \sup\{|f_n(x) - f(x)| : x \in I\}$$

for each n. Then the sequence $\{f_n\}$ converges uniformly to f if and only if the sequence $\{M_n\}$ converges to 0.

Proof. Suppose first that the sequence $\{f_n\}$ converges uniformly to f on I. Let $\varepsilon > 0$ be given. Then there is a positive integer N such that for all $n > N$ and all $x \in I$, $|f_n(x) - f(x)| < \varepsilon$. Then, since for every $n > N$, every element of the set $\{|f_n(x) - f(x)| : x \in I\}$ is less than ε, for all $n > N$, $\sup\{|f_n(x) - f(x)| : x \in I\} < \varepsilon$. But this says that for all $n > N$, $|M_n| = M_n < \varepsilon$ (note that M_n is positive), and so the sequence M_n converges to 0.

Now suppose the sequence $\{M_n\}$ converges to 0. Let $\varepsilon > 0$ be given. Then there is a positive integer N such that for $n > N$, $|M_n| < \varepsilon$. This means that for $n > N$, $\sup\{|f_n(x) - f(x)| : x \in I\} < \varepsilon$. Hence, for each $n > N$, every element of the set must be less than ε, i.e., for each $n > N$ and $x \in I$, $|f_n(x) - f(x)| < \varepsilon$, and so it follows that $\{f_n\}$ converges uniformly to f. $\qquad\qquad\square$

There is also a Cauchy criterion for uniform convergence.

Theorem 12.4.8 (Cauchy Criterion for Uniform Convergence). *A sequence $\{f_n\}$ of functions defined on an interval I converges uniformly on I if and only if for each $\varepsilon > 0$ there is a positive integer N such that for all $x \in I$ and $m, n > N$, $|f_n(x) - f_m(x)| < \varepsilon$.*

Proof. Suppose that for each $\varepsilon > 0$ there exists a positive integer N such that for all $x \in I$ and all $m, n > N$, $|f_n(x) - f_m(x)| < \varepsilon$. The first step in the proof is to find a function f such that the sequence $\{f_n\}$ converges pointwise to f on I. By hypothesis, the sequence $\{f_n(x)\}$ is a Cauchy sequence for each $x \in I$. Hence, for each $x \in I$, $f_n(x)$ converges to a limit; let $f(x)$ be that limit; i.e., for each $x \in I$,

$$f(x) = \lim_{n \to \infty} f_n(x).$$

To show that $\{f_n\}$ converges uniformly to f on I, let $\varepsilon > 0$ be given, and choose a positive integer N such that for all $x \in I$ and all $m, n > N$, $|f_n(x) - f_m(x)| < \varepsilon$. Fix $n > N$ and $x \in I$. Then for each $m > N$, the inequality $|f_n(x) - f_m(x)| < \varepsilon$ is valid, and so the sequence $\{|f_n(x) - f_m(x)|\}$ for $m > N$ (with x and n fixed) converges to $|f_n(x) - f(x)|$. It follows that $|f_n(x) - f(x)| < \varepsilon$. Since $x \in I$ and $n > N$ are arbitrary, this shows that $\{f_n\}$ converges uniformly to f on I.

For the converse, suppose that $\{f_n\}$ converges uniformly to f on I. Then for every $\varepsilon > 0$, there is a positive integer N such that for all $x \in I$ and all $n > N$, $|f_n(x) - f(x)| < \varepsilon/2$. Let $x \in I$ and $m, n > N$. Then we have

$$\begin{aligned}
|f_n(x) - f_m(x)| &\leq |f_n(x) - f(x)| + |f(x) - f_m(x)| \\
&< \frac{\varepsilon}{2} + \frac{\varepsilon}{2} \\
&= \varepsilon.
\end{aligned}$$

Since $x \in I$ and $n, m > N$ are arbitrary, this shows that the condition for each $\varepsilon > 0$ there is a positive integer N such that for all $x \in I$ and $m, n > N$, $|f_n(x) - f_m(x)| < \varepsilon$ is satisfied. $\qquad\qquad\square$

Theorem 12.4.9 (Dini's Theorem [6] **).** *Let $\{f_n\}$ be a sequence of continuous functions that converges pointwise to a function f on a closed interval $[a, b]$, and suppose that $\{f_n\}$ is monotone (i.e., either for each n*

---

[6]Dini was an Italian mathematician and politician from Pisa (1845–1918).

and each $x \in [a,b]$, $f_n(x) \le f_{n+1}(x)$ or for each n and each $x \in [a,b]$, $f_n(x) \ge f_{n+1}(x)$). If f is continuous on $[a,b]$, then $\{f_n\}$ converges uniformly to f on $[a,b]$.

Proof. We do the proof for $\{f_n\}$ increasing (i.e., for each n and each $x \in [a,b]$, $f_n(x) \le f_{n+1}(x)$). (The proof for $\{f_n\}$ decreasing (i.e. for each n and each $x \in [a,b]$, $f_n(x) \ge f_{n+1}(x)$) is similar, simply take $g = f_n - f$ and follow the same steps below.)
For each n, let $g_n = f - f_n$. Then

$$
\begin{aligned}
g_{n+1}(x) - g_n(x) &= (f(x) - f_{n+1}(x)) - (f(x) - f_n(x)) \\
&= f_n(x) - f_{n+1}(x) \\
&\le 0,
\end{aligned}
$$

because $f_{n+1}(x) \ge f_n(x)$. So for each n, $g_n(x) \ge g_{n+1}(x)$. Also, since $\{f_n\}$ converges pointwise to f, $\{g_n\}$ converges pointwise to $g(x) = 0$. Finally, let $x \in [a,b]$, and let $\varepsilon > 0$ be given. Then since f_n and f are continuous on $[a,b]$, there are $\delta_1 > 0$ and $\delta_2 > 0$ such that for all $y \in [a,b]$,
    if $|y - x| < \delta_1$, then $|f(x) - f(y)| < \varepsilon/2$, and
    if $|y - x| < \delta_2$, then $|f_n(x) - f_n(y)| < \varepsilon/2$.
Let δ be the smaller of δ_1 and δ_2. Then if $|x - y| < \delta$, we have

$$
\begin{aligned}
|g_n(x) - g_n(y)| &= |(f(x) - f_n(x)) - (f(y) - g_n(y))| \\
&\le |f(x) - f(y)| + |f_n(x) - f_n(y)| \\
&< \tfrac{\varepsilon}{2} + \tfrac{\varepsilon}{2} \\
&= \varepsilon.
\end{aligned}
$$

Hence, it follows that g_n is continuous on $[a,b]$.

Since $[a,b]$ is a closed interval, it follows that each g_n has a maximum value on $[a,b]$. For each n, let M_n be the maximum value of g_n on $[a,b]$. This means that for each n, there is a real number $x_n \in [a,b]$ such that $g_n(x_n) = M_n$. Since for each n, $x_n \in [a,b]$, the sequence $\{x_n\}$ is bounded. Thus, by the Bolzano-Weierstrass Theorem, there is a subsequence $\{x_{p_n}\}$ of $\{x_n\}$ which converges to a point $z \in [a,b]$. Since g_n is continuous at z, the sequence $\{M_{p_n} = g_{p_n}(x_{p_n})\}$ converges to $g(z) = 0$.

Now let $\varepsilon > 0$ be given. Since $\{M_{p_n}\}$ converges to 0, there is a positive integer N_1 such that for all $n > N_1$, $|M_{p_n}| < \varepsilon$. Let $N = p_{N_1}$ and let $x \in [a,b]$. Then, by the definition of $\{M_n\}$, for all $n > N$, since $\{g_n(x)\}$ is a nonincreasing sequence,

$$
\begin{aligned}
|f_n(x) - f(x)| &= |g_n(x)| \\
&\le |g_{p_N}(x)| \\
&\le |M_{p_N}| \\
&< \varepsilon.
\end{aligned}
$$

It follows that the sequence $\{f_n\}$ converges uniformly to f on $[a, b]$. $\square$

12.4.1 Exercises

Exercise 12.13. For each real number x, the sequence $\{\frac{x}{n}\}$ converges to 0. Given $\varepsilon > 0$ and a real number x, find a positive integer $N(\varepsilon, x)$ such that for all $n > N(\varepsilon, x)$, $|\frac{x}{n}| < \varepsilon$.

Exercise 12.14. For each $x \in (0, 1)$, the sequence $\{x^n\}$ converges to 0. Given $\varepsilon > 0$ and $x \in (0, 1)$, find a positive integer $N(\varepsilon, x)$ such that for all $n > N(\varepsilon, x)$, $|x_n| < \varepsilon$.

Exercise 12.15. Let I be an interval. Suppose that $\{f_n\}$ converges uniformly to f on I. Assume for each n, f_n is bounded (that is, there is M_n such that for each $x \in I$, $|f_n(x)| \le M_n$). Show that:

1. There is a uniform bound for all f_n's. That is, show that there is M such that for each n, for each $x \in I$, $|f_n(x)| \le M$.

2. f is bounded on I (that is, there is M such that for each $x \in I$, $|f(x)| \le M$).

Exercise 12.16. Let I be an interval. Suppose that $\{f_n\}$ converges uniformly to f on I and that $\{g_n\}$ converges uniformly to g on I.

1. Prove that the function $\{f_n + g_n\}$ converges uniformly to $f + g$ on I.

2. Give an example to show that $\{f_n g_n\}$ may not converge uniformly to fg on I.

3. Suppose that f and g are bounded on I. Prove that $\{f_n g_n\}$ converges uniformly to fg on I.

4. Suppose that for all n, f_n and g_n are bounded on I. Prove that $\{f_n g_n\}$ converges uniformly to fg on I.

Exercise 12.17. Determine whether or not the sequence $\{f_n\}$ where $f_n(x) = \frac{n}{x^n + 1}$ converges uniformly on $(0, 1)$.

Exercise 12.18. For each of the sequences of functions given below, determine its pointwise limit on $[0, 3]$ and give a proof whether convergence to this limit is uniform.

1. The sequence $\{f_n\}$, where $f_n(x) = \frac{x^2}{n + 1}$.

2. The sequence $\{f_n\}$, where $f_n(x) = \frac{x}{x - n}$.

3. The sequence $\{f_n\}$, where $f_n(x) = \frac{x}{nx+1}$.

4. The sequence $\{f_n\}$, where $f_n(x) = \frac{x}{nx^2+1}$.

5. The sequence $\{f_n\}$, where $f_n(x) = \frac{nx}{nx+1}$.

6. The sequence $\{f_n\}$, where $f_n(x) = \frac{x^n}{x^n+1}$.

Exercise 12.19. For each of the sequences given below, determine whether there is a function to which the sequence converges pointwise on $[0, 1]$ and if such a function exist:

- Formally show the pointwise convergence.

- Determine whether this convergence is uniform and give a proof for your claim.

1. The sequence $\{f_n\}$, where $f_n(x) = (1 - x^2)^n$.

2. The sequence $\{f_n\}$, where $f_n(x) = x(1 - x)^n$.

3. The sequence $\{f_n\}$, where $f_n(x) = nx(1 - x)^n$.

4. The sequence $\{f_n\}$, where $f_n(x) = nx(1 - x^2)^n$.

- $\{a_n\}$ is nondecreasing (resp. nonincreasing) if $\forall n$, $a_n \leq a_{n+1}$ (resp. $a_n \geq a_{n+1}$). $\{a_n\}$ is monotone if it nondecreasing or nonincreasing.
- A bounded monotone $\{a_n\}$ has a limit. If $\{a_n\}$ is unbounded nondecreasing (resp. nonincreasing) then $\lim_{n\to\infty} a_n = \infty$ (resp. $= -\infty$).
- $\{a_n\}$ is a *Cauchy sequence* if $\forall \varepsilon > 0$, $\exists N$: $\forall m, n > N$, $|a_n - a_m| < \varepsilon$.
- $\{a_n\}$ is Cauchy iff it Converges. Every Cauchy sequence is bounded.
- A sequence of intervals $\{I_n\}$ such that $\forall n$, $I_{n+1} \subseteq I_n$ is called *nested*.
- If $\{[a_n, b_n]\}$ is a nested sequence of closed and bounded intervals, then $\forall n$, $\exists z \in [a_n, b_n]$. And, if $\lim_{n\to\infty}(b_n - a_n) = 0$, then z is unique.
- If $\{p_n\}$ is a strictly increasing sequence of positive integers then $\{a_{p_n}\}$ is a subsequence of $\{a_n\}$.
- If $\{a_n\}$ converges to a, then each of its subsequences converges to a. If $\{a_n\}$ has two subsequences that converge to different limits, then $\{a_n\}$ does not converge.
  Every sequence of real numbers has a monotone subsequence.
- **Bolzano-Weierstrass Theorem** Every bounded sequence has a subsequence that converges to a limit.
- The tail of $\{a_n\}$ determined by N is $T_{a_n,N} = \{a_{N+1}, a_{N+2}, \cdots\} = \{a_n : n > N\}$. When the context is clear, we write T_N.
- $\forall N$, $T_{a_n,N}$ is a non empty subsequence of $\{a_n\}$ and if $\{a_n\}$ converges then $\lim_{n\to\infty} T_{a_n,N} = \lim_{n\to\infty} a_n$
  Define $\limsup a_n = \lim_{N\to\infty} \sup T_{a_n,N}$ and $\liminf a_n = \lim_{N\to\infty} \inf T_{a_n,N}$.
- If $T_1, T_2, \cdots$ are the tails of $\{a_n\}$ then $T_1 \supseteq T_2 \supseteq T_3 \supseteq \cdots$.
  If $\{a_n\}$ is bounded then $\forall N$, $\sup T_N$ and $\inf T_N$ exist, both $\{\inf T_N\}$ and $\{\sup T_N\}$ converge and $\inf T_1 \leq \inf T_2 \leq \cdots \liminf a_n \leq \limsup a_n \cdots \leq \sup T_2 \leq \sup T_1$.
- If $\lim_{n\to\infty} a_n = a$ then $\liminf a_n = a = \limsup a_n$.
- If $\{a_n\}$ is bounded and $\liminf a_n = \limsup a_n$, then $\{a_n\}$ converges and $\lim_{n\to\infty} a_n = \liminf a_n = \limsup a_n$.
- If $\{a_n\}$ is bounded then:
  - $\{a_n\}$ has 2 monotone subsequences with limits $\limsup a_n$ resp. $\liminf a_n$.
  - a is a sequential limit if a subsequence of $\{a_n\}$ converges to a.
  - If S is the set of sequential limits of $\{a_n\}$, then $\inf S \in S$, $\sup S \in S$ and $\liminf a_n = \inf S$ and $\limsup a_n = \sup S$.

Figure 12.1: Concepts on Sequences and Continuity

- **Bolzano Theorem** If f is continuous on $[a, b]$ and $f(a)$ and $f(b)$ have opposite signs then there is $c \in (a, b)$ such that $f(c) = 0$.

- **Intermediate Value Theorem** If f is continuous on $[a, b]$ and $f(a) \leq v \leq f(b)$, then there is $c \in (a, b)$ such that $f(c) = v$.

- A continuous, one-to-one function is strictly monotone.

- If f is one-to-one continuous on I which is strictly increasing (decreasing) then f_{inv} is strictly increasing (decreasing) on $f(I)$.

- If f is a strictly monotone function on interval I and $f(I)$ is an interval then f is continuous on I.

- If f is one-to-one continuous on I then f_{inv} is continuous on $f(I)$.

- Let I and J be intervals and f a continuous one-to-one function on I with values in J. If f is differentiable at c and if $f'(c) \neq 0$, then f_{inv} is differentiable at $f(c)$ and $f'_{\text{inv}}(f(c)) = 1/f'(c)$.

- **Extreme Value Theorem** If f is continuous on $[a, b]$, then $\exists c, d \in [a, b]$ such that $f(c) \leq f(x) \leq f(d) \ \forall x \in [a, b]$.

- If $\{f_n\}$ is a sequence of functions which have the same domain as function f, then $\{f_n\}$ *converges pointwise* to f on the common domain if for each x in the common domain, $\lim_{n \to \infty} f_n(x) = f(x)$.

- If $\{f_n\}$ is a sequence of functions whose domains include I, and the domain of f also includes I, then $\{f_n\}$ converges uniformly to f on I if $\forall \varepsilon > 0, \exists N > 0$ such that $\forall x \in I$ and $\forall n > N$, $|f(x) - f_n(x)| < \varepsilon$.

- Assume the domains of f and f_n for each n include interval I, that $\forall n$, f_n is continuous at c for some $c \in I$, and that $\{f_n\}$ converges to f uniformly on I. Then f is continuous at c.

- Assume the domains of f and f_n for each n include interval I, that $\forall n$, f_n is continuous on I, and that $\{f_n\}$ converges to f uniformly on I. Then f is continuous on I.

- Assume $\{f_n\}$ converges pointwise to f on an interval I and $\forall n$, let $M_n = \sup\{|f_n(x) - f(x)| : x \in I\}$. The sequence $\{f_n\}$ converges uniformly to f if and only if the sequence $\{M_n\}$ converges to 0.

- **Cauchy Criterion for Uniform Convergence** A sequence $\{f_n\}$ of functions defined on I converges uniformly on I iff $\forall \varepsilon > 0, \exists N > 0$ such that for all $x \in I$ and $m, n > N$, $|f_n(x) - f_m(x)| < \varepsilon$.

- **Dini's Theorem** Let $\{f_n\}$ be a sequence of continuous functions that converges pointwise to f on $[a, b]$ where $\{f_n\}$ is monotone. If f is continuous on $[a, b]$, then $\{f_n\}$ converges uniformly to f on $[a, b]$.

Figure 12.2: More on Continuity and Converging Functions of Sequences

Chapter 13

Riemann Integral

In calculus, the definite integral was introduced to solve the problem of finding the area under a curve representing a function. While the Greeks used the method of exhaustion to do such calculation, nowadays we use integration.

To see an example of this, suppose f is a function which is continuous and non- negative on a closed interval $[a, b]$ and we want to find the area under the curve representing the graph of $y = f(x)$ from a to b. The standard procedure is to approximate this area by the areas of rectangles. For a positive integer n, choose points x_i for $0 \le i \le n$ so that

$$a = x_0 < x_1 < x_2 < ... < x_{n-1} < x_n = b.$$

The base of each rectangle will be an interval of the form $[x_i, x_{i+1}]$. The height of each such rectangle will be chosen to be $f(t_i)$ for some value of t_i such that $x_{i-1} \le t_i \le x_i$. The value of t_i can be chosen in various ways:

1. t_i can be chosen so that $f(t_i) = M_i$ is the maximum value that f takes in the interval $[x_{i-1}, x_i]$. This will give an overestimate of the area. See Figure 13.1.

2. t_i can be chosen so that $f(t_i) = m_i$ is the minimum value that f takes in the interval $[x_{i-1}, x_i]$. This will give an underestimate of the area. See Figure 13.2.

3. t_i can be chosen to be some other value. This will presumably give an estimate closer to the true value of the area. See Figure 13.3.

Whichever way t_i is chosen, the sum of the rectangles is an approximation of the area A:

$$A \approx \Sigma_{i=1}^n f(t_i)(x_i - x_{i-1}).$$

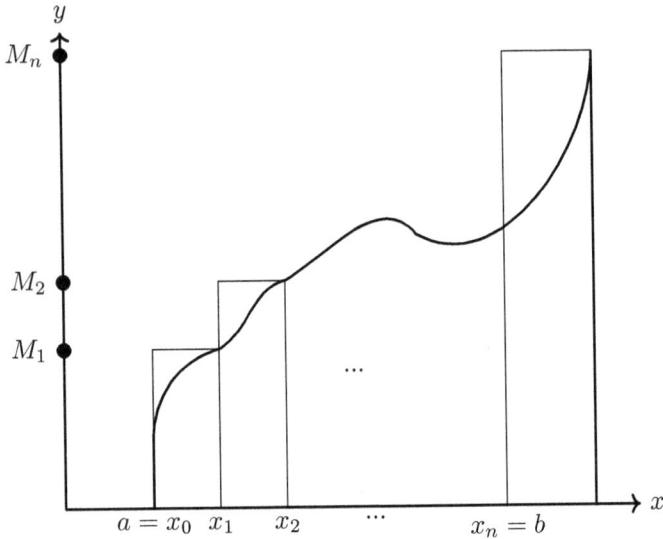

Figure 13.1: Overestimating Areas

Intuitively, as n increases and the rectangles get thinner, we expect these sums to give us better approximations of A. Part of the way we can see this is to note that if t_i is chosen by method 1 above, the result will be an overestimate of A, and if it is chosen by method 2, the result will be an underestimate, so that we have

$$\Sigma_{i=1}^{n} m_i(x_i - x_{i-1}) \leq A \leq \Sigma_{i=1}^{n} M_i(x_i - x_{i-1})$$

As the rectangles get thinner, we expect the two sums to get closer together, thus squeezing A between them. In other words, we expect the sum

$$\Sigma_{i=1}^{n}(M_i - m_i)(x_i - x_{i-1})$$

to get closer to 0. If this sum can be made arbitrarily close to 0 by choosing thin enough rectangles, then any sum obtained by choosing t_i by method 3 above will also get squeezed, and then we would expect the number A to exist by the Completeness Axiom of the real numbers.

Methods of using approximations and summations to calculate the areas and volumes of figures appeared centuries BC. We already saw that Archimedes used the method of Exhaustion in order to give rigorous upper

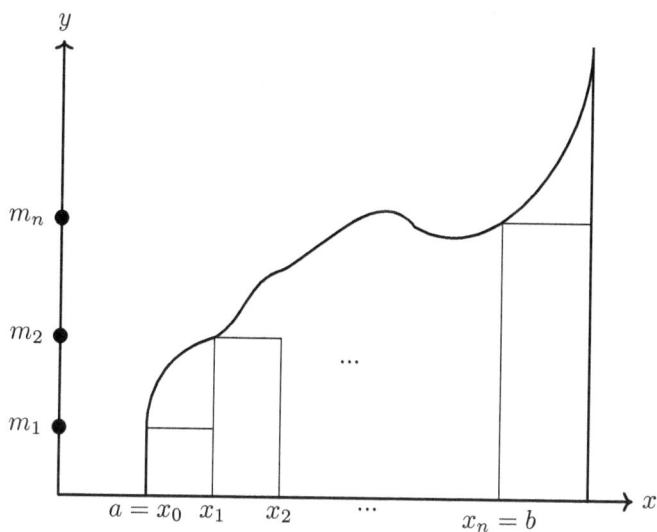

Figure 13.2: Underestimating Areas

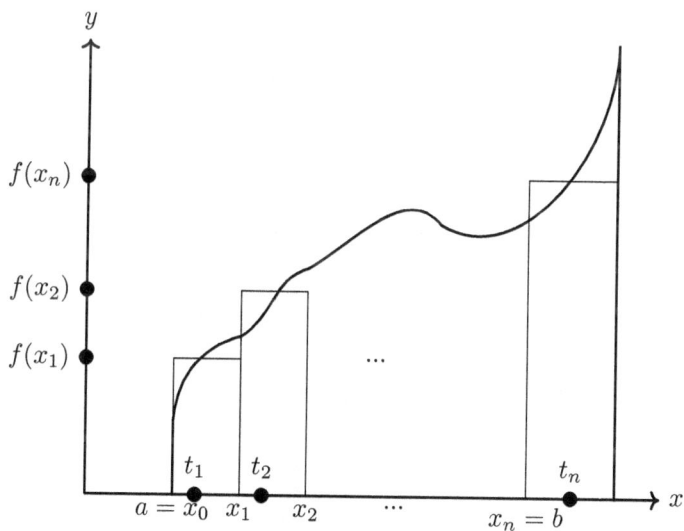

Figure 13.3: Close estimate

CHAPTER 13. RIEMANN INTEGRAL

and lower bounds which both converge towards the area being calculated. Archimedes devised a mechanical method (known as the method of equilibrium) which relies on the law of the lever which pivots around a fixed point. The lever was used to hang thin strips which represented the sub-areas of the area to be measured. If balance is achieved, then the sum of these strips approximates to the whole area. The method was not written in an official publication since Archimedes did not consider it sufficiently rigorous, but he used it to find the areas before writing the rigorous proof. Instead Archimedes described the method in letters. Archimedes manuscript was lost and rediscovered in 1906. Nonetheless, the method was in use in Europe in the 17th century. The purpose of this chapter is to prove that this approach does, indeed, lead to a unique number A representing the area under the curve. This number is called the Riemann[1] integral of f, denoted

$$\int_a^b f(x)d(x)$$

Although the Greeks knew how to compute areas via approximation, summation, infinitesimals, equilibrium and exhaustion, the Riemann integral (which was given in the 19th century as part of Riemann's habilitation manuscript at the University of Göttingen) is the first rigorous definition of integral and is the simplest integral and easiest to understand. All continuous functions and even some discontinuous functions have a Riemann integral. Other integrals (like the Lebesque integral) allow more functions to be integrable but they are more difficult to understand. In this chapter we concentrate on the Riemann integral.

13.1   The Riemann Integral

We need to begin with some terminology. A partition of an interval is just a division of the interval into subintervals.

Definition 13.1.1 (Partition). *A partition of an interval $[a, b]$ is a finite set of points $\{x_i : 0 \leq i \leq n\}$ such that*

$$a = x_0 < x_1 < x_2 < \cdots < x_n = b.$$

The norm of a partition P, denoted by $\|P\|$ is the largest of the numbers $x_i - x_{i-1}$, that is, $\|P\| = max\{x_i - x_{i-1} : 1 \leq i \leq n\}$. If P_1 and P_2 are partitions of $[a, b]$ and $P_1 \subseteq P_2$, then P_2 is a refinement of P_1.

---

[1]Bernhard Riemann, 1826–1866, German mathematician.

The points t_i chosen as in Figures 13.1, 13.2, and 13.1, are called *tags*.

Definition 13.1.2. *A tagged partition* ${}^tP$ *of an interval* $[a, b]$ *consists of a partition* $P = \{x_i : 0 \leq i \leq n\}$ *of* $[a, b]$ *along with a set* $\{t_i : 1 \leq i \leq n\}$ *of values, known as tags, that satisfy* $x_{i-1} \leq t_i \leq x_i$ *for* $1 \leq i \leq n$. *We will express a tagged partition* ${}^tP$ *of* $[a, b]$ *by* ${}^tP = \{(t_i, [x_{i-1}, x_i]) : 1 \leq i \leq n\}$, *say that* ${}^tP$ *is a tagged partition formed from* P, *and refer to* P *as the partition associated with* ${}^tP$. *The norm* $\|{}^tP\|$ *is the norm* $\|P\|$ *of the partition* P *associated with* ${}^tP$.

Tagged partitions of an interval are actually an easy concept. The interval is simply split into subintervals with one point chosen (or tagged) from each interval.

Example 13.1.3. *We can divide the interval* $[0, 1]$ *into 10 subintervals with the midpoint of each interval tagged. This yields the tagged partition*
${}^tP = \{(0.05, [0, 0.1]), (0.15, [0.1, 0.2]), (0.25, [0.2, 0.3]), (0.35, [0.3, 0.4]),$
$(0.45, [0.4, 0.5]), (0.55, [0.5, 0.6]), (0.65, [0.6, 0.7]), (0.75, [0.7, 0.8]),$
$(0.85, [0.8, 0.9]), (0.95, [0.9, 1.0])\}$.
Of course here, $P = \{0, 0.1, 0.2, 0.3, 0.4, 0.5, 0.6, 0.7, 0.8, 0.9, 1\}$ *and the set of tags is*
$\{0.05, 0.15, 0.25, 0.35, 0.45, 0.55, 0.65, 0.75, 0.85, 0.95\}$.
The norm $\|{}^tP\| = \|P\| = 0.1$.

Definition 13.1.4 (Riemann Sum). *Let* f *be a real-valued function whose domain includes the interval* $[a, b]$, *and let* $P = \{x_i : 0 \leq i \leq n\}$ *be a partition of* $[a, b]$ *and* ${}^tP = \{(t_i, [x_{i-1}, x_i]) : 1 \leq i \leq n\}$ *be a tagged partition of* $[a, b]$. *The Riemann sum* $S(f, {}^tP)$ *of* f *associated with* ${}^tP$ *is defined by*

$$S(f, {}^t P) = \Sigma_{i=1}^n f(t_i)(x_i - x_{i-1}).$$

Furthermore, if f *is bounded on* $[a, b]$ *then we also define the upper (resp. lower) Riemann sum* $S^+(f, P)$ *(resp.* $S^-(f, P)$*) as follows:*

$$S^+(f, P) = \Sigma_{i=1}^n \sup\{f(x) : x \in [x_{i-1}, x_i]\}(x_i - x_{i-1}).$$

$$S^-(f, P) = \Sigma_{i=1}^n \inf\{f(x) : x \in [x_{i-1}, x_i]\}(x_i - x_{i-1}).$$

Note that in the above definition, we did not assume f is bounded on $[a, b]$ in order to define its Riemann sum, but we needed this assumption for the upper and lower Riemann sums. We will see in Theorem 13.2.1 that if f is Riemann integrable on $[a, b]$ then f is bounded on $[a, b]$.

Example 13.1.5. *If f is the function defined by $f(x) = x^2$ and if ^tP is the tagged partition of $[0,1]$ of Example 13.1.3, then $S(f, {}^t P) = (0.05)^2(0.1) + (0.15)^2(0.1) + (0.25)^2(0.1) + (0.35)^2(0.1) + (0.45)^2(0.1) + (0.55)^2(0.1) + (0.65)^2(0.1) + (0.75)^2(0.1) + (0.85)^2(0.1) + (0.95)^2(0.1) = 0.3325.$*
Note that $S^+(f, P) = 0.385$ and $S^-(f, P) = 0.285$.

In order for a function to be Riemann integrable, its Riemann sums must approach a limit as the norms of the tagged partitions go to 0. This corresponds to making the rectangles discussed in the introduction very thin, thereby obtaining a better approximation of the area under the curve. This concept is made precise as follows:

Definition 13.1.6 (Riemann Integral, Lower and Upper Riemann Integrals). *A real-valued function whose domain includes a closed interval $[a, b]$ is said to be Riemann integrable on $[a, b]$ if there exists a number L with the following property:*

for any $\varepsilon > 0$ there exists a $\delta > 0$ such that for any tagged partition ^tP of $[a, b]$, such that if $\|{}^t P\| < \delta$ then $|S(f, {}^t P) - L| < \varepsilon$.

That is, $\lim_{\|{}^t P\| \to 0} S(f, {}^t P) = L$.
   The number L is called the Riemann integral of f on $[a, b]$ and is denoted by

$$\int_a^b f(x)d(x),$$

or simply

$$\int_a^b f.$$

It is also usual to speak of the lower and upper Riemann integrals of f on $[a, b]$, where f is bounded on $[a, b]$, which are respectively defined as follows:

$$S^-(f) = \sup\{S^-(f, P) : P \text{ is a partition of } [a, b]\}$$

$$S^+(f) = \inf\{S^+(f, P) : P \text{ is a partition of } [a, b]\}$$

   The adjective "Riemann" will usually be dropped when discussing the integral or integrable functions.

This is a complicated limit process. For every $\delta > 0$, there are many tagged partitions ^tP of $[a, b]$ that satisfy $\|{}^t P\| < \delta$. The subintervals can be chosen in any manner as long as the length of each subinterval is less than δ. Once these subintervals have been chosen, the tag from each subinterval may also be chosen at random. As a result of this variability, it is tedious and/or difficult to prove that a function is Riemann integrable on an interval using the definition unless the function has a very simple form.

Example 13.1.7. *Consider the function defined by $f(0) = 5$ and $f(x) = 1$ for all $x \neq 0$. We will show that f is Riemann integrable on the interval $[0, 1]$ and that $\int_0^1 f = 1$, (The value comes from the area interpretation of the integral: can you sketch the graph?) Let $\varepsilon > 0$ and let $\delta = \frac{\varepsilon}{4}$. Suppose that $^t P = \{(t_i, [x_{i-1}, x_i]) : 1 \leq i \leq n\}$ is a tagged partition of $[0, 1]$ with $\|^t P\| < \delta$. Since $t_i > 0$ for $2 \leq i \leq n$, it follows that $f(t_i) = 1$ for those values of i. It is also clear that $f(t_1)$ is either 1 or 5. Therefore, $(x_1 - x_0) < \delta$, $\Sigma_{i=2}^n (x_i - x_{i-1}) = 1 - (x_1 - x_0)$, and*

$$
\begin{aligned}
|S(f, {}^t P) - 1| &= f(t_1)(x_1 - x_0) + \Sigma_{i=2}^n 1.(x_i - x_{i-1}) - 1 \\
&\leq 5(x_1 - x_0) + [1 - (x_1 - x_0)] - 1 \\
&= 5(x_1 - x_0) - (x_1 - x_0) \\
&= 4(x_1 - x_0) \\
&< 4\delta \\
&= \varepsilon.
\end{aligned}
$$

In general, it is not practical to use Definition 13.1.6 to compute the integral of a function, so we need to develop other methods of doing this. For this, we need to use the definition to obtain general properties of the Riemann integral.

To begin with, note that if f is Riemann integrable on $[a, b]$, it should be clear that $\int_a^b f$ is unique and that its value depends only on the function f and the interval $[a, b]$. The name of the variable is irrelevant. In other words,

$$
\int_a^b f(x)dx = \int_a^b f(t)dt = \int_a^b f(v)dv,
$$

and so on. Variables such as these are known as dummy variables, since they do not play a role in the value of the integral. Thus, the notation

$$
\int_a^b f
$$

contains all the relevant information, namely the function f and the interval $[a, b]$. The reason the variable is so often used follows from the statement

$$
\int_a^b f(x)dx = \lim_{\|^t P\| \to 0} \Sigma_{i=1}^n f(t_i)(x_i - x_{i-1}).
$$

Here, $\int$ is an elongated form of the letter 'S', the first letter of the Latin word for sum, and the product $f(x)dx$ is reminiscent of the product $f(t_i)(x_i - x_{i-1}) = f(t_i)\Delta_i x$, where $\Delta_i x$ is the change in x in the ith subinterval. (This notation for the integral is due to Leibniz.) In applications, the numbers

$f(t_i)$ and $\Delta_i x$ have meanings that are often physical and, if you are willing to accept differentials, $f(x)$ and dx have such meanings as well. Since our purpose here is theoretical, we will use the notation $\int_a^b f$ most of the time here.

The following theorem lists some of the algebraic properties of the Riemann integral. These properties follow easily from the definition, so the proofs are left as exercises (see Exercise 13.2). When proving these statements, also think of the relevant area under the graphs of the respective functions.

Theorem 13.1.8. *Suppose that f and g are Riemann integrable functions defined on $[a, b]$ and that k is a constant. Then*

1. *$\int_a^b f$ is unique.*

2. *Let h be defined on $[a, b]$ such that $h(x) = k$ for all $x \in [a, b]$. Then h is Riemann integrable whose Riemann integral is $k(b - a)$.*

3. *kf is Riemann integrable on $[a, b]$, and $\int_a^b kf = k \int_a^b f$.*

4. *$f + g$ is Riemann integrable on $[a, b]$ and $\int_a^b (f + g) = \int_a^b f + \int_a^b g$.*

In the introduction to this chapter, we discussed the possibility of squeezing the value of the integral between two values as in

$$\Sigma_{i=1}^n m_i(x_i - x_{i-1}) \leq A \leq \Sigma_{i=1}^n M_i(x_i - x_{i-1})$$

where

$$\Sigma_{i=1}^n (M_i - m_i)(x_i - x_{i-1}).$$

is expected to approach 0 as n increases and $\|{}^t P\|$ gets smaller. To consider this approach properly requires the concept of the oscillation of a function on an interval.

Definition 13.1.9. *Let f be a bounded function defined on a closed interval $[a, b]$.*

- *The oscillation of f on $[a, b]$ is defined by*

$$\omega(f, [a, b]) = sup\{f(x) : x \in [a, b]\} - inf\{f(x) : x \in [a, b]\}.$$

- *If $P = \{x_i : 0 \leq i \leq n\}$ is a partition of $[a, b]$, we define the oscillation of f modulo P to be*

$$O(f, P) = \Sigma_{i=1}^n \omega(f, [x_{i-1}, x_i])(x_i - x_{i-1}).$$

The oscillation of a continuous function on an interval is simply its maximum value minus its minimum value on the interval. Since an arbitrary bounded function may not have a maximum or minimum value on an interval, it is necessary to use the least upper bound and the greatest lower bound.

Example 13.1.10. *Consider the function f defined by $f(x) = (1-x)\sin(\frac{1}{x})$ for $x \neq 0$ and $f(0) = 0$. On the interval $[0,1]$, the range of this function is $(-1,1)$, so $\omega(f,[0,1]) = 2$.*

The importance of the oscillation for integrability will become clear in the next section.

The following lemma which is left as an exercise (see Exercise 13.3) gives some properties of oscillations:

Lemma 13.1.11. *Let f be a bounded function whose domain includes the interval $[a,b]$ and let $I = [c,d] \subseteq [a,b]$. Suppose $P_1 = \{z_i : 0 \leq i \leq n\}$ is a partition of $[a,b]$ and that P_2 is a refinement of P_1. Let ${}^tP_1$ and ${}^tP_2$ be tagged partitions of $[a,b]$ formed from the partitions P_1 and P_2 respectively. The following hold:*

1. $\omega(f,I) \geq |f(x) - f(y)| \geq 0$ *for all* $x, y \in I$.

2. $\omega(f,I) = \sup\{|f(x) - f(y)| : x, y \in I\} = \sup\{f(x) - f(y) : x, y \in I\}$.

3. $\omega(f,[c,d]) \leq \omega(f,[a,b])$.

4. *If* $[d,e] \subseteq [a,b]$ *then*
   $\max\{\omega(f,[c,d]), \omega(f,[d,e])\} \leq \omega(f,[c,e]) \leq \omega(f,[c,d]) + \omega(f,[d,e])$.

5. $|S(f,{}^t P_2) - S(f,{}^t P_1)| \leq O(f,P_1)$.

6. $O(f,P_2) \leq O(f,P_1)$.

7. $O(f,P_1) = S^+(f,P_1) - S^-(f,P_1)$.

8. $S^-(f,P_1) \leq S(f,{}^t P_1) \leq S^+(f,P_1)$.

9. $S^-(f,P_1) \leq S^-(f) \leq S^+(f) \leq S^+(f,P_1)$.

10. $S^+(f) - S^-(f) \leq O(f,P_1)$.

13.1.1   Exercises

Exercise 13.1. Let P_1 and P_2 be partitions of $[a, b]$ such that P_2 is a refinement of P_1. Show that $\|P_2\| \leq \|P_1\|$.

Let f be a bounded function defined on a closed interval $[a, b]$ and let $[c, d]$ be a subinterval of $[a, b]$. Show that $\omega(f, [c, d]) \leq \omega(f, [a, b])$.

Exercise 13.2. Prove Theorem 13.1.8.

Exercise 13.3. Prove Lemma 13.1.11.

13.2   Conditions for Riemann Integrability

The goal of this section is to prove that continuous functions are integrable. To prove this we will need some preliminary results, some of which are interesting in their own right. In particular we will prove that all Riemann integrable functions on a closed interval are bounded on that interval (Theorem 13.2.1) and that a bounded function f on a closed interval is Riemann integrable iff one of the following holds:

- f satisfies Cauchy Criterion for Riemann integrability (Theorem 13.2.2);

- for each $\varepsilon > 0$ there is a partition of the closed interval such that the sum of the oscillations of f modulo this partition is less than ε (Theorem 13.2.3).

Theorem 13.2.1 (Riemann Integrable functions are bounded). *If f is a real valued function whose domain includes the interval $[a, b]$, and if f is Riemann integrable on $[a, b]$, then f is bounded on $[a, b]$.*

Proof. Since f is Riemann integrable on $[a, b]$, there exists a positive number δ such that $|S(f, {}^t P) - \int_a^b f| < 0.5$ for all tagged partitions ${}^t P$ on $[a, b]$ that satisfy $\|{}^t P\| < \delta$. We will give a direct proof and leave it as an exercise (see Exercise 13.4) to give a proof by contradiction.

By the Archimedean property of the real numbers (see Page 295), there is an integer q such that $b - a < q\delta$. Let $\beta = \dfrac{b - a}{q} < \delta$. Let $x_j = a + j\beta$ for $0 \leq j \leq q$ and let ${}^t P_0 = \{(x_j, [x_{j-1}, x_j]) : 1 \leq j \leq q\}$. Note that ${}^t P_0$ is a tagged partition of $[a, b]$ with $\|tP_0\| = \beta < \delta$. We will prove that the number M defined by

$$M = \frac{1}{\beta} + max\{|f(x_j)| : 0 \leq j \leq n\}$$

is a bound for f on $[a, b]$.

Let $x \in [a, b]$. Choose an integer j such that $x_{j-1} \leq x \leq x_j$ and define the tagged partition ${}^tP$ from ${}^tP_0$ by replacing $(x_j, [x_{j-1}, x_j])$ by $(x, [x_{j-1}, x_j])$ (in other words, replacing the tagged element x_j in $[x_{j-1}, xj]$ by x). Since ${}^tP$ is simply tP_0 with one tag changed, ${}^tP$ is a tagged partition of $[a, b]$ with $\|{}^tP\| = \|{}^tP_0\| < \delta$. By the definition of ${}^tP$,

$$
\begin{aligned}
S(f, {}^t P) - S(f, {}^t P_0) &= f(x)(x_j - x_{j-1}) - f(x_j)(x_j - x_{j-1}) \\
&= (f(x) - f(x_j))\beta.
\end{aligned}
$$

Note also that

$$
\begin{aligned}
S(f, {}^t P) - S(f, {}^t P_0) &\leq |S(f, {}^t P) - \int_a^b f| + |\int_a^b f - S(f, {}^t P_0)| \\
&< 1.
\end{aligned}
$$

Therefore,

$$
\begin{aligned}
|f(x)| &= |\frac{1}{\beta}(S(f, {}^t P) - S(f, {}^t P_0)) + f(x_j)| \\
&\leq \frac{1}{\beta}|S(f, {}^t P) - S(f, {}^t P_0)| + |f(x_j)| \\
&< \frac{1}{\beta} + |f(x_j)| \\
&< M.
\end{aligned}
$$

Since $|f(x)| \leq M$ for all $x \in [a, b]$, M is a bound for f on $[a, b]$. $\qquad\square$

The converse of the theorem is false. The characteristic function for the rationals f, defined by

$$
f(x) = \begin{cases} 0 & \text{if } x \text{ is irrational;} \\ 1 & \text{if } x \text{ is rational;} \end{cases}
$$

is an example of a bounded function that is not Riemann integrable on any interval. This function is known as the Dirichlet function. Although f is not Riemann integrable (see Exercise 13.11), it is Lebesque integrable.

One of the problems in using the definition of a Riemann integral to prove integrability is that we need to know the value of the integral to use it. It is important to have a way of determining that the integral exists as a limit without knowing its value. The following result is one that does this. It is not very easy to use in practice, but it is useful for the theory we are developing.

Theorem 13.2.2 (Cauchy Criterion for Riemann Integrability). *A bounded function f is Riemann integrable on $[a, b]$ if and only if for each $\varepsilon > 0$ there exists $\delta > 0$ such that $|S(f, {}^t P_1) - S(f, {}^t P_2)| < \varepsilon$ for all tagged partitions ${}^t P_1$ and ${}^t P_2$ of $[a, b]$ with norms less than δ.*

Proof. Suppose that for each $\varepsilon > 0$ there exists a positive number δ such that $|S(f,{}^t P_1) - S(f,{}^t P_2)| < \varepsilon$ for all tagged partitions ${}^t P_1$ and ${}^t P_2$ of $[a, b]$ with norms less than δ. For each positive integer n, choose a positive number δ_n such that $|S(f,{}^t P_1) - S(f,{}^t P_2)| < \frac{1}{n}$ for all tagged partitions ${}^t P_1$ and ${}^t P_2$ of $[a, b]$ with norms less than δ_n. We may assume that the sequence $\{\delta_n\}$ is decreasing; if at any stage we get a value for δ_{n+1} which is greater than or equal to δ_n, then skip that value of δ_{n+1} and go on to the next one. (The decreasing values of $\frac{1}{n}$ indicate that it will be necessary to skip only a finite number of values.) For each n, let ${}^t P_n$ be a tagged partition of $[a, b]$ that satisfies $\|{}^t P_n\| < \delta_n$. If m and n are both greater than some positive integer K, then the tagged partitions ${}^t P_m$ and ${}^t P_n$ have norms less than δ_K (this is where we use the fact that the sequence $\{\delta_n\}$ is a decreasing sequence), and it follows that

$$|S(f,{}^t P_n) - S(f,{}^t P_m)| < \frac{1}{K}.$$

This shows that the sequence $S(f,{}^t P_n)$ is a Cauchy sequence. Let L be the limit of this sequence. We will prove that $\int_a^b f = L$.

Let $\varepsilon > 0$. Since the sequence $\{S(f,{}^t P_n)\}$ converges to L, there is a positive integer N such that[2] $\frac{1}{N} < \frac{\varepsilon}{2}$ and $|S(f,{}^t P_n) - L| < \frac{\varepsilon}{2}$ for all $n \geq N$. Let $\delta = \delta_N$, and suppose that ${}^t P$ is a tagged partition of $[a, b]$ that satisfies $\|{}^t P\| < \delta$. Then we have

$$\begin{aligned}
|S(f,{}^t P) - L| &= |S(f,{}^t P) - S(f,{}^t P_N) + S(f,{}^t P_N) - L| \\
&\leq |S(f,{}^t P) - S(f,{}^t P_N)| + |S(f,{}^t P_N) - L| \\
&< \frac{1}{N} + \frac{\varepsilon}{2} \\
&< \frac{\varepsilon}{2} + \frac{\varepsilon}{2} \\
&= \varepsilon,
\end{aligned}$$

where $|S(f,{}^t P) - S(f,{}^t P_N)| < \frac{1}{N}$ by the definition of δ_n above. It follows that the function f is integrable on $[a, b]$ and $\int_a^b f = L$.

For the converse, suppose f is Riemann integrable on $[a, b]$, and let the value of the integral be L. Let $\varepsilon > 0$ be given. Then there is $\delta > 0$ such that for every tagged partition ${}^t P$ with $\|{}^t P\| < \delta$, $|S(f,{}^t P) - L| < \varepsilon$. Let ${}^t P_1$ and ${}^t P_2$ be any two tagged partitions of $[a, b]$ whose norms are less than δ. Then

$$\begin{aligned}
|S(f,{}^t P_1) - S(f,{}^t P_2)| &= |S(f,{}^t P_1) - L + L - S(f,{}^t P_2)| \\
&\leq |S(f,{}^t P_1) - L| + |L - S(f,{}^t P_2)| \\
&< \frac{\varepsilon}{2} + \frac{\varepsilon}{2} \\
&= \varepsilon.
\end{aligned}$$

---

[2]Let N be the smallest positive integer greater than $\frac{2}{\varepsilon}$.

□

Now we are in a position to prove a major theorem that gives a criterion for Riemann integrability. The proof is tedious, and requires careful reading.

Theorem 13.2.3. *Let f be a bounded function whose domain includes $[a, b]$. Then f is Riemann integrable on $[a, b]$ if and only if for each $\varepsilon > 0$, there exists a partition $P = \{x_i : 0 \le i \le n\}$ of $[a, b]$ such that*

$$O(f, P) = \Sigma_{i=1}^n \omega(f, [x_{i-1}, x_i])(x_i - x_{i-1}) < \varepsilon.$$

Proof. Suppose first that for each $\varepsilon > 0$ there exists a partition $P = \{x_i : 0 \le i \le n\}$ of $[a, b]$ such that

$$\Sigma_{i=1}^n \omega(f, [x_{i-1}, x_i])(x_i - x_{i-1}) < \varepsilon.$$

Let $M > 1$ be a bound for f on $[a, b]$ and let ε be given so that $0 < \varepsilon < 1$. By hypothesis, there exists a partition $P_\varepsilon = \{z_k : 0 \le k \le N\}$ of $[a, b]$ such that $\|P_\varepsilon\| < 1$ (By Lemma 13.1.11.6, we can assume that $\|P_\varepsilon\| < 1$) and

$$\Sigma_{i=1}^n \omega(f, [z_{i-1}, z_i])(z_i - z_{i-1}) < \frac{\varepsilon}{4}.$$

Let ${}^t P_\varepsilon$ be any tagged partition of $[a, b]$ that is formed from the partition P_ε, let $d = min\{z_k - z_{k-1} : 1 \le k \le N\}$, and let $\delta = \frac{d\varepsilon}{8MN}$. Then $d < 1$ and $8MN\delta = d_\varepsilon$ and so $2MN\delta = \frac{d_\varepsilon}{4}$. Now suppose that ${}^t P = \{t_i, [x_{i-1}, x_i]) : 1 \le i \le p\}$ is any tagged partition of $[a, b]$ that satisfies $\|{}^t P\| < \delta$. Since $\delta < d$, the set $(x_{i-1}, x_i) \cap P_\varepsilon$ contains at most one point for any value of i. Let $A = \{i : (x_{i-1}, x_i) \cap P_\varepsilon \ne \emptyset\}$ and let $(x_{i-1}, x_i) \cap P_\varepsilon = \{z_k\}$ for each $i \in A$. Since P_ε contains $N - 1$ points in (a, b), the set A contains at most $N - 1$ elements. We will form a new tagged partition ${}^t P_0$ of $[a, b]$ so that most of the tagged intervals in ${}^t P_0$ are the same as those in ${}^t P$ and the partition associated with ${}^t P_0$ is a refinement of P_ε. Form ${}^t P_0$ using the following criteria:

if $i \notin A$, then $(t_i, [x_{i-1}, x_i])$ belongs to ${}^t P_0$;
if $i \in A$, then $(z_{k_i}, [x_{i-1}, z_{k_i}])$ and $(z_{k_i}, [z_{k_i}, x_i])$ belong to ${}^t P_0$.
Then, since M is a bound for f, and since $x_i - x_{i-1} < \delta$, we have

$$|f(t_i) - f(z_{k_i})|(x_i - x_{i-1}) < 2M\delta$$

and hence, since A has fewer than N elements,

$$\Sigma_{i \in A} |f(t_i) - f(z_{k_i})|(x_i - x_{i-1}) < 2MN\delta.$$

Using Lemma 13.1.11.5,

$$|S(f,{}^t P) - S(f,{}^t P_\varepsilon)| \le$$
$$|S(f,{}^t P) - S(f,{}^t P_0)| + |S(f,{}^t P_0) - S(f,{}^t P_\varepsilon)| =$$
$$|\Sigma_{i \in A}(f(t_i) - f(z_{k_i})(x_i - x_{i-1})| + |S(f,{}^t P_0) - S(f,{}^t P_\varepsilon)| \le$$
$$\Sigma_{i \in A}|f(t_i) - f(z_{k_i})|(x_i - x_{i-1}) + \Sigma_{i=1}^N \omega(f, [z_{i-1}, z_i])(z_i - z_{i-1}) <$$
$$2MN\delta + \frac{\varepsilon}{4} =$$
$$\frac{d\varepsilon}{4} + \frac{\varepsilon}{4} <$$
$$\frac{\varepsilon}{2}.$$

If ${}^t P_1$ and ${}^t P_2$ are any two tagged partitions of $[a, b]$ with norms less than δ, then

$$|S(f,{}^t P_1) - S(f,{}^t P_2)| \le$$
$$|S(f,{}^t P_1) - S(f,{}^t P_\varepsilon)| + |S(f,{}^t P_\varepsilon) - S(f,{}^t P_2)| =$$
$$|\Sigma_{i \in A}(f(t_i) - f(z_{k_i})(x_i - x_{i-1})| + |S(f,{}^t P_0) - S(f,{}^t P_\varepsilon)| <$$
$$\frac{\varepsilon}{2} + \frac{\varepsilon}{2} =$$
$$\varepsilon.$$

By the Cauchy criterion for Riemann integrability (Theorem 13.2.2), the function f is Riemann integrable on $[a, b]$. The proof of the converse is left as an exercise (see Exercise 13.13). $\square$

Corollary 13.2.4. *Let f be a real valued function whose domain includes $[a, b]$. The following hold:*

1. *If f is Riemann integrable on $[a, b]$ then $S^+(f) = S^-(f)$.*

2. *If f is bounded on $[a, b]$ and $S^+(f) = S^-(f)$ then f is Riemann integrable on $[a, b]$.*

3. *If f is bounded on $[a, b]$ then f is Riemann integrable on $[a, b]$ iff $S^+(f) = S^-(f)$.*

Proof.   1. By Theorem 13.2.1, f is bounded on $[a, b]$. Let $\varepsilon > 0$. By Theorem 13.2.3, there exists a partition $P = \{x_i : 0 \le i \le n\}$ of $[a, b]$ such that $O(f, P) < \varepsilon$. By Lemma 13.1.11.10, $S^+(f) - S^-(f) \le O(f, P)$. Hence $S^+(f) - S^-(f) \le \varepsilon$. Since ε is arbitrary, then $S^+(f) = S^-(f)$.

2. Let $\varepsilon > 0$. We will show that there is a partition P of $[a, b]$ such that $O(f, P) < \varepsilon$ and then apply Theorem 13.2.3 to obtain the result.

Since $S^-(f) = \sup\{S^-(f, P) : P \text{ is a partition of } [a, b]\}$ then there is P_1 partition of $[a, b]$ such that $S^-(f) < S^-(f, P_1) + \frac{\varepsilon}{2}$.

Similarly, since $S^+(f) = \inf\{S^+(f,P) : P \text{ is a partition of } [a,b]$ then there is P_2 partition of $[a,b]$ such that $S^+(f) > S^+(f,P_2) - \frac{\varepsilon}{2}$.

But $P_1 \cup P_2$ is also a partition of $[a,b]$ and we can easily show that $S^+(f, P_1 \cup P_2) < S^+(f, P_2)$ and $S^-(f, P_1 \cup P_2) > S^-(f, P_1)$.

Hence, $O(f, P_1 \cup P_2) = S^+(f, P_1 \cup P_2) - S^-(f, P_1 \cup P_2) < S^+(f, P_2) - S^-(f, P_1) < S^+(f) + \frac{\varepsilon}{2} - S^-(f) + \frac{\varepsilon}{2} = S^+(f) - S^-(f) + \varepsilon = \varepsilon$. Hence by Theorem 13.2.3, f is Riemann integrable on $[a,b]$.

3. This is a corollary of items 1. and 2. above.

$\square$

Theorem 13.2.5. *Suppose that f and g are Riemann integrable functions defined on $[a,b]$ and that k is a constant. Then*

1. *If $a < c < b$ then $\int_a^b f = \int_a^c f + \int_c^b f$.*

2. *If $f(x) \leq g(x)$ for all $x \in [a,b]$, then $\int_a^b f \leq \int_a^b g$.*

3. *If $|f(x)| \leq k$ for all $x \in [a,b]$, then $|\int_a^b f| \leq k(b-a)$.*

Proof. 1. Obviously, the area between a and b under the graph of f is the sum of the area between a and c under the graph of f and the area between c and b under the graph of f. The proof is as follows:
Since f is Riemann integrable on $[a,b]$ and $a < c < b$, then by Exercise 13.5, f is Riemann integrable on $[a,c]$ and on $[c,b]$. We will show that for any $\varepsilon > 0$, $|\int_a^b f - (\int_a^c f + \int_c^b f)| < \varepsilon$. Let $\varepsilon > 0$. By definition, since f is Riemann integrable on $[a,b]$ and $[a,c]$ and $[c,b]$, there are $\delta_1, \delta_2, \delta_3 > 0$ such that

- for any tagged partition ${}^tP$ of $[a,b]$ if $\|{}^tP\| < \delta_1$ then $|S(f,{}^t P) - \int_a^b f| < \frac{\varepsilon}{3}$;

- for any tagged partition ${}^tP$ of $[a,c]$ if $\|{}^tP\| < \delta_2$ then $|S(f,{}^t P) - \int_a^c f| < \frac{\varepsilon}{3}$;

- for any tagged partition ${}^tP$ of $[c,b]$ if $\|{}^tP\| < \delta_3$ then $|S(f,{}^t P) - \int_c^b f| < \frac{\varepsilon}{3}$;

Let $\delta = \min\{\delta_1, \delta_2, \delta_3\}$ and let ${}^tP$ be a tagged partition of $[a,b]$ such that $\|{}^tP\| < \delta$. Let ${}^tP_1$ be the subset of ${}^tP$ whose tags are all in $[a,c]$

and let ${}^tP_2$ be the subset of ${}^tP$ whose tags are all in $[c, b]$. Obviously, $S(f,{}^t P) = S(f,{}^t P_1) + S(f,{}^t P_2)$ and $\|{}^tP_1\| < \delta$ and $\|{}^tP_2\| < \delta$. Now,

$$
\begin{array}{ll}
\left|\int_a^b f - (\int_a^c f + \int_c^b f)\right| & = \\
\left|\int_a^b f - S(f,{}^t P) + S(f,{}^t P) - (\int_a^c f + \int_c^b f)\right| & = \\
\left|\int_a^b f - S(f,{}^t P) + S(f,{}^t P_1) + S(f,{}^t P_2) - (\int_a^c f + \int_c^b f)\right| & \leq \\
\left|\int_a^b f - S(f,{}^t P)\right| + |S(f,{}^t P_1) - \int_a^c f| + |S(f,{}^t P_2) - \int_c^b f| & < \\
\frac{\varepsilon}{3} + \frac{\varepsilon}{3} + \frac{\varepsilon}{3} & = \\
\varepsilon.
\end{array}
$$

Since ε is arbitrary, then $\int_a^b f = \int_a^c f + \int_c^b f$.

2. Obviously, if f is smaller than g anywhere on $[a, b]$, then the area between a and b under the graph of f is smaller than the area between a and b under the graph of g. The proof is as follows:
We will show that for any $\varepsilon > 0$, $\int_a^b f < \int_a^b g + \varepsilon$.
Let $\varepsilon > 0$. By definition, there are $\delta_1, \delta_2 > 0$ such that for any tagged partition ${}^tP$ of $[a, b]$,

- if $\|{}^tP\| < \delta_1$ then $|S(f,{}^t P) - \int_a^b f| < \frac{\varepsilon}{2}$;

- if $\|{}^tP\| < \delta_2$ then $|S(f,{}^t P) - \int_a^b g| < \frac{\varepsilon}{2}$;

Let $\delta = \min\{\delta_1, \delta_2\}$ and let ${}^tP$ be a tagged partition of $[a, b]$ such that $\|{}^tP\| < \delta$. Then $|S(f,{}^t P) - \int_a^b f| < \frac{\varepsilon}{2}$ and $|S(g,{}^t P) - \int_a^b g| < \frac{\varepsilon}{2}$.

Hence, since $-\frac{\varepsilon}{2} < S(f,{}^t P) - \int_a^b f$ and $S(g,{}^t P) < \int_a^b g + \frac{\varepsilon}{2}$
we have $\int_a^b f < S(f,{}^t P) + \frac{\varepsilon}{2}$ and $S(g,{}^t P) + \frac{\varepsilon}{2} < \int_a^b g + \varepsilon$.
Note that since $f(x) \leq g(x)$ for all $x \in [a, b]$ then $S(f,{}^t P) \leq S(g,{}^t P)$.
Hence $\int_a^b f < S(f,{}^t P) + \frac{\varepsilon}{2} < S(g,{}^t P) + \frac{\varepsilon}{2} < \int_a^b g + \varepsilon$. Since ε is arbitrary, then $\int_a^b f \leq \int_a^b g$.

3. Obviously, if f is smaller than k anywhere on $[a, b]$, then the area between a and b under the graph of f is smaller than the area of the rectangle whose sides are k and $(b - a)$. The proof is as follows:
First note that k and $b - a$ are positive. Let $h(x) = -k$ and $h'(x) = k$ for all $x \in [a, b]$. By 2. above, h and h' are Riemann integrable on $[a, b]$ and $\int_a^b h = -k(b - a)$ and $\int_a^b h' = k(b - a)$. Since $h(x) \leq f(x) \leq h'(x)$ for all $x \in [a, b]$, by 2. above, $\int_a^b h \leq \int_a^b f \leq \int_a^b h'$. I.e., $-k(b - a) \leq \int_a^b f \leq k(b - a)$. That is, $|\int_a^b f| \leq k(b - a)$.

$\square$

Lemma 13.2.6 (Uniform Continuity). *If f is a real valued function whose domain includes $[a, b]$, and if f is continuous on $[a, b]$, then for each $\varepsilon > 0$ there exists $\delta > 0$ such that for all $x, y \in [a, b]$ that satisfy $|y - x| < \delta$, $|f(y) - f(x)| < \varepsilon$.*

Proof. Suppose f is continuous on $[a, b]$ but it is false that for each $\varepsilon > 0$ there exists $\delta > 0$ such that for all $x, y \in [a, b]$ that satisfy $|y - x| < \delta$, $|f(y) - f(x)| < \varepsilon$. Then there is an $\varepsilon > 0$ such that for each $\delta > 0$ there are values $x, y \in [a, b]$ such that $|y - x| < \delta$ and $|f(y) - f(x)| \geq \varepsilon$. In particular, for each positive integer n there exist values $x_n, y_n \in [a, b]$ such that $|y_n - x_n| < \frac{a}{n}$ and $|f(y_n) - f(x_n)| \geq \varepsilon$, since $\{x_n\}$ is a sequence in $[a, b]$, by the Bolzano-Weierstrass Theorem, it contains a subsequence $\{x_{p_n}\}$ that converges to a value $z \in [a, b]$. As

$$y_{p_n} = x_{p_n} + (y_{p_n} - x_{p_n})$$

and $\{y_{p_n} - x_{p_n}\}$ is a sequence which converges to 0, the sequence $\{y_{p_n}\}$ converges to z as well. Since f is continuous at z, the sequence $\{f(y_{p_n}) - f(x_{p_n})\}$ converges to $f(z) - f(z) = 0$. This contradicts the fact that $|f(y_n) - f(x_n)| \geq \varepsilon$ for all n. This proves the lemma by contradiction. $\qquad\square$

Theorem 13.2.7. *If f is a real valued function whose domain includes $[a, b]$, and if f is continuous on $[a, b]$, then f is Riemann integrable on $[a, b]$.*

Proof. Let $\varepsilon > 0$ be given. Then, by Lemma 13.2.6 and the continuity of f on $[a, b]$, there is a $\delta > 0$ such that for all $x, y \in [a, b]$ satisfying $|y - x| < \delta$,

$$|f(y) - f(x)| < \frac{\varepsilon}{b - a}.$$

Choose a positive integer n such that $\beta = \frac{b - a}{n} < \delta$ and define $x_i = a + i\beta$ for $1 \leq i \leq n$. Since f is continuous, for each $1 \leq i \leq n$, let u_i be a point where f reaches its maximum on $[x_{i-1}, x_i]$ and v_i be a point where f reaches its minimum on $[x_{i-1}, x_i]$. Since $|u_i - v_i| < \delta$ then $|f(u_i) - f(v_i)| < \frac{\varepsilon}{b - a}$.

$$
\begin{aligned}
\Sigma_{i=1}^n \omega(f, [x_{i-1}, x_i])(x_i - x_{i-1}) &\leq \Sigma_{i=1}^n |f(u_i) - f(v_i)|(x_i - x_{i-1}) \\
&= \Sigma_{i=1}^n \frac{\varepsilon}{b - a}(x_i - x_{i-1}) \\
&= \frac{\varepsilon}{b - a} \Sigma_{i=1}^n (x_i - x_{i-1}) \\
&= \frac{\varepsilon}{b - a}(b - a) \\
&= \varepsilon,
\end{aligned}
$$

so f is Riemann integrable on $[a, b]$ by Theorem 13.2.3. $\qquad\square$

Although a continuous function is Riemann integrable, the opposite is not true. For example, the Thomae function g on interval $[0, 1]$ given by

$$g(x) = \begin{cases} 1 & \text{if } x = 0 \\ 0 & \text{if } x \text{ is irrational;} \\ \frac{1}{n} & \text{if } x = \frac{m}{n} \text{ is rational, } n \text{ positive integer} \\ & \text{and } m, n \text{ have no common factors;} \end{cases}$$

is discontinuous at every rational yet it is Riemann integrable. This is an example of a function that is discontinuous at countably infinite number of places, yet it is still Riemann integrable. We can also give examples of functions that are discontinuous at uncountably infinite number of places and remain Riemann integrable.

Theorem 13.2.8. *Let f be a monotone function on $[a, b]$. Then f is Riemann integrable on $[a, b]$.*

Proof. Since for any $x \in [a, b]$, $f(x) \leq \max\{f(a), f(b)\}$, f is bounded on $[a, b]$.

Let $\varepsilon > 0$. By the Archimedean Property, there is an n such that $|f(b) - f(a)|\frac{(b-a)}{n} < \varepsilon$. Let $x_i = a + i\frac{(b-a)}{n}$ for all $0 \leq i \leq n$. Then, $P = \{x_i : 0 \leq i \leq n\}$ is a partition of $[a, b]$ such that $(x_i - x_{i-1}) = \frac{b-a}{n}$ and $\omega(f, [x_{i-1}, x_i]) = |f(x_i) - f(x_{i-1})|$ for all $0 \leq i \leq n$. Moreover, $O(f, P) = \sum_{i=1}^{n}\omega(f, [x_{i-1}, x_i])(x_i - x_{i-1}) = \sum_{i=1}^{n}\omega(f, [x_{i-1}, x_i])\frac{b-a}{n} = \frac{b-a}{n}\sum_{i=1}^{n}|f(x_i) - f(x_{i-1})|$.

Hence, for all $0 \leq i \leq n$,

- if f is increasing, then $\omega(f, [x_{i-1}, x_i]) = (f(x_i) - f(x_{i-1})$ and $O(f, P) = \frac{b-a}{n}\sum_{i=1}^{n}(f(x_i) - f(x_{i-1}) = \frac{b-a}{n}(f(b) - f(a))$.

- else if f is decreasing, then $\omega(f, [x_{i-1}, x_i]) = f(x_{i-1}) - f(x_i)$ and $O(f, P) = \frac{b-a}{n}\sum_{i=1}^{n}(f(x_{i-1}) - f(x_i) = \frac{b-a}{n}(f(a) - f(b))$.

Hence $O(f, P) = \frac{b-a}{n}|f(a) - f(b)| < \varepsilon$.
Hence by Theorem 13.2.3, f is Riemann integrable on $[a, b]$. □

13.2.1   Exercises

Exercise 13.4. Can you give another proof of Theorem 13.2.1? Perhaps a proof by contradiction?

Exercise 13.5. Let f be Riemann integrable on $[a, b]$ and let $[c, d] \subseteq [a, b]$. Show that f is Riemann integrable on $[c, d]$.

Exercise 13.6. Let f be a real-valued function whose domain includes $[a, b]$ and assume that $a < c < b$. Show that f is Riemann integrable on $[a, b]$ iff f is Riemann integrable on $[a, c]$ and on $[c, b]$. Show that if f is Riemann integrable on $[a, b]$ then $\int_a^b f = \int_a^c f + \int_c^b f$.

Exercise 13.7. Suppose that f and g are Riemann integrable functions defined on $[a, b]$. Show that fg is Riemann integrable on $[a, b]$.

Exercise 13.8. Let f be Riemann integrable on $[a, b]$.

1. Define $f_+(x) = \begin{cases} f(x) & \text{if } f(x) \geq 0; \\ 0 & \text{otherwise} \end{cases}$

   and $f_-(x) = \begin{cases} -f(x) & \text{if } f(x) \leq 0; \\ 0 & \text{otherwise} \end{cases}$

   Show that f_- and f_+ are Riemann integrable on $[a, b]$.

2. Show that $|f|$ is Riemann integrable on $[a, b]$ and that $|\int_a^b f| \leq \int_a^b |f|$.

Exercise 13.9. Let f be a function whose domain includes $[a, b]$. Assume that $|f|$ is Riemann integrable on $[a, b]$. Is f is Riemann integrable on $[a, b]$. Justify your answer.

Exercise 13.10. Let f be a function whose domain includes $[a + c, b + c]$ and define on $[a, b]$ the function g by $g(x) = f(x + c)$.
Show that f is Riemann integrable on $[a+c, b+c]$ iff g is Riemann integrable on $[a, b]$ and compare $\int_a^b g$ and $\int_{a+c}^{b+c} f$.

Exercise 13.11. For each of the functions below, show whether it is Riemann integrable or not on the relevant interval.

1. The Dirichlet function f on any interval $[a, b]$.

$$f(x) = \begin{cases} 0 & \text{if } x \text{ is irrational}; \\ 1 & \text{if } x \text{ is rational}; \end{cases}$$

2. The Thomae function g on interval $[0, 1]$.

$$g(x) = \begin{cases} 1 & \text{if } x = 0 \\ 0 & \text{if } x \text{ is irrational}; \\ \frac{1}{n} & \text{if } x = \frac{m}{n} \text{ is rational, } n \text{ positive integer} \\ & \text{and } m, n \text{ have no common factors}; \end{cases}$$

3. The function h on any interval $[0, 2]$.

$$h(x) = \begin{cases} \frac{1}{x} & \text{if } x \text{ is irrational}; \\ 0 & \text{if } x \text{ is rational}; \end{cases}$$

4. Let $[c, d] \subseteq [a, b]$. The function r defined on $[a, b]$.

$$r(x) = \begin{cases} 0 & \text{if } x \in [a, b] \setminus [c, d]; \\ 1 & \text{if } x \in [c, d]; \end{cases}$$

Exercise 13.12. Let f defined on $[a, b]$. Show that f is Riemann integrable on $[a, b]$ iff for every $\varepsilon > 0$, there are two Riemann integrable functions g_ε and h_ε on $[a, b]$ such that $g_\varepsilon(x) \leq f(x) \leq h_\varepsilon(x)$ for all $x \in [a, b]$ and $\int_a^b h_\varepsilon - g_\varepsilon < \varepsilon$.

Exercise 13.13. Prove the remaining part of Theorem 13.2.3.

Exercise 13.14. Suppose that f is a real-valued function whose domain includes $[a, b]$ and that g is a real-valued function whose domain includes $[c, d]$ and which maps $[c, d]$ into $[a, b]$. Prove that:

1. If f and g are continuous, then $f \circ g$ is Riemann integrable on $[c, d]$.

2. If f is continuous on $[a, b]$ and g is Riemann integrable on $[c, d]$ then $f \circ g$ is Riemann integrable on $[c, d]$.

Exercise 13.15. Suppose that f and g are real-valued functions whose domain includes $[a, b]$. Assume $g(x) = f(x)$ for all $x \in [a, b]$ except for a finite number of points. Show that f is Riemann integrable on $[a, b]$ iff g is Riemann integrable on $[a, b]$.

13.3 Relating Integration and Differentiation

As we discussed earlier, the Greeks used integration-like methods to calculate the areas and volumes by breaking them down into smaller units. However, the Greeks did not use fractions to approximate continuous magnitudes and had different kinds of magnitudes for lengths, areas, volumes, etc. They never multiplied two lengths to get another length. The idea of using fractions to approximate continuous magnitudes developed first in the Arab world during the middle ages in Europe, and came to Europe in the 16th and 17th centuries. This idea would have been assumed by both Newton and Leibniz when they introduced the calculus in the second half of the

17th century. Furthermore, in the first half of the 17th century, Descartes used analytic geometry and a ruler-and-compass method to multiply two lengths to get a length. Differentiation as we said before is concerned with measuring the sensitivity of a function to change and this is best done by drawing the tangents to the curve of the function and finding the minimum and maximum of the function. The slopes of the tangents on the graph of the function measure the change of the function. But tangents were also used in Archimedes' equilibrium method to calculate the areas under a graph of a function. This connection of tangents to both the area under the graph of a function, and the slope of the change of the function is what we will discuss in this section. In fact the measure of geometric areas and the measure of the change of functions are intimately connected: they are inverses of each other. This link is given by the Fundamental Theorem of the Calculus which is the subject of this section. This theorem states that the Riemann integral of the derivative is the original function and vice versa, the derivative of the Riemann integral is the original function.

13.3.1 Indefinite Integration/Anti-differentiation

Recall from Definition 11.3.1 that for a function f defined on an interval which includes c, the derivative of f at $x = c$ is defined by

$$f'(c) = \lim_{h \to 0} \frac{f(c+h) - f(c)}{h}$$

if the limit exists and that the *derivative* of f, $f'(x)$ is the function defined by

$$f'(x) = \lim_{h \to 0} \frac{f(x+h) - f(x)}{h}$$

for each value of x for which the limit is defined. Recall also that if f has a derivative at c, then f is said to be differentiable at c and that if f' exists for a set of values x, then f is said to be differentiable for those values of x.

Definition 13.3.1 (Primitive). *Let f be a function defined on an interval I. We call F a primitive (or indefinite integral or anti-derivative) of f iff F is differentiable on I and $F' = f$.*

Note that a function f may have different primitives. For example if $n \in \mathbb{N}$ then for any constant c, the function $f(x) = x^n$ on $\mathbb{R}$ has $F_c(x) = \frac{x^{n+1}}{n+1} + c$ as primitive. We use $\int f$ to denote a primitive of f.

Example 13.3.2. *The first three rows of the following table are by Exer-*

cise 11.13, the rest, are by Exercises 11.23:

f	$\int f$
$4x^3 - 4x$	$x^4 - 2x^2$
$\dfrac{1}{2\sqrt{x}}$	$\sqrt{x}$
$\dfrac{1}{(x^2+1)\sqrt{x^2+1}}$	$x/\sqrt{x^2+1}$
$\cos x$	$\sin x$
$-\sin x$	$\cos x$
$\sec^2 x$	$\tan x$
$-\csc^2 x$	$\cot x$
$\dfrac{1}{\sqrt{1-x^2}}$	arcsin
$\dfrac{1}{1+x^2}$	arctan

The following theorem whose proof is left as an exercise (see Exercise 13.16) shows that indefinite integration is a linear operation and that if F is an indefinite integral of f then $F + c$ is also an indefinite integral of f for any constant c. The theorem below does not show unicity of indefinite integration up to a constant. This will wait until we establish the derivative mean value theorem (see Theorem 13.3.7).

Theorem 13.3.3. *If F resp. G are indefinite integrals of f resp. g on interval I and if c is an arbitrary constant then*

1. $F + c$ is an indefinite integral of f.

2. $F + G$ is an indefinite integral of $f + g$.

3. cF is an indefinite integral of cf.

Theorem 13.3.4 (Integral Mean Value). *Let f and g be continuous functions on $[a,b]$ such that $g(x) \geq 0$ for all $x \in [a,b]$. Then there is a $c \in [a,b]$ such that $\int_a^b fg = f(c) \int_a^b g$.*

Proof. Since f is continuous on $[a,b]$ then by the Extreme Value Theorem 12.3.12, there exist values e and d in $[a,b]$ such that $m = f(e) \leq f(x) \leq f(d) = M$ for all x in $[a,b]$. Since $g(x) \geq 0$ for all x in $[a,b]$, then $mg(x) \leq f(x)g(x) \leq Mg(x)$ for all x in $[a,b]$. By Theorems 13.1.8 and 13.2.5, and Exercise 13.7, mg, fg and Mg are all Riemann integrable and $m\int_a^b g \leq \int_a^b fg \leq M \int_a^b g$.

- If $\int_a^b g = 0$ then $\int_a^b fg = 0$ and any $c \in [a,b]$ will do.

- If $\int_a^b g \neq 0$ then let $L = \dfrac{\int_a^b fg}{\int_a^b g}$. We know that $L \in [m, M]$ and by the

Intermediate Value Theorem 12.3.2, there is $c \in [a, b]$ such that $f(c) = L = \dfrac{\int_a^b fg}{\int_a^b g}$. Hence, there is a $c \in [a, b]$ such that $\int_a^b fg = f(c) \int_a^b g$.

□

Corollary 13.3.5. *Let f is a continuous functions on $[a, b]$ then there is a $c \in [a, b]$ such that $\int_a^b f = f(c)(b - a)$.*

Proof. Let $g(x) = 1$ for all $x \in [a, b]$. Then, g is continuous on $[a, b]$ and also, by Theorem 13.1.8, $\int_a^b g = (b - a)$. Hence by Theorem 13.3.4, there is a $c \in [a, b]$ such that $\int_a^b f = f(c)(b - a)$.

□

13.3.2 Derivative Mean Value Theorem

We start by Rolle's theorem:

Theorem 13.3.6 (Rolle). *Let f be a differentiable function on (a, b) which is also continuous on $[a, b]$. If $f(a) = f(b)$ then there is $c \in (a, b)$ such that $f'(c) = 0$.*

Proof. If f is a constant function then for all $c \in (a, b)$ we have $f'(c) = 0$. Else, if there is $d \in (a, b)$ such that $f(a) \neq f(d)$ then assume that $f(d) > f(a)$ (the case $f(d) < f(a)$ is similar). Since f is continuous on $[a, b]$, then by the Extreme Value Theorem 12.3.12, there exists c in $[a, b]$ such that $f(x) \leq f(c)$ for all x in $[a, b]$. Since $f(b) = f(a) < f(d) \leq f(c)$ then $c \in (a, b)$. But,

- If $x > c$ then $\dfrac{f(x) - f(c)}{x - c} \leq 0$.

- If $x < c$ then $\dfrac{f(x) - f(c)}{x - c} \geq 0$.

Since $f'(c)$ exists then $0 \leq \lim_{x \to c^-} \dfrac{f(x) - f(c)}{x - c} = \lim_{x \to c^+} \dfrac{f(x) - f(c)}{x - c} \leq 0$ and $f'(c) = 0$.

□

Theorem 13.3.7 (Derivative Mean Value). *Let f be a differentiable function on (a, b) which is also continuous on $[a, b]$. There is $c \in (a, b)$ such that $f'(c) = \dfrac{f(b) - f(a)}{b - a}$.*

Proof. Let $g(x) = f(x) - \dfrac{f(b) - f(a)}{b - a} x$. Obviously g satisfies all the preconditions of Theorem 13.3.6 and hence there is a c such that $g'(c) = 0$. I.e., there is $c \in (a, b)$ such that $f'(c) = \dfrac{f(b) - f(a)}{b - a}$. $\qquad\qquad$ $\square$

The next theorem establishes unicity of indefinite integrals up to a constant.

Theorem 13.3.8 (Unicity of indefinite integrals up to a constant).

1. *If F is differentiable on (a, b) and $F'(x) = 0$ for all $x \in (a, b)$ then there is a real constant c such that $F(x) = c$ for all $x \in (a, b)$.*

2. *If F and G are both indefinite integrals of f on interval (a, b) then $F = G + c$ where c is a real constant.*

Proof.   1. Let $e \in (a, b)$. Let $x \in (a, b)$.

  - Assume $x > e$. We have F is differentiable on (e, x) and by Theorem 11.3.5, F is continuous on $[e, x]$ hence by Theorem 13.3.7, there is $c \in (e, x)$ such that $0 = F'(c) = \dfrac{F(x) - F(e)}{x - e}$. Hence $F(x) = F(e)$ for all $x \in (e, b)$.

  - Assume $x < e$. We follow the above steps to show that $F(x) = F(e)$ for all $x \in (a, e)$.

  Hence $F(x) = F(e)$ for all $x \in (a, b)$.

2. Let $H = F - G$. By Theorem 11.3.9, H is differentiable on (a, b) and $H' = F' - G' = f - f = 0$. By 1. above, H is a constant function c and hence $F = G + c$.

$\qquad\qquad$ $\square$

13.3.3 Fundamental Theorem of the Calculus

So far all our discussion of the Riemann integral $\int_a^b f$ has been based on the assumption that $a < b$. Most of this discussion however holds also when $a \geq b$. For this, we will give the following definition:

Definition 13.3.9. *Let I be a closed interval and f be Riemann integrable on I. Let $c, d \in I$ such that $c < d$. We define $\int_c^c f = 0$ and $\int_d^c = -\int_c^d$.*

Now in Theorem 13.1.8.1, we can drop the condition that $a < c < b$. It is easy to show that all the important results in this chapter hold for $\int_a^b$ regardless of whether $a < b$ or $b \leq a$.

Theorem 13.3.10 (Continuity and Differentiation of the antiderivative). *Let f be Riemann integrable on $[a, b]$. For any $x \in [a, b]$, we define $F(x) = \int_a^x f$. The following hold:*

1. *F is continuous on $[a, b]$.*

2. *If f is continuous on $[a, b]$ then F is differentiable on $[a, b]$ and $F' = f$.*

Proof.   1. Let c be an arbitrary point on $[a, b]$. Since f is Riemann integrable on $[a, b]$ then by Theorem 13.2.1, f is bounded on $[a, b]$. Assume that $|f(y)| \leq M$ for any $y \in [a, b]$. Let $\varepsilon > 0$ and let $\delta = \frac{\varepsilon}{M}$. Note that by Theorem 13.2.5.1, $\int_a^x f = \int_a^c f + \int_c^x f$. Also by the extended Theorem 13.2.5.3, $|\int_c^x f| \leq M|x - c|$ for any $x \in [a, b]$.

Now, for any $x \in [a, b]$ such that $|x - c| < \delta$ we have $|F(x) - F(c)| = |\int_a^x f - \int_a^c f| = |\int_c^x f| \leq M|x - c| < M\frac{\varepsilon}{M} = \varepsilon$.

Hence F is continuous on every point c in $[a, b]$.

2. Assume f is continuous on $[a, b]$ and let $x \in [a, b]$. Let $\varepsilon > 0$. For any $y \in [a, b]$ we have: if t is between x and y then $|t - x| \leq |y - x|$ and there is $\delta > 0$ such that $|y - x| < \delta$ then $|f(t) - f(x)| < \varepsilon$.

Now, using the extended version of Theorems 13.1.8 and 13.2.5, freely, when $|y - x| < \delta$ we get:

$$\left| \frac{F(y) - F(x)}{y - x} - f(x) \right| = \left| \frac{\int_a^y f - \int_a^x f}{y - x} - f(x) \right| =$$

$$\left| \frac{\int_x^y f}{y - x} - f(x) \right| = \left| \frac{\int_x^y f - (y - x)f(x)}{y - x} \right| =$$

$$\left| \frac{\int_x^y f(t)dt - \int_x^y f(x)dt}{y - x} \right| = \left| \frac{\int_x^y (f(t) - f(x))dt}{y - x} \right| \leq$$

$$\frac{\int_x^y |f(t) - f(x)|dt}{y - x} \leq \left| \frac{\int_x^y \varepsilon}{y - x} \right| =$$

$$\left| \frac{\varepsilon(y - x)}{y - x} \right| = \varepsilon.$$

Note that the above also applies if $x = a$ or $x = b$ and in that case we also have $F'(a) = f(a)$ and $F'(b) = f(b)$. Hence F is differentiable on $[a, b]$ and $F' = f$ on $[a, b]$.
□

Now we are ready to introduce the important fundamental theorem of calculus:

Theorem 13.3.11 (Fundamental theorem of calculus). *Let f be a continuous on $[a, b]$ and let F differentiable on $[a, b]$ such that $F' = f$. Then, $\int_a^b f = F(b) - F(a)$.*

Proof. Since f is continuous, by Theorem 13.2.7, f is Riemann integrable on $[a, b]$. Define G on $[a, b]$ such that $G(x) = \int_a^x f$. By Theorem 13.3.10, G is differentiable on $[a, b]$ and $G' = f$. Hence $F' = G' = f$ and by Theorem 13.3.8, $F = G + c$ for some real value c.

Now, $F(b) - F(a) = G(b) - G(a) = \int_a^b f - \int_a^a f = \int_a^b f$.
□

13.3.4   Exercises

Exercise 13.16. Prove Theorem 13.3.3.

- A partition P of $[a, b]$ is a finite set $\{x_i : 0 \leq i \leq n\}$ such that $a = x_0 < x_1 < x_2 < \cdots < x_n = b$.
  The norm of P is $\|P\| = max\{x_i - x_{i-1} : 1 \leq i \leq n\}$.
  For partitions P_1 and P_2 of $[a, b]$, P_2 is a refinement of P_1 if $P_1 \subseteq P_2$.

- If $P = \{x_i : 0 \leq i \leq n\}$ is a partition of $[a, b]$ and $\{t_i : 1 \leq i \leq n\}$ is a set of tags such that $x_{i-1} \leq t_i \leq x_i$ for $1 \leq i \leq n$, call $^tP = \{(t_i, [x_{i-1}, x_i]) : 1 \leq i \leq n\}$ a tagged partition ^tP of $[a, b]$. Let $\|^tP\| = \|P\|$.

- Let f be defined on $[a, b]$, and $P = \{x_i : 0 \leq i \leq n\}$ and $^tP = \{(t_i, [x_{i-1}, x_i]) : 1 \leq i \leq n\}$ be a partition resp. a tagged partition of $[a, b]$. The Riemann sum $S(f, {}^tP)$ of f associated with ^tP is:
  $$S(f, {}^tP) = \Sigma_{i=1}^{n} f(t_i)(x_i - x_{i-1}).$$
  The upper (resp. lower) Riemann sum $S^+(f, P)$ (resp. $S^-(f, P)$) is:
  $$S^+(f, P) = \Sigma_{i=1}^{n} \sup\{f(x) : x \in [x_{i-1}, x_i]\}(x_i - x_{i-1}).$$
  $$S^-(f, P) = \Sigma_{i=1}^{n} \inf\{f(x) : x \in [x_{i-1}, x_i]\}(x_i - x_{i-1}).$$

- f is Riemann integrable on $[a, b]$ if there exists L such that for any $\varepsilon > 0$ there exists a $\delta > 0$ such that for any tagged partition ^tP of $[a, b]$ where, if $\|^tP\| < \delta$ then $|S(f, {}^tP) - L| < \varepsilon$.
  I.e., $\lim_{\|^tP\| \to 0} S(f, {}^tP) = L$ for any tagged partition ^tP of $[a, b]$.
  L is the Riemann integral of f on $[a, b]$ written as $\int_a^b f(x)dx$ or $\int_a^b f$.

- A bounded f on $[a, b]$, has *lower* and *upper* Riemann integrals on $[a, b]$:
  $S^-(f) = \sup\{S^-(f, P) : P \text{ is a partition of } [a, b]\}$.
  $S^+(f) = \inf\{S^+(f, P) : P \text{ is a partition of } [a, b]\}$.

- Assume f and g are Riemann integrable on $[a, b]$ and k is a constant.

  1. $\int_a^b f$ is unique.

  2. If $h(x) = k$ for all $x \in [a, b]$, then $\int_a^b h$ exists and $\int_a^b h = k(b-a)$.

  3. kf is Riemann integrable on $[a, b]$, and $\int_a^b kf = k \int_a^b f$.

  4. $f+g$ is Riemann integrable on $[a, b]$ and $\int_a^b (f+g) = \int_a^b f + \int_a^b g$.

- Let f be a bounded function defined on a closed interval $[a, b]$.

  - The oscillation of f on $[a, b]$ is defined by:
    $\omega(f, [a, b]) = \sup(f([a, b])) - \inf(f([a, b]))$.

  - If $P = \{x_i : 0 \leq i \leq n\}$ is a partition of $[a, b]$, the oscillation of f modulo P is defined by:
    $O(f, P) = \Sigma_{i=1}^{n} \omega(f, [x_{i-1}, x_i])(x_i - x_{i-1})$.

Figure 13.4: Main Concepts of Riemann Integrals from Chapter 13

- If f is a function whose domain includes $[a, b]$, and if f is Riemann integrable on $[a, b]$, then f is bounded on $[a, b]$.

- **Cauchy Criterion for Riemann Integrability** A bounded function f is Riemann integrable on $[a, b]$ if and only if $\forall \varepsilon > 0$, $\exists \delta > 0$ such that $|S(f, {}^t P_1) - S(f, {}^t P_2)| < \varepsilon$ for all tagged partitions ${}^t P_1$ and ${}^t P_2$ of $[a, b]$ with norms less than δ.

- A bounded function f defined on $[a, b]$ is Riemann integrable iff $\forall \varepsilon > 0$, $\exists P = \{x_i : 0 \leq i \leq n\}$ a partition of $[a, b]$ such that $O(f, P) = \sum_{i=1}^{n} \omega(f, [x_{i-1}, x_i])(x_i - x_{i-1}) < \varepsilon$.

- Let f be a real valued function whose domain includes $[a, b]$. We have:

  - If f is Riemann integrable on $[a, b]$ then $S^+(f) = S^-(f)$.

  - If f is bounded on $[a, b]$ then:
    * If $S^+(f) = S^-(f)$ then f is Riemann integrable on $[a, b]$.
    * f is Riemann integrable on $[a, b]$ iff $S^+(f) = S^-(f)$.

- Assume Riemann integrable functions f and g defined on $[a, b]$.
  - If $a < c < b$ then $\int_a^b f = \int_a^c f + \int_c^b f$.
  - If $f(x) \leq g(x)$ for all $x \in [a, b]$, then $\int_a^b f \leq \int_a^b g$.
  - For constant k, if $|f(x)| \leq k$ for all $x \in [a, b]$, then $|\int_a^b f| \leq k(b - a)$.

- **Uniform Continuity** If f is continuous on $[a, b]$, then $\forall \varepsilon > 0$, $\exists \delta > 0$ such that $\forall x, y \in [a, b]$ that satisfy $|y - x| < \delta$, $|f(y) - f(x)| < \varepsilon$.

- If f is continuous on $[a, b]$, then f is Riemann integrable on $[a, b]$.

- If f is monotone on $[a, b]$, then f is Riemann integrable on $[a, b]$.

- Take a function f defined on interval I. We call F a primitive (or indefinite integral or anti-derivative) of f iff F is differentiable on I and $F' = f$.

- If F resp. G are indefinite integrals of f resp. g on interval I and c is a constant then $F + c$ (resp. $F + G$, resp. cF) is an indefinite integral of f (resp. $f + g$, resp. cf).

Figure 13.5: More on Riemann Integrals

- **Integral Mean Value** Let f and g be continuous functions on $[a, b]$ such that $\forall x \in [a, b]$, $g(x) \geq 0$. Then $\exists c \in [a, b]$ such that $\int_a^b fg = f(c) \int_a^b g$.

- Let f be continuous on $[a, b]$ then $\exists c \in [a, b]$ such that $\int_a^b f = f(c)(b-a)$.

- **Rolle** Let f be a differentiable function on (a, b) and continuous on $[a, b]$. If $f(a) = f(b)$ then there is $c \in (a, b)$ such that $f'(c) = 0$.

- **Derivative Mean Value** Let f be a differentiable function on (a, b) and continuous on $[a, b]$. There is $c \in (a, b)$ such that $f'(c) = \dfrac{f(b) - f(a)}{b - a}$.

- **Unicity of indefinite integrals up to a constant**
    - If F is differentiable on (a, b) and $F'(x) = 0$ for all $x \in (a, b)$ then there is a real constant c such that $F(x) = c$ for all $x \in (a, b)$.
    - If F and G are indefinite integrals of f on (a, b) then $F = G + c$ for constant c.

- When defined, we let $\int_c^c f = 0$ and $\int_d^c = -\int_c^d$.

- **Continuity and Differentiation of the antiderivative** Let f be Riemann integrable on $[a, b]$ and $\forall x \in [a, b]$, define $F(x) = \int_a^x f$. We have:

    - F is continuous on $[a, b]$.
    - If f is continuous on $[a, b]$ then F is differentiable on $[a, b]$ and $F' = f$.

- **Fundamental theorem of calculus** Let f be a continuous on $[a, b]$ and let F differentiable on $[a, b]$ such that $F' = f$. Then, $\int_a^b f = F(b) - F(a)$.

Figure 13.6: Yet More on Riemann Integrals

Appendix A

Plato's Dialogue of Meno

A.1   Proofs as Reasoning about Diagrams

The following example of proof as reasoning about diagrams is taken from Plato's dialogue *Meno*. In this dialogue, Socrates is trying to convince Meno of something about the nature of mathematical knowledge, and for that purpose questions one of Meno's slave boys. For us, the importance is the way the argument relates to some diagrams.

Soc. Tell me, boy, do you know that a figure like this is a square?

Boy. I do.

Soc. And you know that a square figure has these four lines equal?

Boy. Certainly.

Soc. And these lines which I have drawn through the middle of the square are also equal?

Boy. Yes.

Soc. A square may be of any size?

Boy. Certainly.

Soc. And if one side of the figure be of two feet, and the other side be of two feet, how much will the whole be? Let me explain: if in one direction the space was of two feet, and in other direction of one foot, the whole would be of two feet taken once?

Boy. Yes.

Soc. But since this side is also of two feet, there are twice two feet?

Boy. There are.

Soc. Then the square is of twice two feet?

Boy. Yes.

Soc. And how many are twice two feet? count and tell me.

Boy. Four, Socrates.

Soc. And might there not be another square twice as large as this, and having like this the lines equal?

Boy. Yes.

Soc. And of how many feet will that be?

Boy. Of eight feet.

Soc. And now try and tell me the length of the line which forms the side of that double square: this is two feet-what will that be?

Boy. Clearly, Socrates, it will be double.

Soc. Do you observe, Meno, that I am not teaching the boy anything, but only asking him questions; and now he fancies that he knows how long a line is necessary in order to produce a figure of eight square feet; does he not?

Men. Yes.

Soc. And does he really know?

Men. Certainly not.

Soc. He only guesses that because the square is double, the line is double.

Men. True.

Soc. Observe him while he recalls the steps in regular order. (To the Boy.) Tell me, boy, do you assert that a double space comes from a double line? Remember that I am not speaking of an oblong, but of a figure equal every way, and twice the size of this-that is to say of eight feet; and I want to know whether you still say that a double square comes from double line?

Boy. Yes.

Soc. But does not this line become doubled if we add another such line here?

Boy. Certainly.

Soc. And four such lines will make a space containing eight feet?

Boy. Yes.

Soc. Let us describe such a figure: Would you not say that this is the figure of eight feet?

Boy. Yes.

Soc. And are there not these four divisions in the figure, each of which is equal to the figure of four feet?

Boy. True.

Soc. And is not that four times four?

Boy. Certainly.

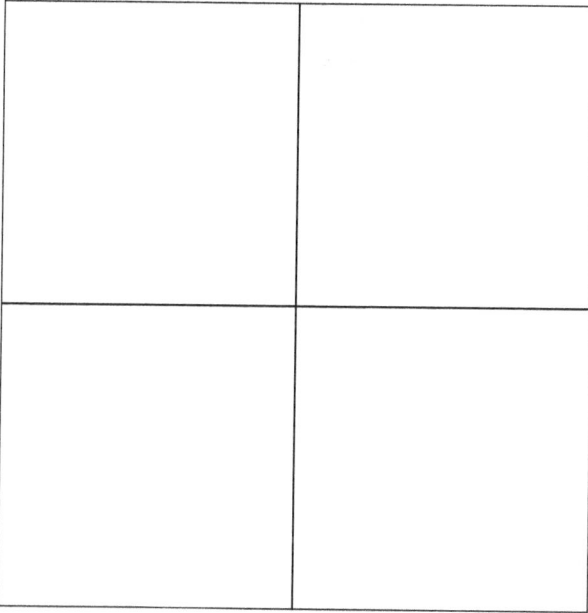

Soc. And four times is not double?

Boy. No, indeed.

Soc. But how much?

Boy. Four times as much.

Soc. Therefore the double line, boy, has given a space, not twice, but four times as much.

Boy. True.

Soc. Four times four are sixteen-are they not?

Boy. Yes.

Soc. What line would give you a space of eight feet, as this gives one of sixteen feet;-do you see?

Boy. Yes.

Soc. And the space of four feet is made from this half line?

Boy. Yes.

Soc. Good; and is not a space of right feet twice the size of this, and half the size of the other?

Boy. Certainly.

Soc. Such a space, then, will be made out of a line greater than this one, and less than that one?

Boy. Yes; I think so.

Soc. Very good; I like to hear you say what you think. And now tell me, is not this a line of two feet and that of four?

Boy. Yes.

Soc. Then the line which forms the side of eight feet ought to be more than this line of two feet, and less than the other of four feet?

Boy. It ought.

Soc. Try and see if you can tell me how much it will be.

Boy. Three feet.

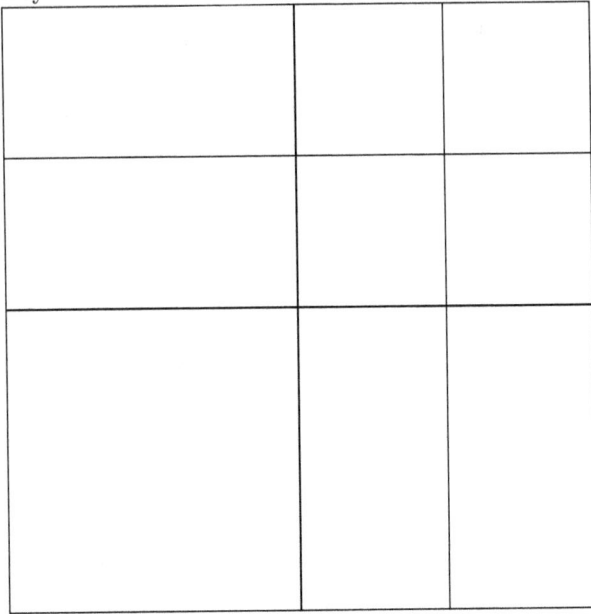

Soc. Then if we add a half to this line of two, that will be the line of three. Here are two and there is one; and on the other side, here are two also and there is one: and that makes the figure of which you speak?

Boy. Yes.

Soc. But if there are three feet this way and three feet that way, the whole space will be three times three feet?

Boy. That is evident.

Soc. And how much are three times three feet?

Boy. Nine.

Soc. And how much is the double of four?

Boy. Eight.

Soc. Then the figure of eight is not made out of three?

Boy. No.

Soc. But from what line?-tell me exactly; and if you would rather not reckon, try and show me the line.

Boy. Indeed, Socrates, I do not know.

Soc. Do you see, Meno, what advances he has made in his power of recollection? He did not know at first, and he does not know now, what is the side of a figure of eight feet: but then he thought that he knew, and answered confidently as if he knew, and had no difficulty; now he has a difficulty, and neither knows nor fancies that he knows.

Men. True.

Soc. Is he not better off in knowing his ignorance?

Men. I think that he is.

Soc. If we have made him doubt, and given him the "torpedo's shock," have we done him any harm?

Men. I think not.

Soc. We have certainly, as would seem, assisted him in some degree to the discovery of the truth; and now he will wish to remedy his ignorance, but then he would have been ready to tell all the world again and again that the double space should have a double side.

Men. True.

Soc. But do you suppose that he would ever have enquired into or learned what he fancied that he knew, though he was really ignorant of it, until he had fallen into perplexity under the idea that he did not know, and had desired to know?

Men. I think not, Socrates.

Soc. Then he was the better for the torpedo's touch?

Men. I think so.

Soc. Mark now the farther development. I shall only ask him, and not teach him, and he shall share the enquiry with me: and do you watch and see if you find me telling or explaining anything to him, instead of eliciting his opinion. Tell me, boy, is not this a square of four feet which I have drawn?

Boy. Yes.

Soc. And now I add another square equal to the former one?

Boy. Yes.

Soc. And a third, which is equal to either of them?

Boy. Yes.

Soc. Suppose that we fill up the vacant corner?

Boy. Very good.

Soc. Here, then, there are four equal spaces?

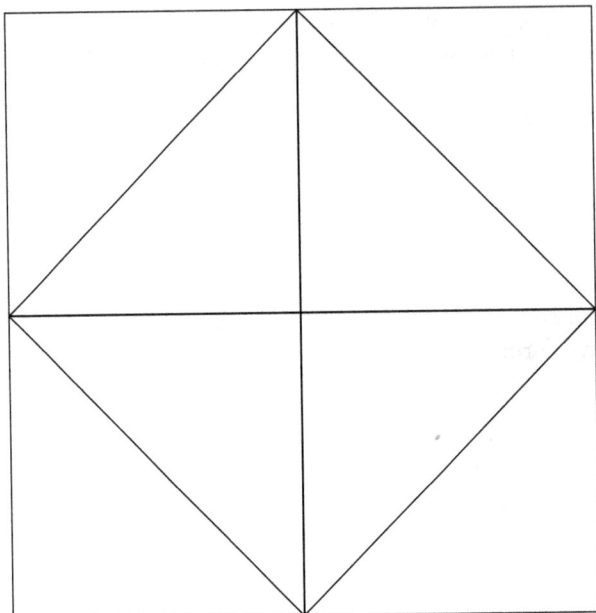

Boy. Yes.

Soc. And how many times larger is this space than this other?

Boy. Four times.

Soc. But it ought to have been twice only, as you will remember.

Boy. True.

Soc. And does not this line, reaching from corner to corner, bisect each of these spaces?

Boy. Yes.

Soc. And are there not here four equal lines which contain this space?

Boy. There are.

Soc. Look and see how much this space is.

Boy. I do not understand.

Soc. Has not each interior line cut off half of the four spaces?

Boy. Yes.

Soc. And how many spaces are there in this section?

Boy. Four.

Soc. And how many in this?

Boy. Two.

Soc. And four is how many times two?

Boy. Twice.

Soc. And this space is of how many feet?

Boy. Of eight feet.

Soc. And from what line do you get this figure?

Boy. From this.

Soc. That is, from the line which extends from corner to corner of the figure of four feet?

Boy. Yes.

Soc. And that is the line which the learned call the diagonal. And if this is the proper name, then you, Meno's slave, are prepared to affirm that the double space is the square of the diagonal?

Boy. Certainly, Socrates.

Appendix B

List of Figures and Tables

B.1 Figures/Tables with Main Concepts

B.2   General Figures/Tables

Bibliography

[1] Aristotle. *Physics*. Hackett Publishing Company, 2018. Translated, with Introduction and Notes, by C. D. C. Reeve.

[2] John L. Bell. Continuity and infinitesimals. In Edward N. Zalta, editor, *The Stanford Encyclopedia of Philosophy*. Metaphysics Research Lab, Stanford University, summer 2017 edition, 2017.

[3] George Berkeley. The Analyst: or A Discourse Addressed to an Infidel Mathematician. Wherein it is examined whether the object, principles, and inferences of the modern Analysis are more distinctly conceived, or more evidently deduced, than religious Mysteries and points of Faith. First printed in 1734. In A. A. Luce and T. E. Jessop, editors, *The Works of George Berkeley Bishop of Cloyne*, volume 4, pages 53–102. Nelson, London, 1951. Full text, edited by David R. Wilkins, available on line at http://www.maths.tcd.ie/pub/HistMath/People/Berkeley/Analyst/Analyst.pdf.

[4] E. Bloch. *Proofs and Fundamentals. A First Course in Abstract Mathematics*. Birkhauser, Boston, 2000.

[5] Eugene Boman and Robert Rogers. *How We Got from There to Here: A Story of Real Analysis*. Milne Open Textbooks, 2013.

[6] David M. Bressoud. *Calculus Reordered: A History of the Big Ideas*. Princeton, 2019.

[7] A. L. Cauchy. Note sur les séries convergentes dont les divers termes sont des fonctions continues d'une veriable réele ou imaginaires, centre des limites données. *Comptes rendus de L'Académie des Sciences*, 36:454, 1853. Also in the Oeuvres Compètes, volume 12 of series 1, pages 30–36.

[8] A. L. Cauchy. Cours d'Analyse de l'Ecole royale polytechnique, 1re partie, analyse algébrique. In *Oeuvres compètes d'Augustin Cauchy*

publiées sous la direction scientifique de l'Académie des sciences et sous les auspices de M. le ministre de l'Instruction publique, volume 3 of series 2. Gauthier-Villars et fils, Paris, 1882–1974. Originally published in 1821.

[9] A. L. Cauchy. Résume des leçons sur le calcul infinitésimal. In *Oeuvres compètes d'Augustin Cauchy publiées sous la direction scientifique de l'Académie des sciences et sous les auspices de M. le ministre de l'Instruction publique*, volume 4 of series 2. Gauthier-Villars et fils, Paris, 1882–1974. Originally published in 1823.

[10] H. B. Curry. The purposes of logical formalization. *Logique et Analyse*, 43:357–366, 1968.

[11] W. DeLong. *A Profile of Mathematical Logic*. Addison-Wesley, Reading, MA & Menlo Park, CA & London & Don Mills, Ontario, 1970.

[12] Keith Devlin. *The language of mathematics, making the invisible visible*. Freeman, 1998.

[13] D. H. Fowler. *The Mathematics of Plato's Academy: A New Reconstruction*. Clarandon Press, Oxford, 1990.

[14] A. A. Fraenkel. *Abstract Set Theory*. North-Holland, Amsterdam, second completely revised edition, 1961.

[15] Alexander Campbell Fraser. *The Works of George Berkeley*. Clarendon Press, Oxford, 1871.

[16] Galileo Galilei. *Dialogues Concerning Two New Sciences*. Macmillan, 1914. Translated from the Italian and Latin into English by Henry Crew and Alfonso de Salvio, with an introduction by Antonio Favoro. Reprinted by Dover Publications, New York.

[17] Kurt Gödel. *The Consistency of the Axiom of Choice and of the Generalized Continuum Hypothesis with the Axioms of Set Theory*. Princeton University Press, Princeton, New Jersey, USA, 1940. Annals of Mathematics Studies, No. 3.

[18] Rod Haggarty. *Fundamentals of Mathematical Analysis*. Prentice Hall, 1993.

[19] T. L. Heath. *The Works of Archimedes*. Cambridge University Press, 1912.

[20] T. L. Heath. *The Thirteen Books of Euclid's Elements.* Cambridge University Press, second edition, 1926. Three volumes. Reprinted by Dover, 1956. The text, along with a Java applet to manipulate the diagrams can be found on line at http://aleph0.clarku.edu/~djoyce/java/elements/elements.html.

[21] Thomas L. Heath. *Diophantus of Alexandria: A study in the history of Greek Algebra.* Cambridge University Press, 1910. Also in Dover 1964.

[22] David Hilbert. *The Foundations of Geometry.* The Open Court Publishing Co, 1902.

[23] B. Jowett. *Dialogues of Plato.* Oxford University Press, 1892.

[24] F. Kamareddine, T. Laan, and R. Nederpelt. *A Modern Perspective on Type Theory.* Kluwer, Dordrecht, Boston, London, 2004.

[25] E. Kamke. *Theory of Sets.* Dover, 1950.

[26] W. R. Knorr. *The Evolution of the Euclidean Elements: A Study of the Theory of Incommensurable Magnitudes and Its Significance for Early Greek Geometry.* Reidel, Dordrecht and Boston and London, 1975.

[27] I. Lakatos. Cauchy and the continuum: the significance of non-standard analysis for the history and philosophy of mathematics. *Mathematical Intelligencer,* 1:151–161, 1978.

[28] G.W. Leibniz. Preface to the general science. In Philip Wiener, editor, *Leibniz: Selections.* Scribner's New York, 1951.

[29] Lillian R. Lieber. *The Education of T. C. MITS.* W. W. Norton & Company, New York, 1944. Drawings by Hugh Gray Lieber. Revised and Enlarged Edition. (First edtion 1942).

[30] Antoni Malet. Renaissance notions of number and magnitude. *Historia Mathematica,* 33(1):63–81, February 2006.

[31] R. R. Middlemiss. *College Algebra.* McGraw-Hill, 1952.

[32] Isaac Newton. *Philosophiae Naturalis Principia Mathematica.* Jussi Societatus Regiae ac typis Josephi Streater; prostat apud plures bibliopolas, Londini, 1687. Translated into English by Andrew Motte, 1729. Incomplete version of translation available on line at http://members.tripod.com/~gravitee/.

[33] Isaac Newton. *Opticks: or a Treatise of the Reflections, Refractions, Inflections & Colours of Light.* Dover, New York, 1952. With a foreword by Albert Einstein, an Introduction by Sir Edmund Whittaker, a preface by I. Bernard Cohen, and an analytical table of contents prepared by Duane H. D. Roller. Based on the fourth edition London, 1730.

[34] Plato. *The Trial and Death of Socrates: Four Dialogues.* Dover, 1992. Four dialogues taken from [23].

[35] G. Pólya. *Induction and Analogy in Mathematics.* Princeton University Press, 1954.

[36] A. Robinson. *Non-Standard Analysis.* North-Holland, Amsterdam, 1966.

[37] J. C. Rolfe (translator). *The Attic Nights of Aulus Gellius, I.* Putham, New York, 1927.

[38] J. P. Seldin. Reasoning in elementary mathematics. In Tessoula Berggren, editor, *Canadian Society for History and Philosophy of Mathematics. Proceedings of the Fifteenth Annual Meeting, Quebec City, Quebec, May 29–May 30, 1989*, pages 151–174, 1990. Available on line from http://www.cs.uleth.ca/~seldin under "Publications.".

[39] Patrick Suppes. *Axiomatic Set Theory.* Dover Publications Inc., 1960.

[40] W.R. Thomas. Moscow mathematical papyrus, no. 14. *The Journal of Egyptian Archeology*, 17(1/2):50–52, 1931.

Index